4TH EDITION
SOIL AND WATER MANAGEMENT SYSTEMS

GLENN O. SCHWAB

Professor Emeritus of Biological and Agricultural Engineering
The Ohio State University
Columbus, Ohio

DELMAR D. FANGMEIER

Professor of Agricultural and Biosystems Engineering
University of Arizona
Tucson, Arizona

WILLIAM J. ELLIOT

Project Leader, U.S. Department of Agriculture, Forest Service
Intermountain Research Station
Moscow, Idaho

JOHN WILEY & SONS, INC.
New York   Chichester   Brisbane   Toronto   Singapore

ACQUISITIONS EDITOR Eric Stano
MARKETING MANAGER Rebecca E. Herschler
PRODUCTION EDITOR Tony VenGraitis
DESIGNER Madelyn Lesure
MANUFACTURING MANAGER Dorothy Sinclair
ILLUSTRATION COORDINATOR Rosa Bryant
PRODUCTION SERVICE Progressive Publishing Alternatives

This book was set in New Baskerville by Digitype, printed by Courier/Stoughton, and bound by Courier/Westford. The cover was printed by Lehigh Press.

Schwab, Glen Orville, 1919–
    Soil and water management systems / Glen O. Schwab, Delmar D.
Fangmeier, William J. Elliot.
        p.   cm.
    Rev. ed. of: Elementary soil and water engineering.   3rd ed.   1993.
    Includes bibliographical references and index.
    ISBN 0-471-10973-8 (alk. paper)
    1. Soil conservation.   2. Water conservation.   3. Hydraulic engineering.   I. Fangmeier, D. D.
(Del D.)   II. Elliot, William J.   III. Schwab, Glen Orville, 1919–   Elementary soil and water
engineering.   IV. Title.
S623.S39   1995   95–25189
631.4'5—dc20   CIP

Printed in the United States of America

10 9 8 7 6 5 4 3 2 1

CONTENTS

PREFACE

As in previous editions, this textbook has been written primarily for classroom and laboratory instruction at the college level for students in agriculture, natural resources, environmental science, and other related nonengineering fields. In this edition the title has been changed from ELEMENTARY SOIL AND WATER ENGINEERING to reflect greater emphasis on a more systematic and integrated approach to solving soil and water problems. The purpose of this book is to present up-to-date information, gathered through experience and research, in a simplified form that will be useful to the beginning student of this subject. The engineering phases of soil and water conservation in agriculture are emphasized with the realization that all aspects must be considered, including agronomic, economic, environmental, biological, and others. Vocational agriculture teachers, instructors for continuing education, engineers, county extension directors, contractors, developers, farm managers, farmers, and others who face rural and urban engineering problems on land may find the information of value.

This text includes subject matter on the management and design of soil and water conservation practices, as well as simple surveying and its application to field problems. The first chapter relates to broader problems, such as population, land use, economics, the environment, food and fiber production, conservation planning, and federal and state environmental legislation; Chapters 2, 3, and 4 cover simple surveying and satellite and computer mapping systems; Chapter 5 deals with rainfall and runoff; Chapters 6 through 9 include channel and upland erosion and control practices for water and wind; Chapter 10 covers water quality and supply; Chapter 11 discusses water storage for farm and domestic uses; Chapters 12 and 13 cover surface and subsurface drainage; Chapters 14, 15, 16, 17, and 18 include irrigation principles and surface, sprinkler, and microirrigation; Chapter 19 covers water measurement in pipes and channels; and a glossary defines many special terms. A selected list of references are included at the end of each chapter for those who wish to pursue a particular subject. Sample instrument survey field notes are given in applicable chapters to illustrate and to standardize the recording of field data. Many examples have been given to explain management and design procedures.

The text is primarily in English units because federal and state agencies and the general public in the United States work predominantly in these units. Most examples in the text show final answers in SI (metric) units. Those who wish to use SI units should refer to other texts, such as SOIL AND WATER CONSERVATION ENGINEERING, (1993).

We are deeply indebted to many individuals and organizations for the preparation of this text. Credit for illustrations and data is given after the figure title or table. Among the many individuals who have made constructive criticisms, provided material, or contributed valuable suggestions are C. E.

Anderson, R. L. Bengston, R. Bonnell, L. C. Brown, D. Conrad, A. B. Daugherty, A. V. Elliot, D. E. Eisenhauer, J. C. Hayes, R. L. Huffman, J. Lyons, J. K. Mitchell, M. L. Palmer, J. A. Replogle, M. R. Reyes, D. C. Slack, M. C. Smith, O. R. Stein, D. E. Storm, G. W. TeSelle, F. R. Troeh, M. Vesterby, L. E. Wagner, R. E. Walker, P. Waller, M. Yitayew, and D. Yoder.

We wish to express appreciation to our wives and families for their sympathetic understanding during the preparation of this edition. We are also grateful to Mrs. Corine Frevert for her support in revising this text and keeping it in print.

We are dedicating this book to the memory of our colleagues Kenneth K. Barnes, Talcott W. Edminster, and Richard K. Frevert, coauthors of previous editions, for their many contributions to the earlier versions.

Columbus, Ohio   Glenn O. Schwab
Tucson, Arizona   Delmar D. Fangmeier
Moscow, Idaho   William J. Elliot

ABBREVIATIONS AND SYMBOLS[a]

a   cross-sectional area; constant

A   watershed area; cross-sectional area; average annual soil loss; apparent specific gravity

ac   acre

ac-ft   acre feet (1 foot depth over 1 acre)

Agr.   Agriculture

Agron.   Agronomy

Ala.   Alabama

Ariz.   Arizona

Ark.   Arkansas

ARS   Agricultural Research Service

ASAE   American Society of Agricultural Engineers

ASCE   American Society of Civil Engineers

ASCS   Agricultural Stabilization and Conservation Service

ASTM   American Society for Testing Materials

AW   available water

b   constant; width; depth

B   outside diameter; width of trench

BHP   brake horsepower

B.M.   bench mark

B.S.   backsight

bu/ac   bushels per acre

Bull.   Bulletin

c   cut

C   Celsius (Centigrade); discharge coefficient; vegetative cover management factor; climate factor; constant; coefficient of variation

Calif.   California

cfd   cubic feet per day

cfm   cubic feet per minute

cfs   cubic feet per second

CFSA   Consolidated Farm Service Agency, formerly ASCS

cir.   circular

cm   centimeter

CN   curve number

Colo.   Colorado

Conn.   Connecticut

CPT   corrugated plastic tubing

d   diameter; depth; distance

D   diameter; depth; distance

D.A.   drainage area

D.C.   drainage coefficient; District of Columbia

Del.   Delaware

dia.   diameter

dS   deciSiemens (unit of electrical conductivity)

DU   distribution uniformity

e   vapor pressure; width

E   efficiency; radiant energy; evaporation; elevation; east; annual soil loss

EDM   electronic distance measurement

EC   electrical conductivity

EL   elevation

EPA   Environmental Protection Agency

ERS   Economic Research Service

ESP   exchangeable sodium percentage

ESSA   Environmental Science Service Administration

[a]Many superscripts may be added in the text.

ET	evapotranspiration	ipd	inches per day
EU	emission uniformity	iph	inches per hour
expt.	experiment	IR	irrigation requirement
f	infiltration rate; fill; water content; monthly evaporation factor	k	monthly evapotranspiration coefficient
F	dimensionless force; fill; safety factor; infiltration amount; Fahrenheit	K	constant; hydraulic conductivity; evapotranspiration coefficient; soil erodibility factor; friction coefficient
FC	field capacity (soil)		
Fla.	Florida		
for.	forestry	Kans.	Kansas
fpd	feet per day	kg	kilogram
fps	feet per second	km	kilometer
F.S.	foresight	kW	kilowatt
ft	feet or foot	Ky.	Kentucky
g	gram; acceleration of gravity	lb.	pound
		L	liter; dimensionless length; length; slope length factor
Ga.	Georgia		
gal.	gallon	La.	Louisiana
GIS	Geographic Information System	LR	leaching requirement
		m	meter; water table height
gph	gallons per hour	MAD	management allowed deficiency
gpm	gallons per minute		
GPO	U.S. Government Printing Office	Mass.	Massachusetts
		Md.	Maryland
GPS	Global Positioning System	mg	milligram
h	hour; height; head; depth	Mg	megagram (metric ton)
H	total head; height	mi.	mile
ha	hectare	Mich.	Michigan
handb.	handbook	min	minute; minimum
H.I.	height of instrument; horizontal interval	Minn.	Minnesota
		misc.	miscellaneous
hp	horsepower	Miss.	Mississippi
i	rainfall intensity; inflow rate; irrigation rate; drainage rate	MJ	megajoule
		mL	milliliter
		mm	millimeter
I	soil erodibility index	Mo.	Missouri
i.d.	inside diameter	mon.	month
Ill.	Illinois	Mont.	Montana
in.	inch or inches	mph	miles per hour
Ind.	Indiana	n	roughness coefficient; drainable porosity; number
info.	information		

	of values; hours of sunshine
N	revolutions per minute; north; hours of sunshine
N.C.	North Carolina
N. Dak.	North Dakota
Neb.	Nebraska
Nev.	Nevada
N.H.	New Hampshire
N.J.	New Jersey
N. Mex.	New Mexico
NOAA	National Oceanic and Atmospheric Administration
NPSH	net positive suction head
NRCS	Natural Resources Conservation Service, formerly SCS
NRI	national resource inventory
N.Y.	New York
o.d.	outside diameter
Okla.	Oklahoma
Oreg.	Oregon
p	wetted perimeter; percent
P	pressure; precipitation; conservation practice factor; power; water content; atmospheric pressure
Pa.	Pennsylvania
PE	potential evapotranspiration
ppm	parts per million
P.R.	Puerto Rico
psi	pounds per square inch
publ.	publication
PVC	polyvinyl chloride
PWP	permanent wilting point
q	flow rate; sprinkler discharge rate
Q	water volume; flow rate
r	radius; rate of application

R	hydraulic radius; radius of influence; rainfall erosivity index; residue cover; range; ridge roughness; solar radiation
RAW	readily available water
rep.	report
res.	research
R.I.	Rhode Island
rpm	revolutions per minute
RUSLE	revised universal soil loss equation
s	slope gradient or percent; second
S	slope gradient or percent; slope steepness factor; sprinkler spacing; drain spacing; south; surface storage
SAR	sodium-adsorption ratio
SC	summation of cuts
S.C.	South Carolina
SCS	U.S. Soil Conservation Service (name changed in 1994 to NRCS, Natural Resources Conservation Service)
S. Dak.	South Dakota
serv.	service
SF	summation of fills
soc.	society
sq	square
sta.	station
t	time; temperature; width
tc	time of concentration
T	dimensionless time; time of concentration; return period; width; temperature; township
t/a	tons per acre
Tenn.	Tennessee
Tex.	Texas
T.P.	turning point

u	monthly evapotranspiration; wind velocity	Wa.	Washington
		WEPP	water erosion prediction project
U	seasonal water use	WHP	water horsepower
UC	uniformity coefficient	Wis.	Wisconsin
U_s	statistical uniformity	W. Va.	West Virginia
U.S.	United States	Wyo.	Wyoming
USBR	U.S. Bureau of Reclamation	x	variable; distance; mean
USDA	U.S. Department of Agriculture	X	variable; distance; horizontal coordinate; climatic constant
USDC	U.S. Department of Commerce	y	variable; depth
USGS	U.S. Geological Survey	Y	variable; distance; vertical coordinate; constant for soil erodibility and cover
USLE	universal soil loss equation		
v	velocity	z	side slope ratio (horizontal to vertical); depth; height
V	volume; vegetative cover factor; velocity	Z	vertical or horizontal distance; depth; vertical coordinate
Va.	Virginia		
V.I.	vertical interval	θ	sideslope angle; slope in degrees; soil water content
Vt.	Vermont		
w	unit weight	φ	rodman
W	width; watt; west; water volume	𝒜	survey instrument person

NOTE: In the text various subscripts to the above symbols may be added. State names not included above are written in full, such as Iowa.

INTRODUCTION

The authors were all raised on farms with somewhat different experiences that motivated them to pursue engineering careers in soil and water conservation. They have had broad and varied experiences in many U.S. states and countries of the world. During these experiences they have seen famine, drought, dust storms, soil erosion, floods, fields too wet to farm, rising water tables and salinization of soils with salt crust on the soil surface, seepage from irrigation canals, inefficient irrigation, and many other soil and water management problems. The authors were also influenced to enter this field because of the many problems they observed during their early childhood.

One field of Dr. Elliot's family farm was a sandy glacial outwash ridge, while the lower slopes were a less permeable glacial till. Where the two soils joined, the field was too wet to support the weight of corn harvesting machinery. After several days of trying to harvest the crop, the effort was given up to wait for the ground to freeze. When cold weather did not come, the corn was harvested by hand. Poor drainage was obviously the problem, which was corrected by a cutoff drain installed on the hill slope (Chapter 13).

In May of the mid 1950s thunder storms following a cold front (Chapter 5) passed through eastern Iowa. During the previous week corn planted on a 20-acre field having a 3 percent slope was just emerging. While inspecting the damage with his dad, Elliot noticed that eroding rills cut into the field at regular intervals (Chapter 6). In some places the rills had followed the row, washing out every seedling in its path. "We are going to have to replant these rows by hand," his dad concluded. That was done in the next couple of days with Elliot helping by dropping two or three kernals of corn in the gash cut by the hoe where the row should have been. Several years later, soil loss in this field was significantly reduced by contour strip cropping (Chapter 7). Thus, Dr. Elliot experienced the impact of soil erosion.

When Dr. Fangmeier was growing up in Nebraska, he recalls seeing heavy rains which caused severe erosion and crop damage by flooding. The eroded soil was deposited on lower land, where crop production was temporarily reduced. Where runoff concentrated in unprotected channels, gullies became larger, wider, and deeper. He helped build terraces and grassed waterways for erosion control. Contouring practices were initiated on other fields while steeper fields were reseeded and converted back to grass. The benefits of these conservation practices were readily apparent.

Fangmeier's visits to irrigated areas brought scenes of erosion on steep slopes and of excess water running along or across roads. His observations of poor practices challenged him to find better methods or management systems. Periods of inadequate rainfall led to his dreams of irrigation and well-watered crops growing lushly on parched fields. He was also concerned about the effects on the environment and how it could be protected. These observations and the satisfaction of working with soils and plants had an impact on his choosing soil and water management as a career.

While Dr. Schwab was growing up in Kansas, he remembers the early 1930s which experienced severe droughts, hot summer temperatures, and dust storms which blew in from western Kansas. He recalls looking out the window and wondering what the future would bring. In an alfalfa field, cracks in the soil were more than one foot wide and as deep as one could see. At the other extreme in another year, heavy rainfall occurred as the wheat was beginning to ripen. The entire crop could not be harvested as the soil was too wet. Along a straightened drainage ditch were many tile outlets which drained the "filled in" old meandering channel. Because the tile did not have a pipe to stabilize the outlet end, these outlets had to be cleaned out each year. He was delegated to perform this messy job, but resolved that there should be a better way. On more rolling land, rills and small gullies formed each time a heavy rain occurred. At the head of a drainage ditch, runoff caused a gully to develop. After a concrete wall failed, brush, trash, and other debris were thrown into the gully, but these also failed to provide adequate control. These observations and experiences had an impact on his decision to work in the field of soil and water management.

The authors were motivated to choose teaching and research careers in soil and water management systems by many of the above described experiences. The opportunity to conserve and manage soil and water has been rewarding because a natural resource will be conserved and crop and fiber production will be enhanced for an ever increasing population.

The principles discussed in this text have broad application within the United States to both rural and urban areas, for both agricultural and nonagricultural land. The same principles have been applied internationally by the authors and many other natural resource managers in developed and developing countries. We hope this text will inspire the reader to appreciate the challenges and responsibilities associated with managing soil and water systems, effectively and efficiently.

CHAPTER 1

CONSERVATION MANAGEMENT AND THE ENVIRONMENT

Soil and water conservation practices are here considered those measures that provide for the management of water and soil in such a way as to ensure the most effective use of each, while preserving or improving the environment. These conservation practices involve the soil, the plant, and the climate, each of which is of utmost importance. The soil phase of conservation requires an inventory of the soil, quantitative measurements of its physical characteristics, and information on soil response to various treatments. The plant phase brings to light (1) questions of adequate, but not excessive, water for plant growth and (2) optimum utilization of the plant as a means of preventing erosion and increasing the rate of movement of water into and through the soil. The plant is also greatly influenced by its biotic environment, which includes all living things and the interrelated actions and reactions that individual organisms directly or indirectly impose on each other. The climate phase is an overwhelming aspect, involving water that can be partly controlled by humans, with appropriate drainage and irrigation practices. Temperature, wind, humidity, chemical constituents in the air, and solar radiation are factors over which humans have little or no control.

This text will emphasize management systems for controlling soil erosion by water and wind, for providing adequate supplies of good quality water for agricultural uses, for removing excess water from the surface and the soil profile, and for supplying water for the irrigation of crops as well as systems that protect and enhance the environment. The objective is the conservation of natural resources while (1) maintaining or increasing food and fiber production for the benefit of society and (2) improving the environment.

1.1   Population, Food and Fiber, Energy, and Pollution Problems

These problems are serious in most countries in varying degrees. Compared to developed countries (Japan, Germany, Great Britain, Australia, United States, etc.), developing countries of the world (Nigeria, India, Pakistan, Mexico, Ecuador, etc.) have about twice the population growth rate, a much lower economic growth rate per capita, and a much higher need for an increase in food and fiber production. These problems are considered *worldwide* because of humanitarian concerns, national security, long-range economic stability, energy supplies, and deterioration of the atmospheric environment. According to the United Nations (1988), the 1984 world population of 4.7 billion is expected to be more than 6 billion by the year 2000. Presently, no critical worldwide food shortage exists, but critical nutritional problems arise from uneven distribution of food among countries, within countries, and among families with different levels of income.

One approach to solving the energy problem is to stabilize or reverse population growth. It is seldom treated as a part of energy policy because the effective means required would offend tradition and religious beliefs. History and social practices favored having many children. This concept is now inappropriate (Davis, 1981). Energy policy has been to produce or save energy for as many people as necessary, not to have fewer people receiving as much energy as needed. Reversing or stopping population growth could play a major role in solving the energy problem.

Thousands of newborns come into a world that can offer them nothing but hunger, disease, and squalor that worsen as the numbers increase. The social costs of overpopulation include overcrowding, unemployment, pollution, energy crises, crime, and so forth. A few private organizations and countries have endeavored to educate the public of the need to control overpopulation, but the results have had little impact.

After government planning policies were adopted in 9 developing countries, birthrates dropped from 10 to 25 percent, indicating that some progress has been made (IIED, 1987). Although the growth rate in the United States has stabilized, about one third of its population growth is due directly to immigration (on the order of one million immigrants annually).

1.2   Conservation Ethics

Increasing world population will dictate the necessity of conserving natural resources now and in the future. Fossil fuels, soils, minerals, timber, and many other materials are being exhausted at a rapid rate. In a lifetime, the average 70-year-old American will use 1 million times his or her weight in water, 10 000 times in fossil fuels and construction materials, and 3000 times in metal, wood, and other manufactured products (Wolman, 1990). Recycling of paper, glass, metals, and so on is being practiced, partly because of the increasing costs of waste disposal and partly because of public support and appreciation for conservation. The decreasing population of wildlife and the disappearance of many species are evidence that much of the problem is related to air and water pollution as well as loss of habitat. In agriculture, soil erosion from farmland is not only one of the major causes of water pollution, but the loss of the soil itself reduces the capacity for production of food and fiber.

Historical evidence shows the rise and fall of many civilizations, particularly in the Middle East and China, have been attributed to soil erosion (Lowdermilk, 1953). After an extensive world tour, Lowdermilk gave a Jerusalem radio broadcast in which he suggested an Eleventh Commandment, "Thou shalt inherit the Holy Earth as a faithful steward, conserving its resources and productivity from generation to generation." Exploitation of the soil and other natural resources for economic benefit has been practiced since the early pioneers developed this country. Future food needs are disregarded as the conversion of prime farmland to urban development and other nonfarm uses continues, because farmland cannot compete with other more profitable uses. Political solutions to these problems are only possible with the promotion and teaching of appropriate conservation ethics.

1.3 Land Use and Crop Production

In the United States, cropland, pastureland, and forests have been diverted to nonagricultural uses or for development. Land for urban uses, roads, and surface-water storage has increased about 69 percent from 1920 to 1987 (USDA, 1989). Land in agriculture may change from one use to another over time, but urban land is not likely to change back to farmland.

The distribution of land use in the 50 states is shown in Fig. 1.1. Private or nonfederal land represents about two thirds of the total. Of the 334 million acres of prime farmland in 1992 (excluding Alaska), 65 percent was cropland and 35 percent was pastureland and forest (SCS, 1994). Prime farmland is our best land for producing high crop yields with minimal hazards or loss by erosion. Prime farmland loss per day by regions is shown in Fig. 1.2 for an 8-year period. The average loss in the United States was 2740

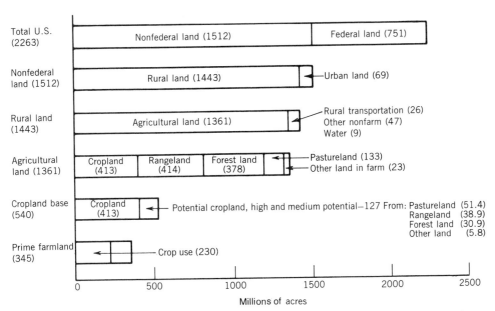

FIGURE 1.1 Distribution of agricultural land in the 50 states. Source: U.S. Department of Agriculture and the President's Council on Environmental Quality (1981).

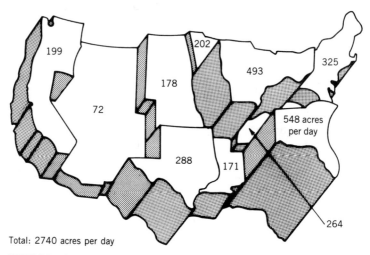

FIGURE 1.2   Average loss of prime farmland in acres per day to urban and water uses by regions for the period 1967–1975. Source: Sampson (1978).

acres per day (about 2 acres per minute), although some later studies showed lower losses. The leading cause of prime farmland loss is poor planning for urban and industrial development. Purchase of development rights is one of the few alternatives for saving prime farmland, but it is expensive. A few private and government agencies are concerned with land preservation, but little is being done.

Land-use changes on rural nonfederal land from 1982 to 1992 are shown in Table 1.1. Acreages were obtained from the National Resources Inventory

TABLE 1.1   Land Use on Rural Nonfederal Land in the United States, 1982 and 1992

	1982	1992	Change from 1982
Land Use	*Million Acres[a]*		
Cropland	421	382	−39
Pastureland	132	126	−6
Rangeland	409	399	−10
Forest land	394	395	+1
Other rural land	53	89	+36[b]
Developed land (urban, etc.)	78	92	+14
Total rural nonfederal land	1409	1391	−18

Source: SCS (1994).

[a]Includes Hawaii, Puerto Rico, Virgin Islands, but not Alaska.

[b]Includes 34 million acres in Conservation Reserve Program (CRP).

(NRI) taken in 1992 (SCS, 1994). Such inventories are conducted at 5-year intervals by the U.S. Soil Conservation Service (SCS) to determine the state and national conditions as well as trends of soil, water, and related resources. Cropland, pastureland, and rangeland decreased by 55 million acres from 1982 to 1992, whereas developed land and other rural land increased by 50 million acres. Of the 36-million-acre increase in other rural land in 1992, 34 million acres represent land under contracts in the Conservation Reserve Program (CRP). Developed land includes land adapted for urban, industrial, transportation, and other nonrural uses.

Crop and livestock production have been gradually increasing in the United States for more than three decades, even though cropland acreage has been nearly constant or decreasing slightly (see Fig. 1.3). Continued increases in yields may be threatened by (1) soil erosion, (2) air and water pollution regulatory constraints, (3) the high cost of fertilizers, water, fuel, and other resource inputs, and (4) the necessity to farm less-productive cropland. In recent years, educational programs have emphasized sustainable agriculture programs as a way to reduce costs and pollution. Increasingly expensive and uncertain energy supplies will make it more difficult to secure high crop yields.

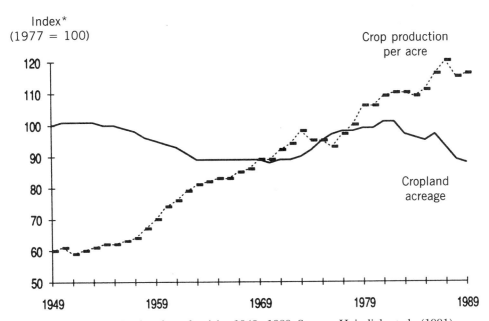

Cropland Productivity, 1949–1989

FIGURE 1.3   U.S. cropland and productivity, 1949–1989. Source: Heimlich et al., (1991).

1.4   Soil and Water Conservation Needs

A 50-state Conservation Needs Inventory (CNI) made in 1967 showed that nearly two thirds of the nonfederal rural land was in need of some kind of conservation practice (SCS, 1969). This inventory showed that the erosion hazard was by far the most serious problem, affecting 49 percent of the land area. About 24 percent of the land had unfavorable soil. Included in this classification are soils with shallow topsoils, stoniness near the surface, low water-holding capacity, high salinity, and high sodium. Land having an excess water problem represents about 19 percent of the total. Poor internal soil drainage, high water table, and overflow from streams are the major causes of wetness. Major concentrations of drainage problem areas occur in the Lake States, Corn Belt, Delta States, Southeast, and in arid irrigated areas in the West.

From the 1992 NRI, land needing conservation treatment is shown in Table 1.2. Land needing treatment in 1982 decreased from 40 percent of the total nonfederal rural land to 31 percent in 1992 (a decrease of 127 million acres). This decrease shows considerable improvement in the adoption of conservation practices, even though 34 of the 127 million acres can be attributed to the CRP (10-year set-aside program).

1.5   Conservation and the Environment

Environmental quality is a relative term measured by some established performance standard. Air, land, and water have a capacity to cleanse or regenerate themselves, biologically or mechanically. Management practices can be beneficial or detrimental to the environment. Erosion is one of the most serious problems, because water is polluted with sediment, plant nutrients, and pesticides, thus increasing environmental damage and water treatment costs. Terracing cultivated land can be beneficial by reducing erosion and sedimentation. Irrigation can be detrimental by increasing drainage water salinity. Minimum tillage,

TABLE 1.2   Rural Nonfederal Land Needing Conservation Treatment in the United States, 1982 and 1992

Land Use	1982	1992	Change from 1982
		Million Acres[a]	
Cropland	270	204	−66
Pastureland	69	58	−11
Forest land	212	158	−54
Other rural land	11	15	+4
Total	563	436	−127
Percent of total rural nonfederal land	40%	31%	

Source: SCS (1994).

[a]Includes Hawaii, Puerto Rico, Virgin Islands, but not Alaska.

which provides greater ground cover but reduces erosion, may increase pesticide requirements. As population increases, greater agricultural production of food and fiber is essential, but increasing attention is being focused on preserving or enhancing the environment. When drinking water becomes contaminated with nitrates, pesticides, sediment, and other materials, the public is—and should be—concerned. The reduction in wildlife and disappearance of some endangered species should confirm any suspicions that the environment is being seriously damaged. More intensive land use can reduce the quantity and quality of the habitat for some wildlife.

Environmental costs are paid either by taxes, when supported by governments, or by higher consumer prices, when supported by the private sector. Many conservation practices can improve the environment with little or no cost to agriculture. An example is sustainable agricultural programs, which promote reduced fertilizer and pesticide use, minimum tillage practices, and other cost-reduction measures. Opponents would argue that increasing fertilizers would increase crop production per acre, which would reduce land requirements and permit taking erodible land out of cultivation. Much depends on who pays the cost.

In general, the impact of drainage on the environment has been beneficial. The drainage of the lake-bed area in the Midwest was instigated by medical doctors, who realized long before the mosquito was the known cause that drainage reduced malaria. Without drainage, much land could not have been occupied by humans. During pioneer days, agricultural interests prevailed and the drainage of swamps and wetlands was done without question. In more recent years, drainage of these naturally wet areas was found to have some adverse effects on migrating wildlife, aquatic life, and other environmental aspects. As much as 50 percent of the wetlands have been drained, mostly in the Southeast, Midwest, and the Pacific Coast states (USDA, 1989). Wetlands vary from permafrost in Alaska to the Everglades in Florida to the desert wetlands in Arizona. In addition to providing a habitat for fish and wildlife, wetlands enhance ground-water recharge, reduce flooding, trap sediment and nutrients, and provide recreational areas.

Irrigation projects, as in the West, have a major impact on the social and political structure of the entire community. Water storage and stream diversion facilities for irrigation, as well as stream flow and ground-water use, influence minimum stream flow, depth of ground water, fish and wildlife populations, natural vegetation and crops grown, recreation activities, roads and public utilities, and other unique and site-specific factors. One example of a serious environmental problem is the chemical pollution (selenium and boron) from irrigation drainage water, such as occurred at the Kesterson National Wildlife Refuge in California. Blockage of the drainage system was mandated by the court. Because of such problems, an environmental impact statement is often mandatory for most conservation projects. It should include an inventory and proposed changes of air, soil, and water quality, as well as all wildlife, people, jobs, transportation, and endangered plant and animal life.

1.6 Economics of Conservation

Soil erosion by wind and water is often considered the most critical conservation problem. In the United States, limiting erosion to tolerable rates would reduce the soil entering our streams by a billion tons and prevent the loss of $5 billion worth of plant nutrients annually (Troeh et al., 1991). Impairment of water quality is one of the most serious off-farm consequences of erosion. Long-term effects of erosion result in destruction and eventual abandonment of the land as well as in the loss of civilization itself (Lowdermilk, 1953).

Failure to drain, irrigate, or conserve soil water usually reduces crop production. Flooding is a naturally occurring problem, which would take place regardless of human activities. Flooding may cause damage to the land, but in some cases soil deposition may be a benefit. Damage to property and structures can often be reduced by proper construction or by avoiding occupation of the floodplain. Government-subsidized flood insurance is available for preapproved areas, but it is not widely used.

Conservation of natural resources implies sustainable utilization without waste, to make possible continuous high crop production while improving the environment. A manager must decide whether to make expenditures for nonproduction goods, machinery, and buildings or for conservation practices, such as drainage, irrigation, or terraces. The financial status of the enterprise, benefit/cost ratios, legal requirements, aesthetic appeal, and conservation awareness of the managers (moral considerations) will largely determine the degree to which they will make specific improvements and land-use adjustments. When land-lease arrangements do not adjust for the length of time between the adoption of conservation practices and the realization of returns, or where off-site or long-term benefits occur, the landowner must pay most of the costs. Owners and operators often share the costs that produce short-term benefits, such as fertilization. Tenancy with absentee owners is generally not conducive to conservation. Governments and other legal entities can encourage conservation practices by making direct payment; providing technical assistance, low-interest loans, tax credits, and other benefits; or by taxing soil loss and passing regulatory laws.

U.S. Department of Agriculture (USDA) programs provide cost sharing as well as technical assistance to land users for protecting natural resources as public benefits. Cost-share payments by the federal government range up to 80 to 100 percent of the cost. Failure to comply may be subject to cross-compliance requirements with the denial of payments to the farmer for other federal programs. Some programs levy fines for poor environmental practices.

Long-term net farm income and intangible benefits to society and the environment are the best bases for evaluating conservation practices. Conservation farms generally require a greater expenditure for conservation improvements, have more land in legumes and grasses, produce higher crop yields, grow better-quality crops, and often have higher livestock production. Conservation costs usually are higher on farms having a high level of conservation than on those having fewer practices. Conservation practices normally increase the net income within 1 to 4 years after establishment. Adoption of erosion-control practices may not increase the net income, except for such practices as minimum tillage, which will reduce energy costs. Strip cropping and terracing costs may outweigh the short-term benefits.

Drainage of wet soils in a humid area is normally a good investment. In the Midwest, the benefit/cost ratio for subsurface drainage is usually greater than 1.5. Because of lower cost and higher economic return, surface drainage generally gives a greater benefit than subsurface drainage.

The economics of irrigation are generally favorable, depending on the reliability, seasonal distribution, and amount of precipitation. In the arid West, where rainfall is less than 15 in./year, irrigation is required for economic crop production. In humid areas where rainfall exceeds 30 in./year, irrigation is usually economical for vegetable and high-value crops, but questionable for field crops, except on low water-holding capacity soils. Where rainfall is from 15 to 30 in./year, irrigation is desirable and usually economical, but benefits and costs must be carefully considered. Dryland farming is often an alternative, where water is not available or is costly. Irrigated acreage in the world since 1980 has nearly stabilized because of low economic returns, limited water supplies, and rising water tables (which cause high salinity and toxicity).

1.7 Conservation Planning

Conservation planning is primarily applicable to private land in farms, cities, and the suburbs, as well as to federal land managed by the Forest Service, Bureau of Land Management, and state and national park agencies. Planning is greatly influenced by land-use zoning, especially in or near densely populated areas. Since the 1930s, farm planning has been done by the Soil and Water Conservation Districts (county size). Technical assistance has been provided by Soil Conservation Service [name changed to Natural Resources Conservation Service (NRCS) in 1994] personnel (federal employees) for cooperating farmers. The following general procedures describe a soil and water conservation farm plan.

BASIC DATA   Data required to develop a conservation farm plan include soil characteristics, topographical data, climatic information, and present farm practices, most of which can be obtained in the field or by consultation with the farmer. The conservation survey should include soil series and type, degree of erosion, length and degree of slope, and present field boundaries, which may be recorded on a base map (see Fig. 1.4). Aerial photographs make a satisfactory base map. Orthophotographic maps (distortion-free aerials) from a Geographic Information System (GIS) are available for direct access by the farmer on his own computer (see Chapter 4).

The degree of erosion is designated by the present depth of topsoil as compared to the depth of virgin topsoil. On the base map, the degree of erosion is shown by a numeral, such as 1 for 0 to 25 percent of the topsoil removed, and

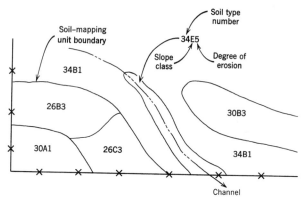

FIGURE 1.4   Soil-mapping units and method of recording soil type, erosion, and land slope for a farm plan.

so on. Erosion on cultivated land is estimated from rainfall, soils, the cropping history, the length and degree of slope, and past conservation practices (such as contouring and terracing).

The slope class is represented by a capital letter indicating a range of slope (i.e. A, nearly level; B, gently sloping, etc.). An area may have the same soil type and degree of erosion, but would be designated separately because of different slope (see soil 26 in Fig. 1.4).

LAND-CAPABILITY CLASSIFICATION   Basic data (as given previously) are necessary to classify the land according to its capability for crop production. As shown in Table 1.3, each of the eight land-capability classes is determined by the general degree of limitations and hazards. The first four classes are suitable for cultivation, but the last four are not suitable. Except for class I land, each capability class may be designated with a subclass, such as II(e), and so on.

In determining the land-capability class, the following are considered: soil type; slope; degree of erosion; extent of drainage; presence of rocks, stones, and other impediments to cultivation; present productivity of the soil; water-holding capacity; and the amount and distribution of rainfall. The land-capability class can be changed to another level (i.e., from class I to class V, if cropland is converted to timber). Several classes of land may be found on the same farm. In Fig. 1.5 all classes are shown.

THE SOIL AND WATER CONSERVATION FARM PLAN   A farm plan is the organized and systematic schedule of all soil and water management practices integrated with a well-balanced farming program. The procedure for developing a farm plan consists of (1) the collection of basic data, (2) the land-capability class determination and field layout, and (3) the farmer's selection of a soil and water management program. Much of the information necessary for a conservation farm plan is shown in Fig. 1.6. On the left is the farm before planning, and in the center map the land is classified according to its capability class (Roman numerals). These

TABLE 1.3 Land-Capability Classification

Land-Capability Class	General Description and Limitations
Suited for Cultivation	
I	Few limitations that restrict its use. No subclasses.
II	Some limitations that reduce the choice of plants or require moderate conservation practices.
III	Severe limitations that reduce the choice of plants or require special conservation practices or both.
IV	Very severe limitations that restrict the choice of plants or require very careful management or both.
Not Suited for Cultivation (except by costly reclamation)	
V	Little or no erosion hazard, but has other limitations (impractical to remove) that restrict its use largely to pastureland, rangeland, woodland, or wildlife food and cover.
VI	Severe limitations that make it generally unsuited to cultivation and restrict its use largely to pastureland, rangeland, woodland, or wildlife and cover.
VII	Very severe limitations that make it unsuited to cultivation and that restrict its use largely to grazing, woodland, or wildlife.
VIII	Limitations that preclude its use for commercial plant production and restrict its use to recreation, wildlife, or water supply or to aesthetic purposes.

Source: Klingebiel and Montgomery (1966).

Note. Except for class I land, the following subclasses are recognized in which the dominant limitations for agricultural use are the result of soil or climate: *(e) erosion,* based on susceptibility to erosion or past damage; *(w) excess water,* based on poor soil drainage, wetness, high-water table, or overflow; *(s) soil limitations within the rooting zone,* based on shallowness, stones, low water-holding capacity, low fertility, salinity, or sodium; and *(c) climate,* based on temperature extremes or lack of water.

land classes are often indicated by color. The final map of the conservation plan as shown on the right in Fig. 1.6 may include field designation, new field boundaries, type of crop and acreage, and supporting conservation practices, such as grassed waterways and terraces. This map shows the over-all picture of the physical changes and the principal features of the farm. If necessary, several closely associated classes of land may be included in one field to obtain a desirable size for efficient operation. For example, field 9 includes both class II and class III land.

Soil loss may be determined by procedures described in Chapter 6, which can help in selecting appropriate conservation practices, such as crop rotations, fertility levels, and erosion-control measures. These practices should reduce soil loss to permissible levels.

The conservation farm plan should include schedules for the application of conservation practices in each field, the planned production for each crop, feed requirements for planned livestock, and the distribution of pasture-carrying capacity for each month of the growing season. Once the plan has been developed, it serves as a ready reference for all farm operations.

LAND CAPABILITY CLASSES			
SUITABLE FOR CULTIVATION		NO CULTIVATION – PASTURE, HAY, WOODLAND, AND WILDLIFE	
I	Requires good soil management practices only	V	No restrictions in use
II	Moderate conservation practices necessary	VI	Moderate restrictions in use
III	Intensive conservation practices necessary	VII	Severe restrictions in use
IV	Perennial vegetation – infrequent cultivation	VIII	Best suited for wildlife and recreation

FIGURE 1.5   Land-capability classes. (Courtesy U.S. Soil Conservation Service.)

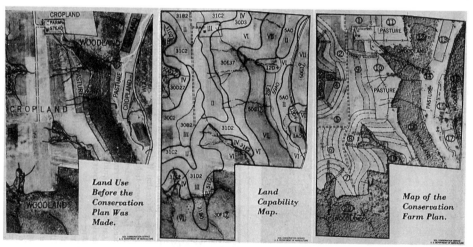

FIGURE 1.6   Development of a soil and water conservation farm plan. (Courtesy U.S. Soil Conservation Service.)

1.8   Watershed and Regional Planning

Over-all systems for flood control, erosion control, irrigation, water supply, and drainage can be accomplished more advantageously by considering watershed or catchment areas rather than political subdivisions, such as a farm or township. Watershed planning is necessary for erosion and flood control. More emphasis has been given to the watershed approach in recent years, especially by federal agencies.

Enabling legislation for the planning and construction of conservation projects that involve several adjacent landowners has been passed and is the law in almost all the states. Where such cooperative projects are undertaken, individuals benefited should and generally do contribute their equitable share of the cost. Because many projects have public benefits, local, state, and federal government funds are often provided.

1.9   Soil and Water Conservation Districts

As a condition to receiving benefits under the Soil Conservation and Domestic Allotment Act, passed by Congress in 1935, the states were required to enact suitable laws providing for the establishment of soil and water conservation districts. The first district was organized in 1937, and all states have now enacted district laws.

State legislation has been patterned largely after the standard state soil conservation districts law prepared by the U.S. Soil Conservation Service (1936). After such districts are established in local communities (usually county size), they may request technical assistance from the Natural Resource Conservation Service, State Cooperative Extension Service, and county officials for carrying out erosion control and other land-use management programs.

Soil and water conservation districts differ from drainage, irrigation, and conservancy districts. Their boundaries are usually the same as a county; operating funds are obtained from state or local appropriations; and in most states, they cannot levy taxes or issue bonds. Election of officials is by all district voters rather than limited to only landowners. These districts have little legal authority and tend to be voluntary in nature. The development of soil and water conservation farm plans, checking cross-compliance for federal payment of conservation measures, and technical assistance in adopting practices have been the principal activities of the district's program.

1.10   Federal and State Legislation

Since the early 1970s, many state and federal statutes and regulations have been created to protect the environment and to promote soil and water conservation (i.e., by controlling erosion, by reducing water pollution, and by protecting wetlands). Governments may provide technical assistance, allow tax credits or other benefits, provide low-interest loans, or establish legal requirements. Most laws have encouraged voluntary practices rather than strict mandatory compliance. Government cost sharing and other federal expenditures have gradually increased since about 1970, partly because of political considerations. Only a few of the many laws will be discussed.

The Federal Water Pollution Control Act (Section 208) 1972, as amended encourages the adoption of best management practices for the reduction of sediment in storm runoff. Also known as the Clean Water Act, it stipulated that all rivers should be clean enough for fishing, swimming, and other recreational uses. It also calls for the elimination of unlawful discharges of pollution into streams, wetlands, and lakes. During the first 15 years after passage, the most effort was made to control point-source pollution (industrial and domestic wastewater). The act as amended in 1987 (Section 319) directed each state to develop and implement programs for the control of nonpoint-source pollution. Nonpoint sources primarily include runoff from agricultural land, forests, and rangeland, which contain sediment and a wide range of chemicals and other pollutants that move with the eroded sediment and the water.

Section 404 of the Clean Water Act stipulates that a permit must be obtained to discharge material into, or fill a wetland. Several government agencies are involved in the permit process. Limited funds have been provided to compensate farmers for restoring and protecting wetlands. Without such funds, many would argue that the taking of wetlands deprives the landowner of his property without due process of law. Much misunderstanding and controversy continues on the definition and use of wetlands (see Chapter 12).

The Food Security Act of 1985, as amended in 1990, requires (1) that highly erodible lands be identified and (2) that farmers using such lands should have conservation plans if they are to receive benefits (Glaser, 1986). The act, also called "the farm bill," requires that significant soil erosion be reduced (referred to as the "conservation compliance" and the "sodbuster" provisions). The act also establishes a CRP for highly erodible land. The "swampbuster" provisions of the act stipulate that some wetlands should not be converted to cropland or pasture. Violations for not actively applying approved practices on highly erodible land or for converting wetland may result in the loss of certain USDA financial benefits. Most states are preparing wetland inventories (GIS maps, see Chapter 4) for purposes of making off-site wetland determinations. The Natural Resource Conservation Service may make site-specific wetland determinations for individual farmers.

A considerable number of state and federal agencies are involved in administering laws relating to soil and water conservation. These agencies are described in detail by Troeh et al. (1991). Local, state, and federal laws are subject to revision by the political process. Resource managers should check with local agencies regarding current regulations pertaining to their activities.

REFERENCES

Davis, K. (1981) "It Is People Who Use Energy" (editorial), *Science*, 211(4481).
Glaser, L. K. (1986) *Provisions of the Food Security Act of 1985*, USDA-ERS, Agr. Info. Bull. 498. U.S. Government Printing Office, Washington, D.C.

Heimlich, R. E., M. Vesterby, and K. S. Krupa (1991) *Urbanizing Farmland: Dynamics of Land Use Change in Fast-Growth Counties,* USDA-ERS, Agr. Info. Bull. 629. U.S. Government Printing Office, Washington, D.C.

International Institute for Environment and Development and the World Resources Institute (IIED) (1987) *Population, Food, and Agriculture.* Basic Books, New York.

Klingebiel, A. A., and P. H. Montgomery (1966) *Land-Capability Classification,* SCS, Agr. Handb. 210. U.S. Government Printing Office, Washington, D.C.

Lowdermilk, W. C. (1953) *Conquest of the Land through Seven Thousand Years,* USDA, SCS, Agr. Info. Bull. 99. U.S. Government Printing Office, Washington, D.C.

Sampson, N. (1978) *Preservation of Prime Agricultural Land, Environmental Comment.* Urban Land Institute, Washington, D.C.

Troeh, F. R., J. A. Hobbs, and R. L. Donahue (1991) *Soil and Water Conservation,* 2nd ed. Prentice–Hall, Englewood Cliffs, New Jersey.

United Nations, Food and Agriculture Organization (FAO) (1988) *FAO Production Yearbook, FAO Statistics Series 70,* FAO, Rome, Italy.

U.S. Department of Agriculture and the President's Council on Environmental Quality (1981) *Final Report, National Lands Study.* U.S. Government Printing Office, Washington, D.C.

U.S. Department of Agriculture (USDA) (1989) *The Second RCA Appraisal. Soil, Water, and Related Resources on Nonfederal Land in the United States.* U.S. Government Printing Office, Washington, D.C.

U.S. Soil Conservation Service (SCS) (1936) *A Standard State Soil Conservation Districts Law.* U.S. Government Printing Office, Washington, D.C.

U.S. Soil Conservation Service (SCS) (1969) "Soil and Water Conservation Needs Inventory," *Soil Conservation,* 35(5):99–109.

U.S. Soil Conservation Service (SCS) (1994) *Summary Report 1992 National Resources Inventory (NRI).* Washington, D.C.

Wolman, M. G. (1990) "The Impact of Man," *EOS, Trans. Am. Geophy. Union,* 71(52):1884–1886.

CHAPTER 2
DISTANCE AND AREA MEASUREMENT

Measurement of field distances and areas, building layouts, locations of drainage, irrigation, and other conservation facilities is often necessary. Simple surveying procedures and equipment described in this chapter can be applied to achieve the needed accuracy. The discussions in this and other sections of the book should not be applied to property boundary surveys or other applications involving legal aspects. For these applications, a licensed professional surveyor should be retained. Surveyors are licensed in each state.

MEASUREMENT OF DISTANCE

2.1 Pacing

Distances may be measured by pacing when the desired accuracy is not greater than 2 ft in 100 ft (1 in 50). Although the length of a pace is approximately 3 ft, each individual should check his pace with an accurately measured distance. A common way of remembering the length of pace is to determine the number of paces per 100 ft. It must be kept in mind that the length of a pace may vary when walking through short or tall vegetation, uphill or downhill, on wet or dry ground, or on plowed or firm soil.

2.2 Measuring Equipment

The 100-ft steel tape is commonly used for measuring distances in the field. Fiber glass or metallic woven tapes are also suitable, but they are not as durable as the steel tape. Most tapes 50 to 100 ft in length can be rolled up in a pocket-size case. For ease of reading, some tapes are graduated in tenths and hundredths of a foot (or meter) for their entire length. For most agricultural purposes, a 100-ft steel tape, $\frac{5}{16}$ in. wide, with markings for each foot, and with the last foot divided into tenths, is quite satisfactory. Other equipment needed for measuring distances includes a set of taping pins, range poles, and a plumb bob. A set of taping pins consists of 11 pins with a suitable carrying ring (Fig. 2.1). Each pin is about 12 in. in length with a loop on one end. Taping pins may

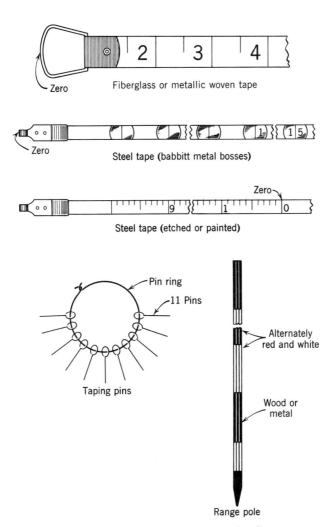

FIGURE 2.1 Taping and sighting equipment for distance measurement.

be purchased or made from heavy gage wire. Range poles are about 8 ft in length with a steel point on one end and painted alternately red and white. They may be seen from a considerable distance. Four-foot lath stakes may serve as good substitutes when the vegetation is small and the distances are short.

An odometer is a simple device that measures distance by registering the number of revolutions of a wheel. It is convenient to make the wheel circumference 10 ft and to divide the circumference into 10 equal parts. Such a device, as shown in Fig. 2.2, may be pushed by hand or attached to a vehicle. On smooth ground the accuracy may be within 1 percent, but on rough ground or in rank vegetation measurements are less accurate. The distance indicated by the odometer or by taping on sloping land is always greater than the true horizontal distance (see Table 2.1).

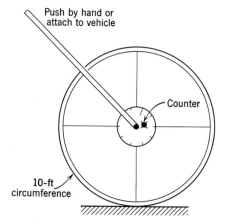

FIGURE 2.2   Odometer for one-person distance measurement.

2.3 Taping

Taping (chaining) refers to the operation of measuring the distance between two points. Land surveys, building layouts, and most other applications require horizontal distances. On sloping land, the horizontal measurements can be obtained with a plumb bob, so the tape can be held level (Fig. 2.3), or by taping the slope distance, taking the slope angle, and converting to a horizontal distance. On land that has a slope less than 2 percent (2 ft in 100 ft) slope distance may be sufficiently accurate (error less than 0.02 percent, see Table 2.1). If the slope is not more than 5 percent, the plumb bob may be used with the full 100-ft length of the tape. On slopes greater than 5 percent, it is necessary to "break tape" (Fig. 2.3). The extra pins required when breaking tape for distances less than the length of the tape should be returned to the head tapeperson so as not to cause the pin count to be in error (see Section 2.5).

TABLE 2.1   Conversion of Slope Distance to True Horizontal Distance[a]

Slope (percent or ft per 100 ft)	Error in Slope Distance (ft per 100 ft)	True Horizontal Distance (ft)
1	0.005	99.995
2	0.020	99.980
3	0.045	99.955
4	0.080	99.920
5	0.125	99.875
10	0.50	99.50
15	1.13	98.87
20	2.02	97.98
30	4.61	95.39

[a]Horizontal distance $= [100^2 - (\text{percent slope})^2]^{1/2}$

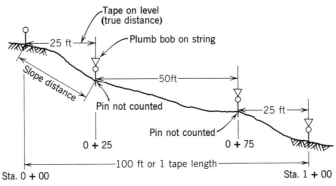

FIGURE 2.3   Measurement of sloping land by the *breaking tape* technique.

When a series of points are set along a prescribed line, a standard system of numbering helps to distinguish these distances from other horizontal or vertical measurements. The zero point is designated $0 + 00$ and each 100-ft point is called a station, which is the number before the plus sign (see Fig. 2.3). For example, $4 + 00$ is station 4. A point 550 ft from the starting point is written $5 + 50$, a point at 646.2 ft is $6 + 46.2$, and so forth. The $+50$ and $+46.2$ are called pluses.

Example 2.1.   On a uniform slope of 5 percent (5 ft/100 ft), the slope distance was measured to be 420.20 ft. Determine the true horizontal distance between the two points.

Solution.   From Table 2.1 read an error of 0.125 ft/100 ft. The total error to be subtracted from the slope distance = $(420.20 \times 0.125/100) = 0.53$ ft. The true horizontal distance = $420.20 - 0.53 = 419.67$ (127.92 m). An alternate procedure assuming a right triangle with a slope length of 100 ft and a 5-ft rise gives the

$$\text{True horizontal distance}/100 \text{ ft} = (100^2 - 5^2)^{1/2} = 99.875 \text{ ft}/100 \text{ ft}$$

$$\text{Total distance} = 99.875/100 \times 420.20 = 419.67 \text{ ft } (127.92 \text{ m})$$

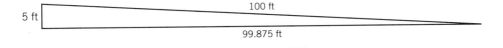

2.4   Handling Measuring Equipment

The following precautions should be observed when handling measuring equipment.

    1. When unrolling the tape, be careful not to leave kinks, as stretching the tape may easily break it.

2. If a tape becomes wet, it should be wiped dry with a cloth before storing on a reel or in a case. Some tapes will rust and should be oiled.

3. Taping pins should not be left in the ground or lying around loose, but should be placed on the ring. A small strip of cloth tied to the ring end of each pin will make them more noticeable.

2.5   Taping Procedure

The proper procedure for taping (chaining) over nearly level ground between two points follows.

1. The head tapeperson with the zero end of the tape unrolls it while walking toward the distant point. The rear tapeperson holds the 100-ft end of the tape at the starting point.

2. For the first 100-ft measurement, the head tapeperson should have 10 pins and the rear tapeperson 1 pin, regardless of whether the 1 pin is needed as the starting point.

3. The rear tapeperson should align the head tapeperson by sighting on the distant point. The tape should be straight and stretched with a tension of about 15 lb.

4. When the rear tapeperson has the 100-ft mark exactly on the starting point (or on a pin in the ground), he/she calls *"stick."* The head tapeperson places a pin exactly at the "0" mark on the tape and calls *"stuck."*

5. The rear tapeperson picks up the pin and walks forward, while the head tapeperson pulls the tape to the next point. When the end of the tape is about 5 ft from the pin in the ground, the rear tapeperson calls *"chain"* or other verbal command to signal the head tapeman to stop.

6. The procedure in 3, 4, and 5 is repeated until the head tapeperson has no more pins (or the distant point is reached). The rear tapeperson hands the head tapeperson the 10 pins. Both tapepersons should count the pins. The distance to the eleventh pin, which is in the ground, is 10 tape lengths or 1000 ft. The number of pins held by the rear person is always the distance (in hundreds of feet) from the starting point to the pin in the ground.

7. Taping is continued until the distant point is reached. Care should be taken in measuring the distance from the last 100-ft pin to the final point, especially with tapes that have subdivisions of a foot only at the end foot.

2.6   Units of Measurement

In surveying, tenths and hundredths of a foot are preferred rather than inches. Since there are 12 in. or 10 tenths in a foot, it is quite confusing to convert from one system to the other. The two scales are shown in Fig. 2.4 which may be used

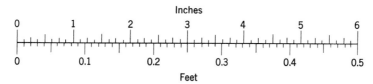

FIGURE 2.4 Conversion scale for inches to tenths of a foot and vice versa.

to convert tenths to the nearest $\frac{1}{8}$ in. and vice versa. To convert values above 6 in., add 0.5 ft or 6 in., for example 0.75 (0.5 plus 0.25) converts to 9 in. (6 plus 3).

2.7 Common Errors in Measurement

Some of the more common errors in measuring distance with a tape follow.

1. Tape not pulled tight enough.
2. Tape not in proper alignment.
3. Pins not carefully placed at proper marks on tape.
4. Mistake in counting pins or lost pins.
5. Mistake in determining the number of feet less than 100. This happens when measuring the fractional tape length at the end of the line.
6. Plumb bob not used when measuring on a slope.
7. Wrong points on the tape used for the zero or the 100-ft mark.
8. Reading or recording the wrong numbers.

2.8 Stadia

To measure distance by stadia, a surveying instrument must be equipped with stadia hairs. In addition to the leveling cross hair, such an instrument (Fig. 2.5) has two additional cross hairs called *stadia hairs*. The stadia cross hairs are so placed that when the level rod is 100 ft from the instrument, the interval between the stadia hairs as read on the level rod is 1 ft. (This conversion factor is correct only when the correction constant in the instrument is zero and the telescope is level.) The accuracy of stadia measurement depends on the accuracy of the instrument, the distance from the instrument to the level rod, and the ability of the user. Normally, an accuracy of $\frac{1}{2}$ ft per 100 ft can be obtained with a good instrument. Levels do not ordinarily come equipped with stadia hairs; however, most manufacturers will install them.

2.9 Electronic Distance Measurement

Electronic instruments for distance measurement (EDM) consist of a transmitter-receiver that sends out a beam of light or high-frequency microwaves to a prism (reflector) located at the point to be measured. The prism reflects the

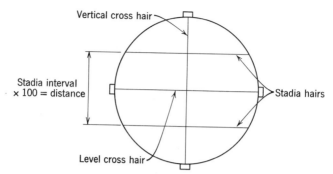

FIGURE 2.5   Stadia distance and leveling cross hairs in a telescope.

light or microwaves back to the instrument where they are analyzed electronically to give a digital readout of the distance between the instrument and the reflector (Fig. 2.6). EDM instruments are portable, highly accurate for long distances, and operate at night or through fog or rain; but they require a clear line of sight. Both EDM and stadia methods are especially suitable for measuring across water, rough terrain, or other land features where taping would be difficult or impossible. EDM instruments are much more accurate than stadia and are available commercially for measuring short-to-long distances. A good EDM instrument has an accuracy of 0.01 ft at 40 miles. They measure slope distance, but some will give a direct readout in horizontal length and vertical difference. For topographic mapping, EDM instruments can be set up next to a transit or can be mounted directly on a standard transit. They can also be combined with data recorders and computer software to produce quickly boundary or topo-

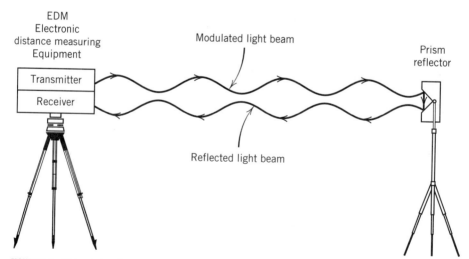

FIGURE 2.6   Principle of operation for electronic distance measuring (EDM) equipment. (Courtesy Keuffel and Esser.)

graphic maps. EDM devices are becoming more affordable for the occasional user, but are standard equipment for professional surveyors.

2.10 Laying Out Right Angles

The steel tape is quite convenient for laying out right angles without the use of a compass or a transit.

THREE-FOUR-FIVE METHOD To lay out a right angle with this method it is necessary to lay out a triangle with sides in the proportion 3, 4, and 5. Convenient lengths are 30, 40, and 50 or 60, 80, and 100 ft. From some base line AB (Fig. 2.7a) and a starting point D, measure the distance DC and swing an arc on the ground at C. Next measure the distance DE and then from E swing another arc near C that will intersect the previous arc at that point and locate the point C. It is always desirable to check the distances to be sure that no mistake has been made. A simple, approximate method is shown in Fig. 2.7b. A person stands at the intersection D, faces in the 90° direction, and moves both arms as shown.

CHORD METHOD If a perpendicular is to be erected through some point A (Fig. 2.8a) within 100 ft of the base line EF, swing an arc from A toward C until it crosses the base line EF, likewise swing the same length arc so that it will intersect the base line again at D. Measure the distance CD along the base line and locate B so that it is midway between C and D. This will establish the point B. The arc length AC must be greater than the distance AB.

When a base line EF is given (Fig. 2.8b) and when the perpendicular line must be constructed through D, measure off two equal distances from D to A and from D to B. As shown in Fig. 2.8b, swing an arc from B to C and then the same length arc from A to C. AC and BC should be greater than the distance AD or DB.

2.11 Laying Out Acute Angles

Angles other than right angles can be laid out with a tape if reference is made to trigonometric functions. These functions exist on most hand calculators, are available in mathematical tables, or are explained in some field books. The tan-

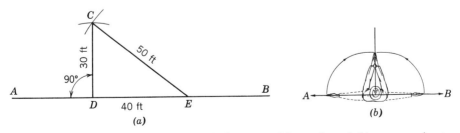

FIGURE 2.7 (a) Three-four-five method of laying out a right angle and (b) an approximate method, the person standing at the intersection D.

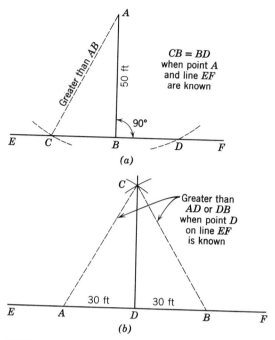

FIGURE 2.8   Chord methods of laying out a right angle *(a)* when a line and point *A* are known and *(b)* when a line and point *D* on a line are known.

gent of an angle is defined as the opposite side of a right triangle divided by the adjacent side and is written as "tan." Two other useful functions are the sine and cosine, written "sin" and "cos," respectively. The sin is the opposite side divided by the hypotenuse. In Fig. 2.9, sin $A = 57.74/AC = 0.5$, thus $AC = 57.74/0.5 = 115.48$. The cos is the adjacent side divided by the hypotenuse. In Fig. 2.9, cos $A = 100/115.48 = 0.866$. For example, suppose a 30° angle is to be established at point *A* in Fig. 2.9. The tangent of 30° is 0.5774. Measuring off *AB* (equal to 100 ft), erecting a perpendicular at *B*, and setting *BC* as 57.74 ft will establish the angle at *A* as the 30° angle with a tangent of 57.74/100. Any lengths of *AB* and *BC*

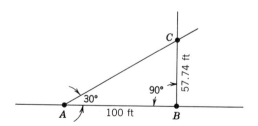

FIGURE 2.9   Laying out an acute angle from point *A* on a line.

may be used so long as BC/AB is 0.5774. Longer lengths will give greater accuracy.

2.12   Extending Straight Lines Through Obstacles

PERPENDICULAR OFFSETS   When an obstacle such as a building lies in a tape line, it is possible to establish an auxiliary line to measure the portion of the taped distance blocked by the building and to permit the extension of the taped line beyond the obstacle, as shown in Fig. 2.10a. This may be accomplished by laying out right angles at A and B and measuring $AA' = BB'$ of sufficient length to establish line $A'B'$ clear of the obstacle. Line $A'B'$ is then extended and points C' and D' are established. Right angles are laid out at C' and D'. Measurement of $CC' = DD' = BB' = AA'$ establishes points C and D, which lie on the extension of the original line AB.

Since distance $B'C'$ equals distance BC, that portion of the taped distance obstructed by the obstacle has been determined. Accuracy is improved if distances AB and CD are 100 ft or more.

EQUILATERAL TRIANGLES   A second method for extending a line through an obstacle is illustrated in Fig. 2.10b. Equilateral triangle AEB is established by swinging arcs with a tape. Line AE is extended to X, and equilateral triangle XYZ is established to

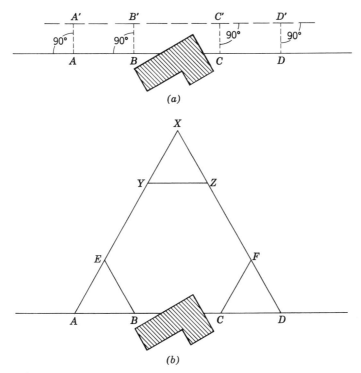

FIGURE 2.10   Extending a straight line through an obstacle (a) by a perpendicular offsets and (b) by equilateral triangles.

locate line *XZ*. Distance *XD* is laid off equal to *XA*. Equilateral triangle *CDF* is then established to define *CD* as the extension of *AB*. *BC* = (*AX* or *XD*) − *AB* − *CD*.

SIGHT METHOD   A straight line can be laid out between two points separated by a ridge (even if the points cannot be seen from the other point) by setting range poles or tall stakes between the points and by aligning them by sight. Starting from one point, set two poles in approximate alignment. Proceed to the first pole and set a third pole in line with the first two. Continue in this manner until the second point can be seen. Reset the poles until they are all in alignment.

MEASUREMENT OF AREAS

On a farm, the area of fields is frequently desired. In general, most irregular-shaped areas may be measured with the steel tape by dividing the field into smaller divisions forming triangles, trapezoids, or rectangles.

2.13 Polygons

The field shown in Fig. 2.11 is an irregular-shaped area subdivided into smaller areas. It is not necessary to measure any of the angles to determine the shape and size of the area. If the lengths of the three sides of a triangle are known, the area may be computed by the formula:

$$A = \sqrt{s(s-a)(s-b)(s-c)} \qquad (2.1)$$

where *a*, *b*, *c* = lengths of the three sides and

$$s = \frac{a+b+c}{2}$$

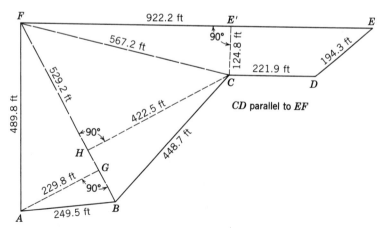

FIGURE 2.11   Measurements to obtain the area of an irregular-shaped field with straight boundaries.

The areas of triangles *ABF* and *BCF* may be found in this manner. Since the formula requires considerable calculations, it may be desirable to measure the perpendicular distances *AG* and *HC* from the line *BF* and thus compute the area of the triangles by the following formula:

$$A = \frac{(\text{base} \times \text{height})}{2} \qquad (2.2)$$

Since the lines *CD* and *EF* are parallel, the figure *CDEF* is a trapezoid. The area may be computed by using the formula:

$$A = \frac{h\,(a+b)}{2} \qquad (2.3)$$

where h = perpendicular distance between the two parallel sides and a, b = lengths of the two parallel sides.

Example 2.2.   Determine the area of the field ABCDEF in Fig. 2.1 in acres.

Solution.   Subdivide the field into two triangles and a trapezoid and compute the areas as follows:

Area			Square Feet
ABF	$\dfrac{FB \times AG}{2}$	$\dfrac{592.2 \times 229.8}{2}$	60805.1
BCF	$\dfrac{FB \times HC}{2}$	$\dfrac{529.2 \times 422.5}{2}$	111793.5
CDEF	$\dfrac{(CD + EF)}{2} \times \text{CE}'$	$\dfrac{(221.9 + 922.2)}{2} \times 124.8$	71391.8
		Total area	243990.4

Since one acre is 43560 sq ft, the total area is

$$\left(\frac{243990.4}{43560}\right) = 5.60 \text{ ac } (2.27 \text{ ha}).$$

2.14   Straight and Curved Boundary Areas

If a field is bounded on one side by a straight line and on the other by a curved boundary, as shown in Fig. 2.12, the area may be computed by the trapezoidal rule:

$$A = d\left(\frac{h_0}{2} + \Sigma h + \frac{h_n}{2}\right) \qquad (2.4)$$

where d = distance between offsets, h_0, h_1, etc.

h_0, h_n = end offsets

Σh = sum of the offsets, except end offsets

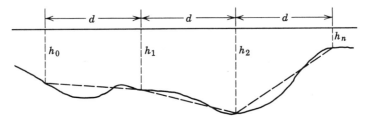

FIGURE 2.12   Measurements to obtain the area of a field with three straight sides and one irregular boundary.

In the preceding formula, the curved boundary is assumed to be a straight line between offsets h_0 and h_1, h_1 and h_2, etc.

If the end offsets h_0 and h_n were zero, which would occur when the curved boundary crosses the base line at the end points, the preceding formula would become

$$A = d \times \Sigma h \tag{2.5}$$

Example 2.3.   Determine the cross-sectional area of a 10-ft-wide stream in which water depths at each 2-ft interval are 0, 0.8, 1.1, 1.5, 0.3, and 0 ft.

Solution.   Substitute in Eq. 2.5 with d = 2 ft,

$$A = 2(0 + 0.8 + 1.1 + 1.5 + 0.3 + 0) = 7.4 \text{ ft}^2 \ (0.69 \text{ m}^2)$$

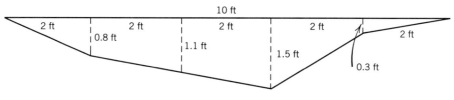

2.15   Irregular Boundary Areas

Natural watersheds and soil-type boundary areas are good examples of irregular areas. When drawn to scale on maps, these areas can be measured with a mechanical planimeter, digitizer, or other electronic device. By manually tracing the boundaries of these irregular areas with a pointer, the area can be computed from the scale of the map with a hand calculator or computer, or read directly from the instrument. Another method is to draw the boundary on a marked grid map or to use an overlay grid. Areas can be approximated by counting the grid squares and multiplying by the area per grid square to obtain the total area.

RECORDING FIELD NOTES

An important part of surveying is the proper recording of field data in a note-book. Standard forms for notes have been devised to provide a systematic procedure for recording field data.

Field notes are those recorded in the field at the time the work is done. Notes made later, from memory or copied from other field notes, may be useful, but they are not field notes. It is *not easy* to take good notes. The recorder should realize that field notes are regarded as a permanent record and are likely to be used by other persons unfamiliar with the locality who must rely entirely on what has been recorded. For this reason the field book should contain all the necessary information along with a good sketch showing the locations of the various points.

2.16   Field Book

The standard field book has horizontal lines extending across the double page as shown in Fig. 2.13. The left-hand side of the double page has six columns for tabulating numerical data, and the right-hand side has vertical columns about $\frac{1}{8}$ in. wide for recording explanatory notes and sketches. These books are made with flexible or hard back covers and may have loose-leaf, spiral, or permanent bindings. Field books will conveniently slip into a coat pocket. The hard-covered field book with the permanent binding is generally desired for

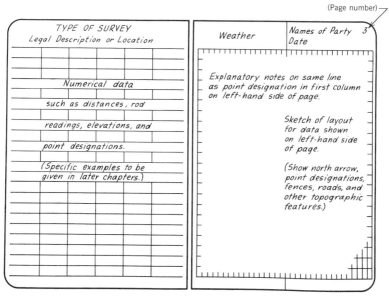

FIGURE 2.13   Suggested format for recording field information in a surveying field book.

field records. For student work, the flexible-covered field book may be satisfactory. Hand-held electronic field books (similar to a computer) are available, from which the data may be transferred to an office computer.

2.17 Suggestions for Keeping Good Field Notes

The following is a list of suggestions for keeping a set of good field notes (see Fig. 2.13):

1. The first few pages of the field book should be reserved for an index, which should be kept up-to-date as the work is completed.
2. The double sheet is considered one page. Pages should be numbered in the extreme upper right-hand corner of the right-hand side of the page.
3. A descriptive title should be printed at the top of the first page for each job. It should show the type of survey, the legal description, and the name of the job.
4. At the top of the right-hand side of the page the names of the survey party should be recorded along with the job assigned to each. For example, the instrument person is designated by $\overline{\wedge}$ the rodman by ϕ, the note recorder by REC, the head tapeperson by H.C., and the rear tapeperson by R.C. The date and weather conditions at the time the data were taken should be recorded at the top right-hand side of the page as shown in Fig. 2.14.
5. A sharp 3H or 4H hard lead pencil should be used in recording data. Ink should *never* be used, as it will smear if the field book gets wet.
6. Titles, descriptions, and words should be lettered in the best form possible, usually capital letters for headings and titles, and lowercase letters for other information.
7. Numbers recorded should be neat and plain, and one figure should never be written over another. In general, numerical data should not be erased; if a number is in error, a line should be drawn through it, and the corrected value written above. Portions of sketches and explanatory notes may be erased if there is a good reason.
8. In tabulating numbers, all figures in the tens column, for example, should be in the same vertical line. Where decimals are required, the decimal point should never be omitted. The number should always show to what degree of accuracy the measurement was taken; thus a rod reading to the nearest 0.01 ft would show 7.40 rather than 7.4 without the zero.
9. Sketches should be neat and large enough to show details without crowding the figures together. Rod readings, distances, and other numerical data are generally not shown on the sketch, provided they are recorded on the left-hand side of the page; however, reference points, a north arrow, names of streams, roads, landowners, and property lines

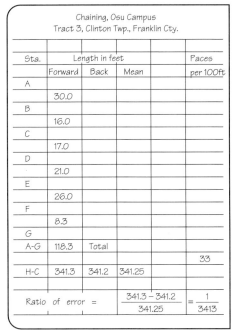
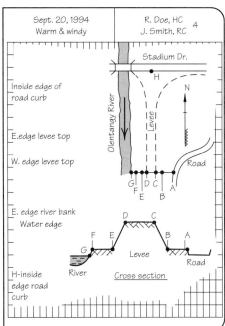

FIGURE 2.14 Sample surveying field book page for recording distances.

are shown. The sketch is drawn on the right-hand side and should correspond to the data recorded on the left-hand side or on succeeding pages.

10. Explanatory notes, such as the approximate location of reference points, should be shown on the same horizontal line, but on the right-hand side. They are needed to clarify what the numerical data and sketches fail to bring out.

11. If a page of notes becomes illegible or erroneous, the data should be retained and usable notes reentered in the book before writing the word "void" in large letters diagonally across the page. The page number of the continuation of the notes should be indicated. Voiding portions of a page may be done in the same manner as voiding a full page.

12. Scribbling should not be done in the field book. A piece of scratch paper held in the field book with a rubber band is convenient for making calculations. If this is not available, the back pages of the field book will suffice. Particular attention should be paid to neatness and arrangement of the data.

2.18   Field Notes for Recording Distances

Example field notes are shown in Fig. 2.14. In the left-hand column, the abbreviation "Sta." refers to the station or the point from which the measurement is taken. The number of paces per 100 ft is recorded for later use and is not a part of the survey. The distances are recorded on a line between the stations from which the measurements were taken. This method is often preferred so that each station can be easily described directly across the page on the right-hand side. An alternate method (Sta. A–G) of recording is shown at the bottom of Fig. 2.14.

2.19   Accuracy in Surveys

The degree of accuracy in any survey will depend on the nature of the survey and the use that will be made of the data. There is little reason for reading the rod to the nearest 0.01 if the work can be done only to the nearest 0.1 ft. It is necessary for the surveyor to use good judgment and common sense. The degree of accuracy will depend on the nature of the work, but it should be consistent. No measurement should be considered correct until it is verified. This verification should be done by an alternate method, and the figures should never be juggled to make the data check.

When justified, distances are measured twice and the mean distance computed as shown in Fig. 2.14. The actual or desired accuracy of taping is often expressed as a ratio of error. It is the difference between the two measurements divided by the mean or average distance. For convenience, the ratio of error is converted to a fraction with a numerator of 1. As shown in Fig. 2.14, the ratio of error is 0.1/341.25 or 1/3413, meaning that the error is 1 unit in 3413 units. Units may be inches, feet, or meters.

REFERENCES

McCormac, J. C. (1985) *Surveying*, 2nd ed. Prentice–Hall, Englewood Cliffs, New Jersey.

Moffitt, F. H., and H. Bouchard (1992) Surveying, 9th ed. Harper Collins Publishers, New York.

Nathanson, J. R., and P. Kissam (1988) Surveying Practice, 4th ed. McGrawHill, New York.

PROBLEMS

2.1 In taping between two points, the slope distance was 1840.3 ft and the average slope was 1 percent. Compute the horizontal distance. If the average slope was 10 percent, what would be the horizontal distance?

2.2 Convert 3, 5, and 10 in. to a decimal part of a foot.

2.3 Convert 0.25, 0.1, and 0.85 ft to inches.

2.4 In laying out a right angle by the 3-4-5 method, what are the distances *DC* and *CE* in Fig. 2.7*a* if *DE* is 96 ft?

2.5 Compute the distance *AB* in Fig. 2.9 if *BC* is 100 ft and the angle *CAB* is 40°. Compute *AC*.

2.6 Determine the area in acres of block *CDEFH* in Fig. 2.11 if *CD* is 200 ft, *CE'* is 112.5 ft, *FE* is 840 ft, *FH* is 400 ft, and *HC* is 380 ft.

2.7 Determine the area in acres for Fig. 2.12 if *d* is 100 ft, h_0 is 300 ft, h_1 is 500 ft, h_2 is 800 ft, and h_n is 50 ft.

2.8 A distance between two points was measured twice and found to be 820.11 ft and 819.95 ft. What is the ratio of error?

2.9 If the desired accuracy for a survey is such that the ratio of error should not exceed 1 in 4000 (1/4000), what is the maximum error in feet for a distance of one-half mile (2640 ft)?

2.10 In taping with a 100-ft steel tape, pins were exchanged twice with the head tapeperson, and the rear tapeperson had 6 pins in hand and one pin was in the ground. If the last pin was 38.7 ft from the point to be measured, how far was the total distance measured?

2.11 Determine the area of a triangular field in square feet if the lengths of the three sides are 500, 600, and 900 ft.

2.12 Compute the acreage of a contour strip of wheat if the up and down slope distance (*d* in Fig. 2.12) between consecutive parallel offsets is 100 ft. The offset distances, which are the widths of the strip, are 100, 180, 210, 190, 141.2, 100, and 0 ft.

2.13 If the watershed area on a map measured with a plainmeter is 36.2 square inches and the scale of the map is 100 ft/in., compute the area of the watershed in acres.

2.14 Determine the area of a field from a sketch and measurements supplied by your instructor.

CHAPTER 3 _____
LEVELS AND LEVELING

_____

Much of the work in soil and water management is devoted to control of water movement. Water must be moved in desired directions at controlled velocities, and to accomplish this it is necessary to accurately measure differences in elevation. The surveying level is an instrument used to determine such differences. It is essential for satisfactory work in drainage, erosion control, and irrigation.

TYPES OF LEVELS

There are many types of levels, ranging in cost from a few dollars to many hundreds of dollars. The type selected for a given job depends on the accuracy required.

3.1 Hand Levels

Instruments of this type are suitable for rough leveling, such as the running of guidelines for contour farming or for approximations of land slope. The simplest type, represented by Fig. 3.1, consists of a metal tube 6 in. long, on which is mounted a level vial. A prism in the tube reflects the image of the level vial into the eyepiece, so that the position of the bubble may be observed as a sight is taken through the tube. Since most hand levels do not have magnification, sight distance is limited to about 100 ft. When held in the hand without stabilizing, readings at 50 ft cannot be taken more accurately than about 0.2 ft. For this reason, hand levels are not suitable for accurate surveys, such as tile drainage and terrace layout.

The Abney level shown in Fig. 3.2 is a modified hand level with a level vial attached to a vertical arc, which gives slope in degrees or percent directly. When sighting on the desired point, the moveable arm is rotated until the reflection of the bubble in the eyepiece is centered. The line of sight must be parallel to the land slope. This slope can be obtained by sighting on a second person at a point equal to the eye height of the observer.

A clinometer is another type of hand-held instrument for measuring vertical angles. It is useful for special purposes, especially for forestry and military applications.

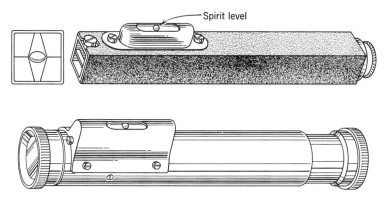

FIGURE 3.1 Two types of hand levels. (Courtesy Keuffel and Esser.)

3.2   Tripod Levels

Levels designed for use with a tripod are available in a wide range of accuracies. They are given the names farm level, builder's level, architect's level, or engineer's level, according to the use for which they are primarily designed. Instruments of this type are well suited for the layout of conservation and water management systems. Based on their construction, these instruments may be referred to as dumpy, tilting, or self-leveling levels. The dumpy level is shown in Fig. 3.3. Its telescope is attached rigidly to the frame of the instrument. This level was originally shorter than other types, hence the name "dumpy." The name has little present significance. Builder's and architect's levels have a circular horizontal scale for reading angles.

The tilting level and the self-leveling level (Fig. 3.4) often use only three screws for rough leveling of the instrument on the base plate. In the tilting level, a bubble tube affixed to the telescope tube is viewed through a prism system with an eyepiece adjacent to the telescope eyepiece. The prisms permit both ends of the bubble to be viewed simultaneously in a split image. The instrument is leveled for each reading. It is level when the images of the two bub-

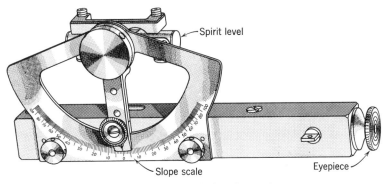

FIGURE 3.2 Abney level. (Courtesy Keuffel and Esser.)

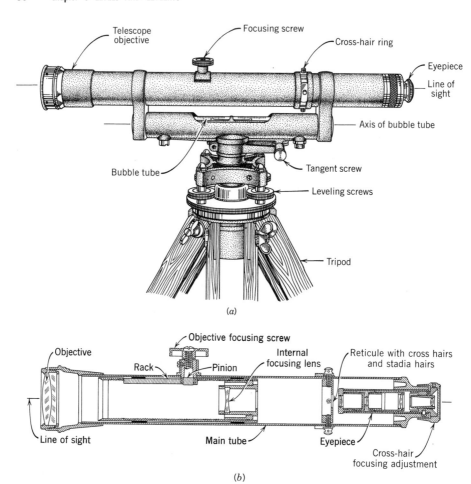

(a)

(b)

FIGURE 3.3 *(a)* Engineer dumpy level and *(b)* cross section of an internal focusing telescope. (Courtesy Keuffel and Esser.)

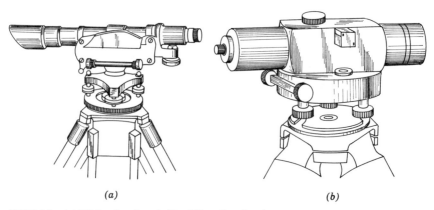

(a)   (b)

FIGURE 3.4 *(a)* Tilting level and *(b)* self-leveling level.

ble ends are brought into coincidence by adjusting a micrometer screw that tilts the telescope. The self-leveling level automatically sets its own level line of sight through a prism and pendulum system. In the three-screw levels, a spherical bubble vial is used for approximate leveling. This bubble vial is independent of the line of sight. The leveling screws may be rotated individually or in pairs to move the bubble into the target.

The cross section and parts of a telescope and level are shown in Fig. 3.3b. The cross hairs are focused by rotating the eyepiece or by moving it, and the focusing screw is turned to focus the telescope on the rod.

3.3 Rotating Beam Levels

Rotating beam levels may have an electronic or a laser beam not visible to the naked eye. A rotating electronic beam unit is shown in Fig. 3.5. It has a range of about 400 ft. The beam is projected as a level plane by a rotating prism. The detector unit on the level rod is moved up or down until the beam intersects the detector, which makes an audible signal. Only one person (the rodperson) is required, and the unit will shut off automatically if it is out of level.

The laser beam level developed for earthwork with heavy equipment is more accurate, heavier, and more expensive than the electronic beam level. Readings at distances up to 1500 ft with the laser are accurate to within a few hundredths of a foot. The laser beam may be adjusted to the desired height and tilted to produce a sloping plane. Such a slope is desirable when it is used with

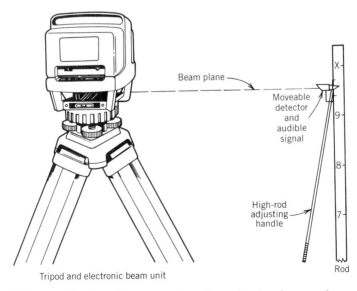

FIGURE 3.5 Rotating electronic or laser beam level and survey detector rod. (Courtesy Spectra Physics, Construction and Agricultural Division.)

trenching or other earth-moving equipment. A special detector unit on the machine automatically maintains the same slope as the laser plane. Any number of machines can be operated from one laser unit. Surveys can be made at night with either the electronic or laser instruments.

3.4 Electronic Total Stations and Transits

These instruments are essential for the professional surveyor. Both the transit and electronic instruments are more versatile than a tripod level and can obtain any measurements possible with a level and many more. They can measure vertical and horizontal angles as well as distances. Although the cost of the total station instrument may be several thousand dollars, it may be justified by decreasing time in the field and in the office (thus lowering labor costs), and by increasing accuracy and recording of data. The transit shown in Fig. 3.6 has been in existence for over 200 years, but it is becoming obsolete since the development of the electronic total station shown in Fig. 3.7. The transit has a telescopic sight and a bubble tube similar to those on tripod levels. The sight and bubble tube are fastened rigidly together and rotate in the vertical plane. If the

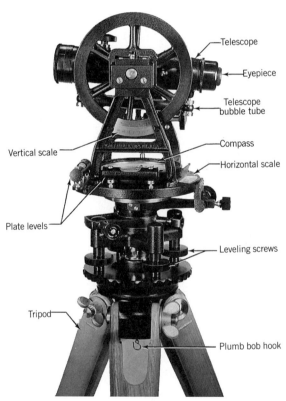

FIGURE 3.6 Transit or theodolite. (Courtesy W. & L. E. Gurley.)

bubble tube is centered, the transit serves as a tripod level. At most distances, especially more than a few hundred feet, the electronic instrument is far superior in accuracy and ease of measurement than stadia distances obtained with a level or transit. Level readings, angles, and stadia distances with a level or transit are mechanical measurements that rely on sight readings and are more subject to error than digital data. The total station incorporates the electronic distance measurement (EDM) system described in Chapter 2. Slope distances are easily converted to horizontal distances, which can be read on the screen or recorded in an electronic field book. Some of the more elaborate instruments have a tape deck for storing the data. The information obtained can thus be easily stored in the field and can be transferred manually or electronically to an office computer for further analysis.

The name, total station, is somewhat of a misnomer, but it likely was con-

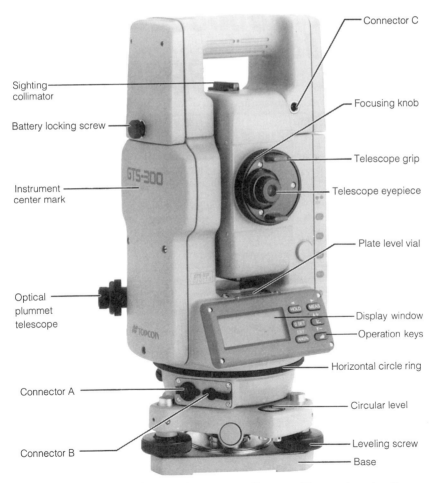

FIGURE 3.7 Electronic total station instrument. (Courtesy Topcon America Corporation.)

ceived because the instrument will perform other functions not possible with a transit. Further description can be found in commercial literature and surveying texts, such as Davis et al. (1981) and Moffitt and Bouchard (1992).

3.5   Care of Instruments

To secure continued reliable service, instruments must be used properly and carefully handled. The following suggestions for the care and handling of surveying instruments should be observed.

1. When transporting instruments, protect them from impact and vibration. Place them on a firm base in vehicles rather than on top of other equipment.
2. Place the lens cap and tripod cap in the instrument box while the instrument is in use.
3. Leave the instrument box closed and in a place where it will not be disturbed while the instrument is in use.
4. Avoid running or otherwise taking a chance on falling while carrying an instrument.
5. Never force screws or other moving parts of an instrument.
6. If provided, use the sunshade, regardless of the weather.
7. Cross fences with an instrument by spreading the tripod legs and placing the instrument on the far side of the fence before climbing.
8. Bring leveling screws to a snug bearing, but do not jam them.
9. Rub lenses only with soft tissue, not with fingers or rough cloth. Do not remove lenses.

LEVELING

3.6   The Hand Level

The hand level is simply held to the eye with the hand. Resting the hand against the cheek aids in holding the instrument steady. Some users rest the instrument on a staff of convenient length. Its operation is basically the same as that of the tripod level, though measurements are not as accurate. Sight distances should be limited to about 100 ft.

3.7   Tripod Levels

SETTING UP THE TRIPOD LEVEL   In setting up a tripod level, the instrument is screwed to the head of the tripod, the lens cap is removed, and the sunshade is put in its place over the objective lens. The instrument is set up in the following steps:

1. Loosen the thumb nuts that fasten the tripod legs to the head. (Most modern levels have friction joints.)

2. Spread the legs 3 or 4 ft, push them firmly into the ground, and adjust the legs so that the tripod head is approximately level in both directions.

3. Tighten the thumb nuts holding the legs to the tripod head.

4. Make sure that the telescope clamp screw is loose and swing the telescope to a position directly over an opposite pair of leveling screws. With a three-screw level, place the bubble vial over one of the screws.

5. Move the leveling screws as shown in Fig. 3.8 and turn them simultaneously to level the bubble tube. These screws should be turned so that the thumbs move toward one another or away from one another. The bubble will follow the left thumb. Keep the leveling screws working firmly against one another, but not tight enough to jam.

6. Rotate the telescope 90° to a position across the other pair of leveling screws and again level the bubble.

7. Repeat over each opposite pair of screws until the bubble stays level or nearly level throughout a 360° circuit. With a three-screw level and circular bubble vial (Fig. 3.9), leveling screws may be turned one at a time or in pairs to bring the bubble into the target circle on the vial.

8. Turn the telescope to bring the rod into the field of vision.

9. Focus on the cross hairs by adjusting the eyepiece.

10. Focus on the rod by turning the objective focusing knob.

11. When using a level, check the bubble immediately before and after each reading and make any necessary adjustment of the leveling screws. Do this while standing in position to sight through the telescope, as movement that shifts body weight may affect the level. With a tilting level, adjustment of the telescope-tilt micrometer screw should be made for each observation. No adjustment need be made between observations with a self-leveling level so long as the bubble remains within the circle. Some levels will display a red light when the level is not within the automatic range.

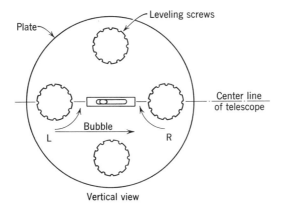

FIGURE 3.8 Leveling an instrument with four leveling screws. Note that the bubble follows the left thumb.

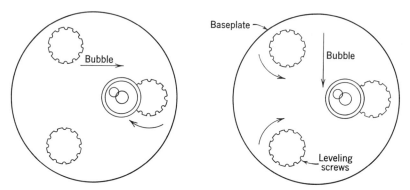

FIGURE 3.9   Leveling an instrument with three leveling screws. Note that when using two screws the bubble follows the left thumb.

TYPES OF LEVEL RODS   Many types of direct reading level rods with different scales are available. The Philadelphia rod has two sections about 7 ft long as shown in Fig. 3.10*a*. It can be extended to read 13 ft and has a target. The target is convenient for laying out a contour line, a grade line as for a terrace, or for precise leveling. The Frisco rod shown in Fig. 3.10*b* has three sections 4 ½ ft long. Either English or metric scales are available.

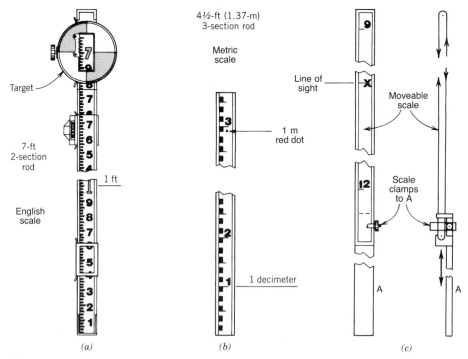

FIGURE 3.10   Three types of leveling rods: *(a)* Philadelphia, *(b)* Frisco, and *(c)* Lenker (the scale is moveable on the rod and inverted to read elevations directly).

The Lenker rod with an inverted moveable scale is shown in Fig. 3.10*c*. Elevations are read directly from the rod. Low numbers are at the top and high ones are at the bottom. An index elevation on the rod, for example, 10 ft, is moved to the line of sight having the same elevation. It is especially suitable for mapping relatively flat fields and for rotating beam levels.

READING THE ROD   Most rods in English units are graduated in feet and tenths of feet as shown in Fig. 3.11*a*. The width of one black line is 1/100 (0.01) ft, and the width of a white space between black lines is 1/100 ft. Several rod readings are indicated on the figure. The reading gives the distance in feet from the lower end of the rod.

Figure 3.11*b* illustrates a common source of error in rod readings. Notice that if the rod is slanted in any direction, the reading will be too large. The rod-person should hold the rod lightly, letting it balance in the vertical position. The fingers should not obscure the face of the rod, for in this position they will frequently obscure the instrument user's view of the rod. Many surveyors find it helpful if the rodperson waves the rod back and forth through the true vertical position. If this is done, the instrument user can read the minimum reading, which is the true reading.

To save time in leveling, the instrument user and the rodperson should communicate with each other by means of hand signals or two-way radios. Some standard signals are given in Fig. 3.12.

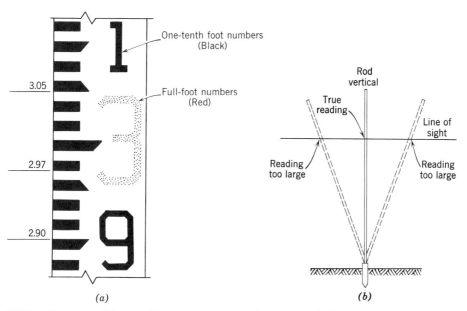

FIGURE 3.11   (*a*) English unit divisions on a rod and (*b*) "waving" the rod to obtain a true reading.

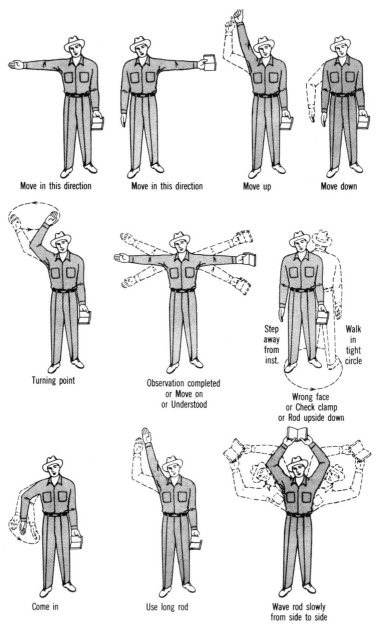

Move in this direction   Move in this direction   Move up   Move down

Turning point   Observation completed
or Move on
or Understood

Step
away
from
inst.

Walk
in
tight
circle

Wrong face
or Check clamp
or Rod upside down

Come in   Use long rod   Wave rod slowly
from side to side

FIGURE 3.12 Hand signals to communicate with the rod-person. (Courtesy U.S. Soil Conservation Service.)

3.8 Differential or Bench Mark Leveling

The operation of leveling to determine the relative elevations of points some distance apart is known as differential leveling. It consists of making a series of instrument setups along the general route between the points. From each setup a rod reading is taken back to a point of previously determined elevation and a reading is taken forward to a point of unknown elevation. These points at which elevations are known or determined are called *bench marks* or *turning points*. A bench mark is a permanently established reference point, the elevation of which is assumed or is accurately measured. A turning point is a temporarily established reference point having its elevation determined as an intermediate step in a differential leveling traverse.

For example, referring to Figs. 3.13 and 3.14, if the elevation of bench mark number 1 (B.M. 1) is known to be 100.00, the elevation of B.M. 2 can be found by differential leveling. The first setup of the instrument is made at some point a convenient distance away from B.M. 1 and along the general route to B.M. 2. This distance will depend on the magnifying power and accuracy of the instrument, but in most instances it will be from 75 to 200 ft. The rod is held on B.M. 1, and the rod reading (in this case 5.62) is noted in the field book (Fig. 3.14). This reading is a *backsight* (B.S.); it is a reading taken on a point of *known* elevation.

Addition of the backsight on B.M. 1 to the elevation of B.M. 1 gives the elevation of the line of sight. (In this case 100.00 + 5.62 = 105.62.) This is the *height of instrument* (H.I.) and is entered in the H.I. column of the field notes. Notice that in the field notes the H.I. is entered on the line between the B.M. 1 line and the *turning point* line (T.P. 1), indicating that the instrument is set up between B.M. 1 and T.P. 1. This form of notes is convenient, although it is not necessary to leave a line for the H.I. between the B.M. and T.P. lines.

After the backsight on B.M. 1 has been recorded, the rodperson moves to turning point number 1 (T.P. 1) and sets a hub stake or finds some other firm object on which to rest the rod. With the instrument still at setup 1, a rod reading is taken on T.P. 1. This reading is a *foresight* (F.S); it is a reading taken on a

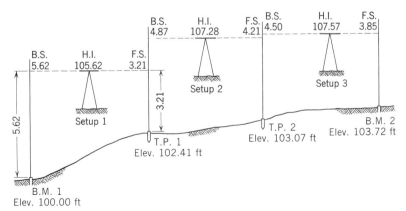

FIGURE 3.13 Differential leveling procedure.

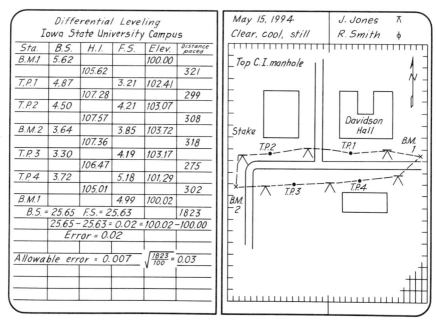

FIGURE 3.14   Field notes for differential leveling.

point of *unknown* elevation. The foresight (in this case 3.21) is entered in the field book opposite T.P. 1. It is good practice to make the line of sight distance for the foresight essentially equal to the line of sight distance of the backsight by pacing. The reason for this will be explained under the discussion of adjustment of the instrument.

The elevation of T.P. 1 can now be computed by subtracting the foresight on T.P. 1 from the instrument height (in this case $105.62 - 3.21 = 102.41$), and this elevation is entered in the notes opposite T.P. 1. Note that T.P. 1 now becomes a point of known elevation.

The rodperson remains at T.P. 1 while the instrument user moves to setup 2. From this position a backsight is taken on T.P. 1, the new H.I. is determined, and T.P. 2 is established and its elevation is determined as was previously done for T.P. 1. Careful study of Figs. 3.13 and 3.14 should make this procedure clear. The same sequence of events is carried through (i.e., determine H.I. from a backsight on a T.P. of known elevation and determine the elevation of a new T.P. by foresight) until B.M. 2 is reached.

The standard form for keeping level notes is shown in Fig. 3.14, but for the beginner they may be written out as follows and kept on a separate sheet:

Elevation B.M. 1	100.00
B.S. on B.M. 1	+5.62
H.I.	105.62
F.S. on T.P. 1	−3.21
Elevation T.P. 1	102.41 etc.

The standard format takes less time to record the data and shows the information more clearly.

As a check on the accuracy of the work, a line of differential levels is then run from B.M. 2 back to B.M. 1, and the difference between the original elevation of B.M. 1 and its calculated elevation is the error. In the example given, this is $100.02 - 100.00 = 0.02$. A reasonable allowance for the error is given by

$$\text{allowable error} = 0.007 \sqrt{\frac{\text{length of traverse in feet}}{100}}$$

In the sample given, the length of the traverse is 1823 ft and the allowable error is 0.03. Thus, 0.02 is within reasonable accuracy.

The length of the traverse shown in Fig. 3.14 should be obtained by the rodperson. The distance of 321 ft from B.M. 1 to T.P. 1 should be the sum of the foresight and the backsight distances rather than the straight-line distance between the two points. Normally, the traverse length and ratio of error are not required, unless the accuracy of the survey is specified.

A check on the arithmetical accuracy of differential leveling notes may be made by subtracting the sum of all the foresights from the sum of all the backsights and comparing this difference with the difference between the final and initial elevations. If the two differences are equal, the notes are arithmetically correct. However, this does not check errors in rod readings or in instrument adjustment. A thorough understanding of differential leveling is basic to the use of the level for other work.

3.9   Profile Leveling

Profile leveling is the process of determining the elevations of a series of points at measured intervals along a line. This is particularly important in drainage and terrace layouts. Essentially, it is a process of differential leveling with a number of intermediate foresights between turning points. As illustrated in Figs. 3.15 and 3.16,

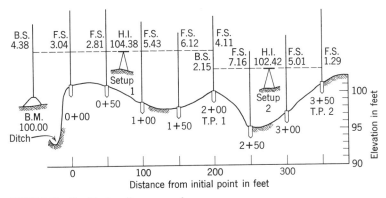

FIGURE 3.15   Profile leveling procedure.

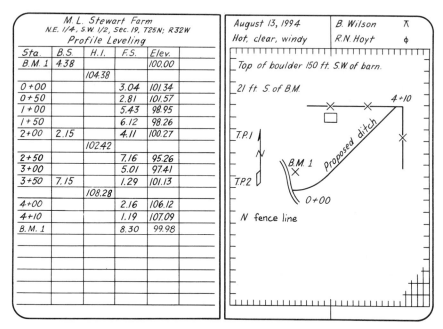

FIGURE 3.16   Field notes for profile leveling.

the foresights taken from setup 1 are each subtracted from the height of instrument at this setup to determine the elevations of the intermediate points between B.M. 1 and T.P. 1. The elevations of points between T.P. 1 and T.P. 2 are determined in a similar manner. The location of the instrument at setups 1 and 2 may be at any convenient point so long as the rod can be read accurately. In Fig. 3.15 the elevation of each of the stations is plotted and a line is drawn between the points, thus giving a profile along the proposed ditch. From such a profile, the depth of the ditch can be determined, as well as the amount of soil to be moved. Further applications of profile leveling will be given in later sections of this book.

In surveying, stations along a line are generally designated as $0 + 00$, $3 + 50$, etc. (see Fig. 3.15), in which the numeral preceding the plus sign is the number of hundred feet from the starting point and the two numerals following the plus sign represent the additional number of feet (which must be less than a hundred). Thus, $3 + 50$ station means that it is 350 ft from the initial point designated as $0 + 00$. The advantages of this system of marking are that station numbers cannot be confused with rod readings and that designation of stations may be simplified. For example, a station 300 ft from an initial point is referred to as station 3 rather than station 300.

CHECKING OF LEVELS

Accurate measurements with a level are possible only if the instrument is checked and adjusted properly. It is thus essential that the surveyor be thoroughly familiar with the methods of field checking the instrument prior to use. Adjustment should be made only by an experienced surveyor or at an instrument repair shop.

3.10   The Horizontal Cross Hairs

The following steps should be taken to ensure that the horizontal cross hairs are horizontal when the instrument is level (Fig. 3.17). For optical telescopic instruments, set up and level the bubble tube and sight on some well-defined point, such as a tack head on a tree or post. Turn the telescope about its vertical axis so that the point appears to traverse the field of view. If the point does not remain on the horizontal cross hair, the instrument should be adjusted.

3.11   The Axis of the Bubble Tube and the Vertical Axis

The following steps should be taken to ensure that the axis of the bubble tube is perpendicular to the vertical axis. Set up and level the bubble tube perfectly with the telescope parallel to two level screws. Rotate the telescope 180° about its vertical axis. If the bubble does not remain centered, adjustment is necessary. This adjustment is not essential for accurate work, but it enables the user to rotate the instrument without having to center the bubble between level readings.

With a self-leveling level, center the bubble tube with the leveling screws and then rotate the telescope 180°. If the bubble does not remain centered, adjustment is required. Relevel and repeat the check and adjustment until the bubble remains centered through 360° of rotation of the telescope.

3.12   The Line of Sight and the Axis of the Bubble Tube

Take the following steps to ensure that the line of sight is parallel with the axis of the bubble tube (2-Peg Test). Set up the instrument exactly midway between two hub stakes that are 150 ft apart, as shown in the field notes of Fig. 3.18. From the midway position take a rod reading on hub A and on hub B, and enter these readings in the notes as *a* and *b*. The difference between these readings is the true difference in elevation, whether or not the instrument is in adjustment. Because fore-

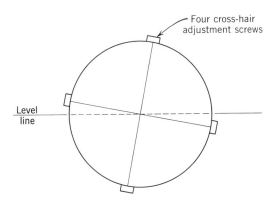

FIGURE 3.17   Cross-hair ring and adjusting screws.

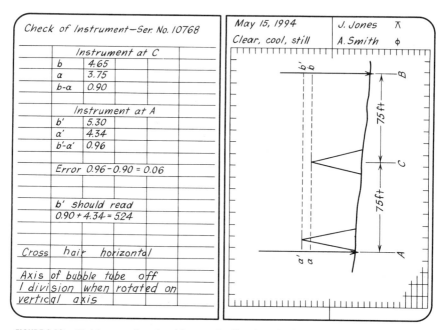

FIGURE 3.18   Field notes for checking and adjusting the level.

sight and backsight distances are of equal length, any error in a will be the same for b. Thus, $(a - b)$ is always correct, i.e. 0.90 for the example in Fig. 3.18.

Move the instrument to A (or B) as in Fig. 3.18 and set up so that the eyepiece nearly touches the rod held on the hub. Sighting backwards through the telescope, take reading $a' = 4.34$, and then sighting in the normal manner take reading $b' = 5.30$. Thus, $b' - a' = 0.96$, which is 0.06 greater than the correct difference of 0.90. Adjust the level until b' reads $5.24(0.90 + 4.34)$, which gives the correct difference of 0.90. After the adjustment is made, the checking procedure should be repeated. Go through the entire checking procedure twice and adjust the cross hairs only after the same error is observed on two consecutive checks. This double check will prevent errors in operation of the instrument from being construed as errors in adjustment. With the self-leveling level, this check is made to center the circular bubble. With the tilting level, it is made to establish a level line.

REFERENCES

Davis, R. E., F. S. Foote, J. M. Anderson, E. M. Mikhail (1981) *Surveying Theory and Practice,* 6th ed. McGrawHill, New York.

McCormac, J. C. (1985) *Surveying,* 2nd ed. Prentice–Hall, Englewood Cliffs, New Jersey.

Moffitt, F. H., and H. Bouchard (1992) *Surveying,* 9th ed. Harper Collins Publishers, New York.

Nathanson, J. R., and P. Kissam (1988) *Surveying Practice,* 4th ed. McGrawHill, New York.

PROBLEMS

3.1 Determine the elevation of B.M. 2 from the following notes. Check arithmetic by adding F.S.s and B.S.s.

Sta.	B.S.	H.I.	F.S.	Elevation
B.M.1	1.21			50.00
T.P. 1	6.20		4.65	
T.P. 2	4.82		3.11	
T.P. 3	3.03		5.22	
B.M. 2			3.16	

3.2 An error of 0.08 ft was made in leveling a distance of 4900 ft. Is this within the allowable error?

3.3 Rod readings when adjusting a level were as follows, using the notation in Fig. 3.18: a was 4.13, b was 6.14, a' was 3.85, and b' was 5.90. What was the error and what should the reading b' have been?

3.4 If B.S.s of 3.71, 4.36, and 6.13 and F.S.s of 5.68, 6.50, 5.23, 5.09, 5.02, 4.03, 3.42, 3.04, 5.34, and 5.21 were recorded in Fig. 3.16 instead of those shown, what would be the new elevation of each station? Tabulate the data in the standard form for field notes.

3.5 From the following rod readings, compute the correct elevation of B.M. 2 if all sight distances were 100 ft and the level had an error of 0.02 ft per 100 ft (rod reads too high). Assume that the level bubble was always centered and that no errors were made in reading the rod. *Hint:* Correct each rod reading before computing.

Sta.	B.S.	H.I.	F.S.	Elevation
B.M.1	3.20			50.00
T.P. 1	0.92		4.60	
B.M. 2			8.14	

3.6 If both B.S. sight distances in Problem 3.5 were 200 ft and both F.S. sight distances were 100 ft, compute the correct elevation of B.M. 2.

3.7 From the survey in Fig. 3.16, check the arithmetic by summing the B.S. and the F.S. *Hint:* Select only F.S. applicable to the differential survey. What is the error?

3.8 Compute the allowable error from the survey in Fig. 3.16 if the total length of the traverse back to B.M. 1 was 802 ft. Is the actual error within the allowable?

3.9 In a differential level survey from B.M. 1 to B.M. 2, the sum of all B.S. was 34.62 ft and the sum of all F.S. was 39.66 ft. If the elevation of B.M. 1 was 50.00 ft, what is the elevation of B.M. 2? What procedure should be followed to verify that the elevation of B.M. 2 is correct?

3.10 Determine the elevation of all T.P.s and other points of unknown elevation from data supplied by the instructor. Record the data according to the standard form in a field book. Prove that no mistake has been made in the arithmetic. What is the error in the survey?

CHAPTER 4

LAND SURVEYS AND MAPPING SYSTEMS

Location and legal description of specific tracts of land are based on the original U.S. public land surveys. Topographic maps showing ground elevations and many other special maps showing natural features and works of man have become increasingly available for all habitable land, but these maps are usually not in sufficient detail to layout or plan most natural resource management practices. Remote sensing, such as aerial photographs and satellite imagery, has enhanced the detail on maps and increased their accessibility over the entire earth. Aerial photographs became readily available in the 1930s and satellite imagery in about 1990. With these many advances in technology and the rapid development of computer software, tremendous improvements in mapmaking have occurred and will continue. These developments will likely have a great impact on conservation and environmental planning and other activities of humans, such as land location, property records, transportation facilities, land use, water quality, and so forth.

PUBLIC LAND SURVEYS

4.1 Metes and Bounds

A tract of land identified by the direction and length of its sides is said to be described by *metes and bounds*. The location of property corners may be described by map coordinates and other terrain features. Tracts of land so described may be relocated, provided at least one of the original corners can be identified, and the true direction of one of the sides can be determined. This method of land survey, which originated prior to 1785, is found in the eastern states and in a few isolated areas in other parts of the United States.

4.2 Rectangular System of Public Land Survey

In 1785, the Continental Congress passed a law that provided for the subdivision of public lands into townships, sections, and quarter sections. As shown in Fig. 4.1, these townships are located with respect to some *initial point* through which passes a true north-south line, called the *principal meridian,* and an east-west line, a

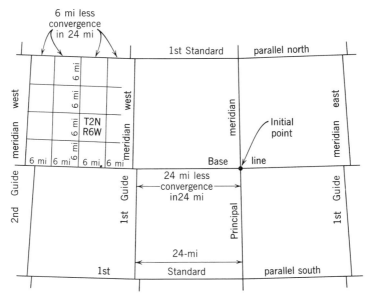

FIGURE 4.1   Rectangular system of U.S. public land survey showing quadrangles and townships.

true parallel of latitude, called the *base line*. Standard parallels of latitude are located at intervals of 24 miles north and south from the base line, and they intersect the principal meridian at right angles. Guide meridians are located at intervals of 24 miles east and west of the principal meridian measured along the base line or along one of the standard parallels. Because the guide meridians converge, the north side of each 24-mile quadrangle is less than 24 miles in length. Likewise, the north side of each township (6 miles square) is less than 6 miles. In Fig. 4.1 only the upper left 24-mile quadrangle is divided into townships. The true meridian lines that subdivide the 24-mile quadrangles into townships are called *range lines,* and the latitudinal lines at 6-mile intervals are called *township lines.* Townships are designated by numerals corresponding to the number of townships (also called tiers) from the initial point. For example, township T2N, R6W is in the second row of townships north of the base line and in the sixth column of townships west of the principal meridian.

The subdivision of a township into 1-mile square sections (640 acres) is shown in Fig. 4.2. Sections are numbered by starting in the northeast corner and continuing east and west across the township as shown. If the survey were made without error, all sections would be 1 mile square except those along the west boundary of the township. These fractional sections are less than 1 mile in width because of convergence of the range lines. If errors were made in the survey, sections 1 through 6 may be less than or more than 1 square mile.

Each section may be further subdivided into as small tracts as necessary, usually into fourths or halves of quarter sections and of sections (40, 80, 160, or 320 acres). The legal description of these subdivisions, as shown in Fig. 4.3*a*, al-

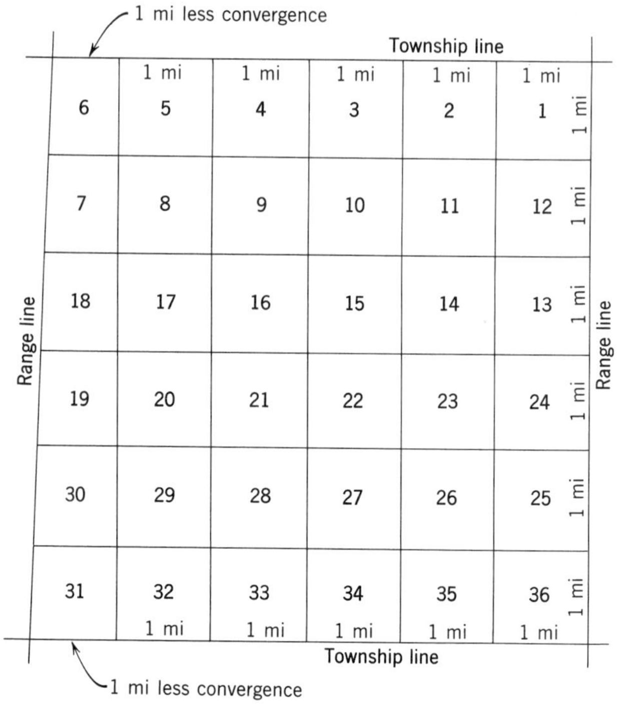

FIGURE 4.2 Subdivision of a township into sections.

ways begins with the smallest unit. For example, the $NE\frac{1}{4}$ Sec. 23, T2N, R6W, "Any county" contains 160 acres in the second tier of townships north of the base line and in the sixth range of townships west of the principal meridian. Section 23 would be the ninth mile $(6+3)$ north and the thirty-second mile $(30+2)$ west from the initial point (see Figs. 4.1 and 4.2). As shown in Fig. 4.3a, the $SE\frac{1}{4}$, $SW\frac{1}{4}$, $SW\frac{1}{4}$, Sec. 23 contains 10 acres or $\frac{1}{64}$ of a section.

Distances between two tracts of land within the same survey can be determined from the legal description, as in the following example.

Example 4.1. Find the distance between the southeast corner of Section 33, T2N, R6W and the southeast corner of Section 36, T2N, R4E.

Solution. Since the two points are in T2N, they are directly east and west.

SE corner Section 33 to SE corner Section 36, R6W = 3 miles
SE corner Section 36, R6W to principal meridian = 30 miles
principal meridian to SE corner Section 36, R4E = <u>24 miles</u>
Total distance = 57 miles

Corrections for errors made in the survey or caused by convergence of the range lines occur in the west and north rows of sections in the township. In these sections, corrections fall in the west or north rows of quarter sections, as

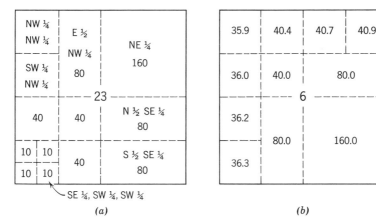

FIGURE 4.3 Subdivision of a section: (*a*) standard section and (*b*) fractional section.

shown in Fig. 4.3*b* for Section 6. Since Section 6 is the northwest section of the township, the corrections occur both in the west and north quarter sections. Sections 1 to 5 have corrections on the north row of quarter sections, which may result in areas greater than or less than the standard 40 acres. Sections 7, 18, 19, 30, and 31 have corrections on the west row of quarter sections, and these areas are usually less than 160 acres because of convergence.

TYPES OF MAPS

A map is a useful, and frequently an essential, means of recording information for selecting soil and water management systems and for recording the practices themselves. Several types of available maps are described in the discussion that follows.

4.3 Aerial Photographs

Effective use of aerial photographs requires understanding of their characteristics. In taking the pictures, 40 percent to 60 percent overlap between adjacent pictures is allowed. The center portion of each photograph generally has a usable area of 4 square miles. Outside the 4-square mile area, the scale is generally too distorted for accurate measurements.

The scale on aerial photographs is only approximate because of slight distortion in taking the picture (except where the camera is directly above) and because of variation in shrinkage of photographic paper. The 10 by 10 contact prints of USDA aerial photos have a scale of about 1667 ft per in. Enlargements can be obtained with scales of 1320, 1000, 660, and 400 ft per in. If the scale of an aerial photograph is not known, it can be obtained by measuring both the ground and map distance between two points that can be easily defined and identified. Preferably, the two points should be at least 500 ft apart. For example, if the ground distance is 600 ft and the map distance is 1.5 in., the scale is 400 ft per in. Where accurate distances are to be obtained from aerial photographs, the scale should be verified with a field measurement as just described. In reading an aerial photograph, it should be held so that shadows of objects fall toward the reader, otherwise valleys appear as ridges and vice versa. Identifying objects on the photograph can best be achieved by noting (1) tone or shade of gray, (2) shape, (3) shadow, and (4) relative size. Tone depends on the amount of light reflected by the object, and it is affected by the color and texture of the surface and by the angle at which the sun's rays strike the surface. In fields, the color and texture change with the seasons, which means that the date the picture was taken is important. Fields, roads, bridges, buildings, and contoured fields can be identified principally by shape. Shadows of trees and buildings not only help in recognition, but they also indicate the height of such objects. Unknown objects may be identified by comparing their relative size to objects of known size.

Much progress is being made by applying remote sensing techniques to aerial mapping. With infrared and special color films with various combinations of color filters, vegetation and soil features can be detected and differentiated on the ground. Such techniques are being developed for mapping and inventorying crop cover; wet, dry, and other soil conditions; tile drains; insect damage; incidence of plant disease; and certain types of polluted water. Although still in the process of development, such techniques promise to provide powerful new tools for large-scale and timely specialized surveys. On a much grander scale, mapping can be done with orbiting earth satellites.

Up-to-date aerial photographs provide details that could not be obtained by ground survey except at prohibitive cost. Government aerial photographic maps are available for most sections of the United States from the Cartographic Division, Natural Resource Conservation Service, Federal Center Building, Hyattsville, Maryland 20782. Prints are available in sizes up to 40 by 40 in. for a 9-square-mile area. Local offices may have prints that can be enlarged to 1 in. = 200 ft and retain good detail. When ordering prints, include the identification, such as that shown at the top of the photo in Fig. 4.4. Each negative and print will have such data as "6-30-49; DE-1F-140"; indicating the date flown, the photograph identifying symbol (DE), the photograph roll number (1F), and the exposure number in the roll (140).

Aerial photographs may be obtained from the State Department of Transportation, other state and local agencies, and private surveying companies. For maps of drainage systems, photographs may be taken with hand-held cameras from low-flying light aircraft. Although the scale may be distorted, these maps are useful for the location of drains at a later time. Satellite and space photography may be obtained from the U.S. Geological Survey, EROS Data Center, Sioux Falls, S. Dak. 57198.

Another source of aerial photographs is the 2 by 2 in. color or infrared slides taken for the U.S. Department of Agriculture, Consolidated Farm Service Agency (CFSA), formerly ASCS, for checking the compliance and acreage of

FIGURE 4.4   Government aerial photograph flown June 30, 1949.

conservation practices on individual farms. These slides are taken once and sometimes twice each year and may be purchased from the local offices of that agency at nominal cost. Although prints from slides may not be as accurate or as high quality as those obtained from high altitude black-and-white aerial photos, they are likely to be more up-to-date and may be adequate for many purposes.

By viewing adjacent pairs of photographs of the same object, a three-dimensional image may be seen with a stereoscope. These devices range from simple lens or mirror stereoscopes to large complex stereoscopic plotting instruments, from which contour maps can be prepared. Unless the area is greater than a few hundred acres, ground surveys and hand plotting are more economical than stereophotographic mapping techniques.

4.4   Soil Survey Maps

Most counties have soil survey maps, which have been prepared by soil scientists of the Natural Resource Conservation Service and the state experiment stations. These have been prepared from aerial photographs, usually with a scale of 1667 ft per in., on which the following information may be added.

1. Soil series, type, and phase, including land slope classes and extent of erosion. (Slope classes and erosion are not shown in Fig. 4.5.)
2. Roads, streams, houses, and cities.
3. Section numbers, legal descriptions, and township names.

These maps do not show contour lines. An example of such a map is shown in Fig. 4.5. Each soil mapping unit is indicated by a designation, such as BfA2, in which Bf is the soil series and type, A is the slope class, and 2 is the degree of erosion. Several mapping units may be grouped together and called a "Soil Management Group." These are useful for making general recommendations for lime and fertilizer, for forestry plantings, for septic disposal systems, and for the design of irrigation and drainage systems. Further grouping into eight land capability classes is discussed in Chapter 1.

These maps are available from the Superintendent of Documents, U.S. Gov-

FIGURE 4.5 Soil survey map from Madison County, Ohio (1981), showing soil series and types. *(Courtesy Soil Conservation Service.)*

ernment Printing Office, Washington, D.C., and other local sources. Some maps may not have all of the preceding information, and in some counties, maps have not yet been prepared.

4.5   Topographic Maps

A topographic map shows the relief or the topography of an area. Relief is usually shown by a contour line, which passes through all points of the same elevation. It is the best and simplest method of accurately representing the three-dimensional surface of an area on a two-dimensional drawing. The irregular lines shown on the farm map in Fig. 4.6 are contour lines. Such a map is useful for planning erosion control, water supply, drainage, irrigation, and other conservation systems.

Large-scale quadrangle topographic maps for 79 percent of the contiguous United States have been prepared by the U.S. Geological Survey (Fig. 4.7). These usually have a scale of 2000 ft per in. (1:24 000) and a contour interval of 5 to 20 ft. Location is by latitude and longitude rather than by counties. Roads, buildings, streams, water areas, forests, and many other features are shown. They are multicolored for easy interpretation. Although quadrangle maps are valuable for large-scale applications, the contour interval is too wide for detailed planning. A portion of such a map (as shown in Fig. 4.7) includes the same area as Fig. 4.5, the soil survey map. Some property lines are shown, but green forested areas are not reproduced. This area in Ohio was surveyed by metes and bounds.

Quadrangle maps are available from U.S. Geological Survey and more than

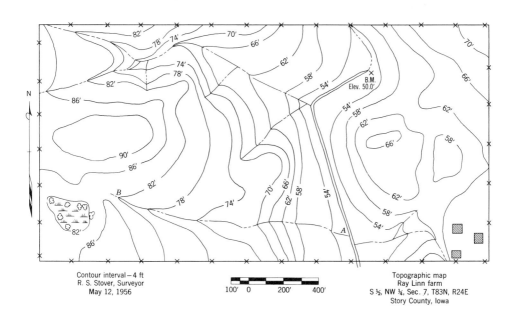

FIGURE 4.6   A topographic map of an 80-acre farm.

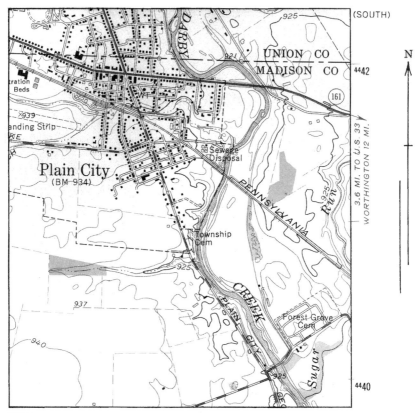

FIGURE 4.7   U.S. Geological Survey topographic map from Plain City, Ohio, quadrangle with a 5-ft contour interval. (Covers about the same area as the soil survey map in Fig. 4.5.)

2500 private retailers. For maps east of the Mississippi River, order from 1200 S. Eads St., Arlington, Virginia 22202; and for maps west of the Mississippi River, from Box 25286, Building 41, Federal Center, Denver, Colorado, 80225.

GROUND SURVEYS

The type of map survey selected will depend on the equipment available and the experience of the surveyor.

4.6   Level and Steel Tape Grid Survey

A good topographic map can be made by the grid method illustrated in Fig. 4.8. The field shown in the sketch is laid out in a square grid pattern. In the case shown, the grid lines are run parallel to the south and west boundaries of

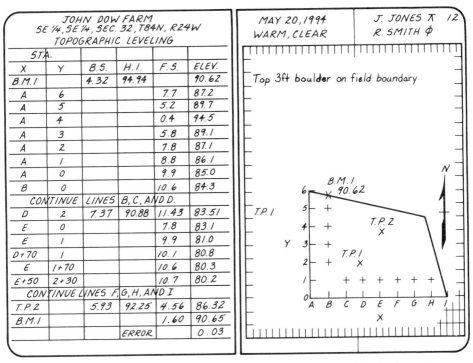

JOHN DOW FARM
SE ¼, SE ¼, SEC. 32, T84N, R24W
TOPOGRAPHIC LEVELING

STA. X	STA. Y	B.S.	H.I.	F.S.	ELEV.
B.M. 1		4.32	94.94		90.62
A	6			7.7	87.2
A	5			5.2	89.7
A	4			0.4	94.5
A	3			5.8	89.1
A	2			7.8	87.1
A	1			8.8	86.1
A	0			9.9	85.0
B	0			10.6	84.3
CONTINUE LINES B, C, AND D.					
D	2	7.37	90.88	11.43	83.51
E	0			7.8	83.1
E	1			9.9	81.0
D+70	1			10.1	80.8
E	1+70			10.6	80.3
E+50	2+30			10.7	80.2
CONTINUE LINES F, G, H, AND I					
T.P. 2		5.93	92.25	4.56	86.32
B.M. 1				1.60	90.65
			ERROR		0.03

MAY 20, 1994
WARM, CLEAR
J. JONES ⊼
R. SMITH ϕ
12
Top 3ft boulder on field boundary
B.M. 1
90.62
T.P. 1
T.P. 2
T.P. 1
Y
X
N
6
5
4
3
2
1
0
A B C D E F G H I

FIGURE 4.8 Field notes for topographic survey using the grid method for horizontal control.

the field. Some persons prefer to lay out the grid system on perfect squares for easy plotting on graph paper. Grid points may be readily identified by the scheme noted in the sketch. For example, point *E* 5 is the intersection of the east-west line through 5 with the north-south line through *E*. It is not necessary to stake each grid point. If a tall stake is placed at each point of the east-west lines through 0 and through 1 and at each point of the north-south lines through *A* and through *B*, it is possible to locate oneself at any other grid point in the field by lining up on the appropriate pairs of tall stakes.

The dimensions of the grid will depend on the topography and the detail desired. For gentle uniform slopes, a 100-ft grid pattern is usually satisfactory. The sides of the field are taped and the internal angles at the corners are measured. A check on the accuracy of the measurement of angles may be made by applying the rule stating that the sum of the internal angles of any closed figure bounded by straight lines is $(n-2)180°$, where n is the number of sides bounding the figure. If equipment is not available for measuring angles, the shape of the field may be determined by chaining the diagonals in addition to the sides. With the grid system located on the ground, the elevations of the various grid points are determined by profile leveling. That is, a foresight is taken to each grid point, and the elevations are calculated as indicated in Fig. 4.8. In addition to all grid intersections, intermediate points should be taken at high or low elevations and where slope changes occur. For example, a point 70 ft from D to E on line 1 is shown as $(D+70)1$ in Fig. 4.8. Likewise $(E+50)(2+30)$ is 50 ft from E and 30 ft from line 2. This system of identifying points is also convenient when the computer is used to make a contour map.

4.7   Electronic Total Station and Transit Surveys

As discussed in Chapter 3, a transit can measure accurately both vertical and horizontal angles as well as stadia distances. Points in the field can be located by taking azimuth angles and stadia distances. Elevations, other than level readings, can be obtained by calculating vertical distances from vertical angles and stadia distances. Some transits have a special (Beaman) arc scale that simplifies these conversions. Formats for keeping field notes can be found in surveying texts, such as McCormac, 1985.

Electronic total station instruments can perform most of the functions more accurately and quickly than a transit. As discussed in Chapter 2, distances are measured with an electronic distance measurement system. The principal disadvantage of these instruments is their high cost.

4.8   Rotating Beam Level Surveys

As described in Chapter 3, the electronic or laser beam level establishes a horizontal plane from which rod readings are taken to obtain elevations. The elevation of the beam is obtained from a bench mark as with a level. The Lenker rod with a light beam detection unit allows reading the elevations directly. When the detector unit is on the beam, a signal is given. Using an aerial photograph or suitable map, points are located, and the rodperson records the elevation directly on the map. Only one person is needed to make a laser survey. A special detector used on earth-moving equipment can also be mounted on a vehicle for surveying. With a suitable recorder, a record of distance and elevation can be obtained. Such a detector follows the beam automatically. Some contractors survey the ground surface prior to drain installation with this type of equipment. Land-grading machines equipped with a laser can make surveys of the land surface prior to leveling or grading to determine the cut or fill.

SATELLITE AND COMPUTER MAPPING SYSTEMS

Spatial data from satellites and other sources can be entered into a computer to produce an accurate map in a shorter time than older land-based and manually produced maps. Spatial data include physical dimensions, such as points, lines, areas, elevations, and geographic location on the ground. The computer is essential to the development of these systems because of the large amount of data collected. Improved maps will be of great assistance in conservation planning, management of natural and human resources, decision-making by governmental agencies, and many other uses. Several examples of these systems were evident during the Middle East Gulf War. Many systems are being developed and applied, and much research is being conducted worldwide.

4.9 Geographic Information Systems (GIS)

A geographic information system (GIS) is a comprehensive information management tool that includes aspects of surveying, mapping, cartography, photogrammetry, remote sensing, landscape architecture, and computer science (TeSelle, 1991). It allows resource managers to input, manage, analyze, manipulate, and display geographic data. Although there are some differences of opinion as to the definition of a GIS, a true GIS can be distinguished from other cartographic systems through its capacity to conduct spatial searches and to make overlays that actually generate new information. A project can be large in scope, such as the analysis of global data, or can involve the analysis of a single farm field for recommending a conservation practice.

Input data for the GIS can be obtained from satellites, aerial photographs, ground-based data, or other sources. It must have a digital base. One of the advantages of satellite data is that geographic location is by latitude and longitude, but other coordinates may be specified. The GIS data base may include land ownership, acreage, soil properties, distances, buildings, assessed value, topography, land cover, roads, streams, field boundaries, and other desired information. As shown in Fig. 4.9, some of these data base layers are used in a GIS to produce interpretative maps and their related data. For example, the GIS system can create a map depicting highly erodible cropland areas.

The Natural Resource Conservation Service (NRCS) has adopted a long-

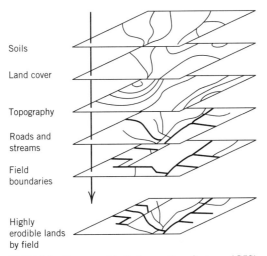

Soils

Land cover

Topography

Roads and streams

Field boundaries

Highly erodible lands by field

FIGURE 4.9 Geographic Information System (GIS) maps. (From TeSelle, 1991.)

term strategy of implementing GIS at the local (3000 offices), state, and national levels within the agency. By working with one or more map layers, a GIS can quickly and accurately perform the following: (1) compute acreages (soil types by fields), (2) measure distances, (3) combine maps with different data, (4) import photo and satellite imagery, (5) compute data relationships among map layers, (6) generate distance zones (e.g., areas within a given distance from a stream), (7) change maps to a desired scale, (8) generate slope, aspect, and three-dimensional maps, (9) print or plot and tabulate reports, (10) digitize existing maps for GIS use, and (11) provide analytical tools for addressing water quality issues.

The NRCS and the U.S. Geological Survey (USGS) are changing their traditional aerial photographs to digital orthophotos, which are very accurate photo image maps (Rohaley, 1992). NRCS began orthophoto production in 1993 and plans completion by 1998. Since soil, farm plan maps, and others have been in use for over 50 years by local and federal agencies, the user should be aware that these maps are not as accurate as the orthophotos being developed for the GIS.

NRCS selected the Geographic Resources Analysis Support System (GRASS) as the support software for GIS. This software was developed by the U.S. Army Corps of Engineers and is used and supported by several federal agencies, academia, and private industry. GIS software packages range in price from less than $1000 to $100 000, with a median of about $20 000 (TeSelle, 1991). A user-friendly system is being developed to allow NRCS staff to generate GIS outputs with little knowledge of GRASS or GIS.

Two examples that describe NRCS applications of GIS are for conservation farm planning and for national-level strategic planning. In farm planning, the planner would key in on the farmer's name and farm from the geographic data base computer files, as illustrated in Fig. 4.10. Conservation plan maps (Chapter 1) and alternative conservation treatment maps would be generated from the GIS to help arrive at an appropriate solution. A few examples are given in Fig. 4.10. In the future, it may be possible for farmers to access the data bases on their computers. At the national level, some aspect or component regarding the implementation of the federal farm bill may be desired, such as the effectiveness of conservation tillage for complying with erosion control on a county, state, or national basis. Other federal agencies could use GIS to track the spread of diseases in animals or to analyze the relationships among water quality problems, natural resources, and human activities. The Forest Service is using GIS in support of pest, wildlife habitat, and fire management.

City, township, and county governments in many states are rapidly converting to GIS maps for roads, utilities, property records, real estate appraisals, and other uses because of their greater accuracy and ease of retrieval and storage. In a few years, most map records will be available on computers. A large number of map layers (Fig. 4.9), each with different information, can be combined to produce a map with almost any scale, size, or detail desired.

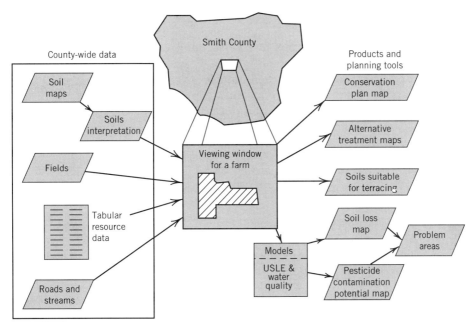

FIGURE 4.10 Example scenario using GIS for conservation planning. (From TeSelle, 1991.)

4.10 Global Positioning System (GPS)

The GPS space satellite system was developed in the late 1980s by the Department of Defense primarily to locate planes, ships, submarines, and other military items. Several orbiting satellites located at a fixed distance above the earth send out radio signals from which precise horizontal and vertical positions on the earth's surface can be obtained. When all satellites have been placed, an observer will be able to receive signals from nearly any point on the earth's surface. The accuracy is so good that the present permanent horizontal-control monuments established by the National Geodetic Survey may not be needed (McCormac, 1985). With proper equipment, horizontal control can be within one inch. Inexpensive equipment is available with less accuracy.

Global Positioning System (GPS) has many applications for soil and water management, such as land surveys and the control of moving equipment in the field. Land-leveling equipment may be developed to control automatically the depth of cut or fill and to deposit the soil at the depth and location desired. Tractors could be controlled to till, plant, cultivate, and harvest the crop, as well as to apply fertilizer on the field with variable amounts as specified by soil conditions and crop requirements. Some of these systems have already been developed using lasers and other equipment, but not necessarily requiring GPS input. Landing of aircraft by GPS rather than by radar appears to have great potential. These and many other possible uses of GPS will likely be developed.

4.11   Other Computer-Based Systems

Many computer-generated maps for special purposes can be prepared from ground-based data. Such systems are generally more economical than GIS. A plotter is required that will take the output from the computer and place it at the appropriate location on the map. For example, a pipe drainage map may start with a contour map, the location of outlets, and the design layout of the drainage system. As the work progresses, the map is corrected for changes made during installation. Changes or corrections are easily and quickly made so that the final map is accurate. Profiles of the drain lines can also be prepared if desired.

PREPARATION OF TOPOGRAPHIC MAPS

4.12   Characteristics of Contour Maps

A contour is an imaginary line of constant elevation on the surface of the ground. The shoreline of a lake is a contour readily seen in nature. A contour line on a map is a line connecting points on the map that represent points on the surface of the ground with the same elevation. The elevation of a contour line is given by a number that appears on the contour line. The following characteristics of contour lines are useful guides in drawing and interpreting contour maps:

1. The horizontal distance between contour lines is inversely proportional to the slope. Hence, on steep slopes the contour lines are spaced closely.
2. On uniform slopes the contour lines are spaced uniformly.
3. On plane surfaces the contour lines are straight and parallel to one another.
4. As contour lines represent level lines, they are perpendicular to the lines of steepest slope. They are perpendicular to ridge and valley lines where they cross such lines.
5. All contour lines must close on themselves, either within or outside the borders of the map.
6. Contour lines cannot merge or cross one another except in the rare cases of vertical or overhanging cliffs.
7. A single contour line cannot lie between two contour lines of higher or lower elevations, except in very rare instances.

A profile of the soil surface (elevations along a given line) may be plotted from a contour map as illustrated in Fig. 4.11. This profile along waterway *AB* in the south portion of the farm (Fig. 4.6) shows graphically the slope in the channel. Such information could be used for waterway design as discussed in Chapter 8.

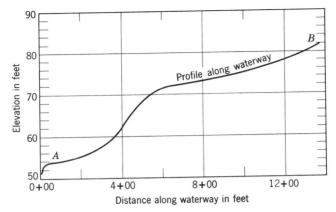

FIGURE 4.11   Ground-surface profile along line *AB* in Fig. 4.6.

4.13   Map Construction

Where the grid system of topographic mapping is used, the field boundaries and the grid are reproduced on paper. In any system of topographic mapping, the observed elevations are noted on the preliminary drawing as shown in Fig. 4.12.

MAP SYMBOLS   Map data gathered from a survey do not attain full usefulness until they are clearly presented on paper to form the finished map. Some features of the landscape occur so frequently that standard symbols have been adopted to facilitate their representation on maps. Figure 4.13 shows a number of common map symbols.

SCALE   The relationship between distances on the map and distances on the ground is the scale of the map. If 1 in. on the map represents 100 ft on the ground, then the scale will appear on the map as 1 in. = 100 ft. This is often conveniently shown on the map in graphic form, as indicated in the sample title block of Fig. 4.14. The graphic presentation of the scale is particularly desirable in that if the map is reduced or enlarged, the graphic presentation of the scale is changed accordingly; whereas a written scale becomes incorrect. A scale of 1 in. = 100 ft may also be written 1:1200, a ratio of map to ground distance in any units.

Selection of the scale is based on the use of the map, the area covered, and the detail desired. In general, the scale should conform to one of those found on a standard engineer's scale. These are 1 in. equals 10, 20, 30, 40, 50, or 60 ft, or any multiple of 10 of these values. A 160-acre farm can be conveniently represented on a 17-in. by 22-in. sheet to a scale of 1 in. = 200 ft.

TITLE BLOCK   The following information should appear on the map or on a neatly arranged title block such as Fig. 4.14.

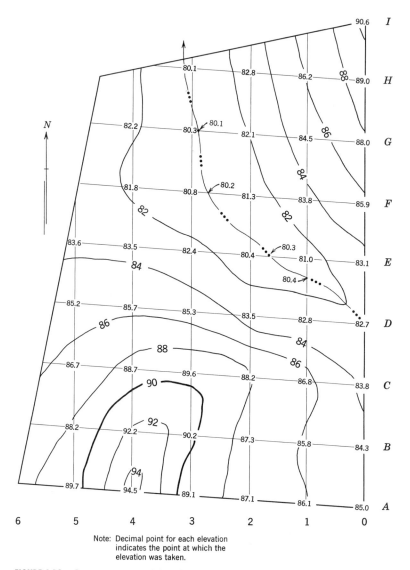

FIGURE 4.12 Contour map with a 2-ft contour interval. (Construction details included to show survey points for elevation.)

1. An arrow indicating true north, magnetic north, or both.
2. A legend or key to symbols used if other than conventional symbols (Fig. 4.13).
3. A graphic scale.
4. A statement of the map type or the purpose of the map.
5. The name of the tract mapped.

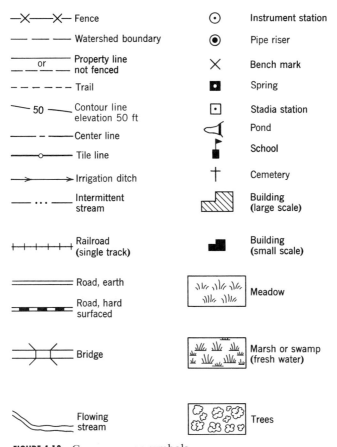

FIGURE 4.13 Common map symbols.

6. The legal description of the tract mapped.
7. The name of the project or purpose of the map. (This may be combined with item 4 in many instances.)
8. The names of the surveyor and draftsman.

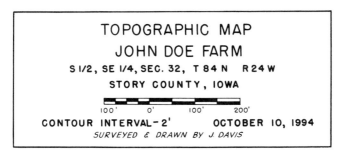

FIGURE 4.14 Example title block for a map.

CONTOUR LINES On a contour map, the vertical distance represented by the spacing between adjacent contour lines is the contour interval. The same contour interval should be maintained throughout a given map to avoid confusion and misinterpretation. Selection of the contour interval will depend on the slope of the ground, the scale of the map, the detail of the survey, and the purpose of the map. The contour interval together with the scale of the map should be selected to avoid excessive crowding of the contour lines.

Seldom does a contour line pass directly through a point of observed elevation. The locations of the contour lines are determined by interpolation between two adjacent points of known elevation. It is assumed that the ground has a uniform slope between such points. For example, in Fig. 4.12 the difference in elevation between points $H1$ and $H2$ is 3.4 ft. From $H1$ to the 86-ft contour line there is a difference in elevation of 0.2 ft. Thus, the 86-ft contour line will fall 0.2/3.4 or $\frac{1}{17}$ of the horizontal distance from $H1$ to $H2$ away from $H1$. The 84-ft contour is 1.2 ft above $H2$; it will fall 1.2/3.4 or $\frac{6}{17}$ of the horizontal distance from $H2$ to $H1$ away from $H2$. This type of reasoning is applied to find the location of contour lines between each two adjacent points of known elevation.

Since the slope is assumed to be uniform between points, the rodperson should keep this in mind when the field survey is made. Elevations should always be taken at the low points and at the high points. For example, in Fig. 4.12, the point F (2 + 70) is a low point (elevation 80.2) and is necessary to locate the flow channel on the map.

The contour lines are drawn in as smooth freehand lines of uniform weight. Usually, every 10-ft, a contour line (see 90-ft contour in Fig. 4.12) will be drawn in heavier than the intermediate lines. This enhances the readability of the map.

Numerous computer programs are available to generate contour maps from grid or transit data. Users should be careful in using units, scales, and correct specifications of the *x-y* coordinate system. All computer-generated maps should be compared to field notes to ensure data have been entered correctly.

REFERENCES

McCormac, J. C. (1985) *Surveying*, 2nd ed. Prentice–Hall, Englewood Cliffs, New Jersey.

Moffitt, F. H., and H. Bouchard (1992) *Surveying*, 9th ed. Harper Collins Publishers, New York.

Nathanson, J. R., and P. Kissam (1988) *Surveying Practice*, 4th ed. McGraw–Hill, New York.

Onsrud, H. J., and D. W. Cook (eds.) (1990) *Geographic and Land Information Systems for Practical Surveyors, a Compendium.* Amer. Congress on Surveying and Mapping, Bethesda, Maryland.

Rohaley, G. M. (1992) "Orthophotos to Cover the Continent," *Soil and Water Conservation News*, 13(2):15–16.

TeSelle, G. W. (1991) "Geographic Information Systems for Managing Resources." In *U.S. Department of Agriculture, Yearbook of Agriculture, 1991.* U.S. Government Printing Office, Washington, D.C.

PROBLEMS

4.1 Give the complete legal description of 40 acres located in the extreme north-west corner of a section $23\frac{3}{4}$ miles north of the base line and $12\frac{1}{4}$ miles east of the principal meridian.

4.2 What should be the sum of all the internal angles of a field with five straight sides?

4.3 Determine the average land slope (in percent) between points A and B, which are 300 ft apart, with each point on different contour lines. The contour interval is 2 ft and 2 other contour lines lie between points A and B.

4.4 What is the scale of an aerial photograph if the map distance between 2 points is 2.5 in. and the ground distance is 1650 ft?

4.5 Two points with elevations of 43.2 and 45.6 ft are 100 ft apart. How far should the 44-ft contour line be from the lower point?

4.6 The shortest horizontal distance between two consecutive contour lines is 40 ft. If the contour interval of the map is 2 ft, what is the land slope in percent?

4.7 On a sketch of Section 18, T10N, R4E locate the E $\frac{1}{2}$, SE $\frac{1}{4}$, NE $\frac{1}{4}$. How many acres does it contain?

4.8 Determine the distance in miles between the center of Section 36, T84N, R14W and the center of Section 1, T5S, R14W.

4.9 A topographic map having a scale of 200 ft per in. is reduced to 75 percent of its original size by photographic processes. What is the scale of the reduced map in feet per inch? What is the advantage of a graphical scale?

4.10 The map distance between the 40-ft and the 60-ft contour lines is 1.5 in. If the scale of the map is 1 in. = 100 ft, compute the land slope in percent.

4.11 In a field with a 1.6 percent slope, how far is the 42-ft contour line from a point with an elevation of 41.52 ft?

4.12 Describe three useful applications of your department or college computer GIS or topographic software package.

CHAPTER 5

RAINFALL AND RUNOFF

Water problems, whether involving too much, too little, or poorly distributed water, limit plant growth. The problems of reducing soil erosion, conserving water, increasing infiltration, managing irrigation, and providing drainage cannot be solved without consideration of rainfall-runoff relationships.

5.1 The Hydrologic Cycle

Water circulation on the earth consists of continual movement of water over, through, and beneath the surface. This pattern, depicted in Fig. 5.1, is commonly referred to as the hydrologic cycle. Precipitation may occur in many forms, including rain, snow, and sleet. Some precipitation evaporates partially or completely before reaching the ground, and some changes from one form to another before reaching the earth's surface. Precipitation reaching the surface can (1) be intercepted by vegetation, (2) infiltrate into the soil, (3) flow over the surface as runoff, or (4) evaporate. Evaporation can be from the surface of the soil, from free water surfaces, or from the leaves of plants through transpiration. Some of the total rainfall moves over the earth's surface as runoff to streams and rivers. Another portion of rainfall infiltrates the soil surface to water vegetation, recharge ground-water aquifers, or seep slowly to streams or the ocean.

5.2 Air Masses

The characteristics of air masses are controlling factors in the development of precipitation. These characteristics are determined by the association of the air mass with the surface (land or water) and seasonal radiation conditions. Fig. 5.2 shows the predominant air masses affecting the continental United States.

The air mass that contributes the largest amounts of water to the central and eastern portions of the United States is the warm moist air from the Gulf of Mexico. By contrast, air masses from Canada result in cold dry winters. On the West Coast, most of the precipitation comes from the Pacific Ocean. Other important air masses shown in Fig. 5.2 are dry winds from the Southwest and the cool damp air masses formed over the northern parts of the Atlantic and Pacific Oceans. The portion of precipitation originating from continental evaporation

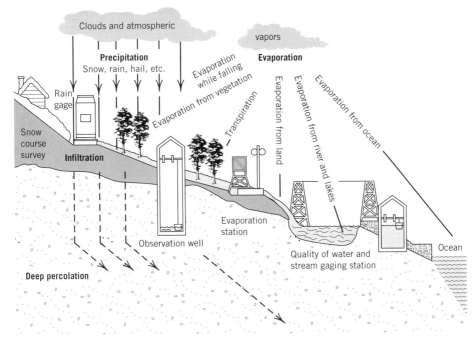

FIGURE 5.1   The hydrologic cycle.

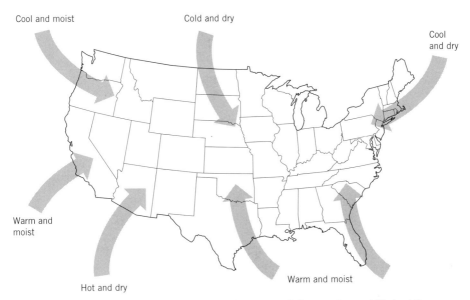

FIGURE 5.2   Characteristics of the dominant air masses of the continental United States.

is small compared to the oceans. Most of the water removed by evaporation is carried away by cool dry continental air masses.

5.3 Weather Maps and Forecasting

Current weather conditions and forecasts are commonly depicted by weather maps that show barometric pressure with the position of isobars, the ground position of warm and cold fronts, temperatures, locations of high- or low-pressure centers, and areas of precipitation. A portion of such a weather map is shown in Fig. 5.3*a*. Maps may also show additional information, such as dew-point temperature, wind direction and velocity, cloud cover, and changes in barometric pressure. Fig. 5.3*b* is a diagram interpreting the barometric pressures and wind patterns shown on the weather map in Fig. 5.3*a*.

A weather forecaster predicts the movement of air masses and associated precipitation. Weather forecasts depend on observations made at stations throughout the world, on observations from weather balloons, and on data collected by meteorological satellites. A forecast is relatively accurate for a short time, but the accuracy decreases as the prediction time increases.

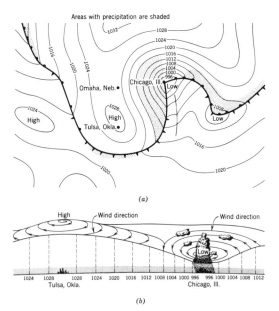

FIGURE 5.3   *(a)* Portion of a weather map in April showing cloudy weather in the East, rain in the Midwest, and clear skies in the Southwest. *(b)* Wind circulation around a high-pressure center at Tulsa and a low center at Chicago. (Courtesy of World Book Encyclopedia, 1961, and copyright by Rand McNally & Co., R.L. 70-S-35.)

STORM PATTERNS

There are three major types of storms: frontal, convective, and orographic (Fig. 5.4). In all three types, precipitation results from the lifting of warm moist air masses to cooler elevations. As the air cools, the water vapor in the air condenses and forms droplets or ice crystals. If the droplets or crystals become sufficiently large, they will fall as precipitation.

5.4 Frontal Storms

Frontal storms occur at the boundaries of warm moist air and relatively dry cold air. The warmer air rises above the denser cool air, leading to precipitation on both sides of the front. The most common frontal combination in the central and eastern United States is between tropical air from the Gulf of Mexico and polar air from central Canada. Frontal storms are more likely to occur during the fall, winter, and spring months when the temperature variations across frontal boundaries are 10° to 30°F, compared to differences of only a few degrees in the summer.

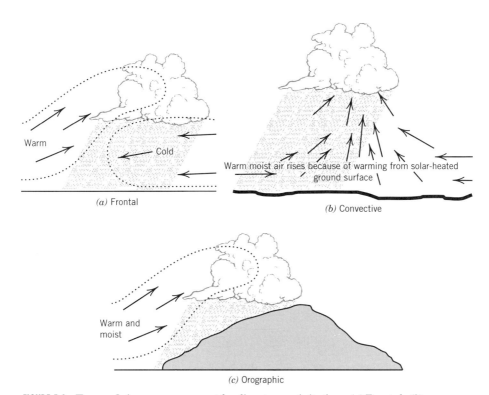

FIGURE 5.4 Types of air mass movement leading to precipitation: *(a)* Frontal; *(b)* Convective; and *(c)* Orographic.

5.5 Convective Storms

During the summer, the air in the central and eastern states has a high water content and, near the earth's surface, is subjected to considerable radiation heating. The heated air with water vapor rises upward, being cooled both by the surrounding air and by the expansion process. When it is cooled to its condensation point, it forms a cloud that may develop into a thunderstorm. Any source of ground heating, even a large fire, can start circulation of this type. The variation of temperature with height and moisture conditions in the atmosphere determine whether convective circulation develops into a thunderstorm. Though these storms generally cause precipitation over smaller areas, they may result in very intense precipitation and are particularly important causes of floods from small watersheds during the summer.

5.6 Orographic Storms

The influence of topography on precipitation is especially important. As air masses move over high elevations, such as mountain ranges, the air is pushed upward, cooled, and often reaches the condensation point. Thus, there is more precipitation on the windward side of the mountain range. This process causes the highest annual precipitation in North America along west coast mountain ranges and the high precipitation in the Appalachian Mountains. Conversely, as the air moves downslope, it is warmed and, having had most of the water condensed from it, deposits little precipitation. The mountain ranges along the western coast are largely responsible for the arid areas further inland. As these dry air masses reach the Central States, they are combined with moist air masses from the Gulf of Mexico, leading to increased precipitation as the East Coast is approached.

RAINFALL

When planning soil and water management systems, the expected rate and depth of rainfall, as well as the frequency of occurrence, need to be estimated. In the United States, recording rain gages, which give the amount and rate of rainfall, have been available only since about 1890. Daily precipitation records were kept in many areas several years before this time. Statistical analyses of these records allow the prediction of a storm, with a given probability of occurrence.

5.7 Intensity, Duration, and Frequency of Rainfall

One of the most important rainfall characteristics is rainfall rate or intensity, commonly expressed in inches per hour. The duration is the length of time from the beginning of a storm until its end. Very intense storms are not necessarily more frequent in areas having high total annual rainfall. Usually, storms of high intensity have fairly short durations and cover small areas. Storms covering large areas are seldom of high intensity, but may last for several days.

The combination of relatively high intensity and long duration occurs infrequently, but when it does occur, a large total amount of rainfall results. These infrequent storms cause much of our soil erosion damage and may result in devastating floods. They are frequently associated with warm-front precipitation or extreme conditions, such as hurricanes.

Storms with greater rainfall occur less frequently than smaller storms. For a given frequency of occurrence, rainfall intensity decreases with duration, whereas the depth increases with duration. This relationship is shown in Table 5.1. Frequency or probability of occurrence can be expressed as the *return period*, which is defined as the average number of years within which a given rainfall event will be equaled or exceeded. The data in Table 5.1 are for a 10-year return period. Rainfall intensity and amount increase with the return period, as shown in Table 5.2. Rainfall intensities vary greatly with geographic location, but the relative intensities for different durations (Table 5.2) are similar.

In planning any water-control facility, managers must balance cost of construction with economic and environmental costs of failure and safety risks. The desired return period storm is generally selected that will provide the most acceptable compromise to these costs and risks. Planning for a smaller storm with a periodic failure may be more economical than planning for a larger storm with a corresponding high investment cost, if the risks of failure can be tolerated.

Large amounts of rainfall data are available from many stations in every state. Rainfall maps showing lines of equal rainfall have been prepared by the Weather Service for 1- and 24-h durations and return periods of 2 and 100 years (Hershfield, 1961). By using these maps and a procedure developed by Weiss (1962), the intensities shown in Table 5.2 were developed for St. Louis, Missouri. Similar tables can be prepared for any location in the United States for varying return periods and/or durations. An alternative procedure is to use an estimate for St. Louis rainfall and then adjust that value with a geographic rainfall factor from Fig. 5.5. An example of this method is presented in Example

TABLE 5.1 Storm Duration and Rainfall Intensity and Depth at St. Louis, Missouri, for a Return Period of 10 years

	Duration of Rainfall					
	10	30	60	120	360	144 min
	(0.17)	(0.5)	(1)	(2)	(6)	(24 h)
Intensity, in./h	7.7	3.7	2.3	1.4	0.6	0.2
Depth, in.	1.3	1.8	2.3	2.7	3.6	4.9

TABLE 5.2   Rainfall Intensity, Total Rainfall Amount, and Relative Intensity for Different Return Periods, for a 24-h Duration Storm for St. Louis, Missouri

	Return Period in Years					
	2	*5*	*10*	*25*	*50*	*100*
Intensity, in./h	0.16	0.18	0.20	0.24	0.28	0.30
Amount, in.	3.84	4.32	4.90	5.76	6.72	7.20
Relative Intensity[a]	0.8	0.9	1.0	1.2	1.4	1.5

[a]Compared to a 10-year return period storm.

5.1. It is an approximate method and should be used only when local data are not available.

Example 5.1.   Determine the rainfall intensity and total rainfall for a 24-h duration storm that will occur once in 10 years at Raleigh, North Carolina.

Solution.   From Table 5.2, read an intensity of 0.2 in./h for a 24-h storm. By interpolation from Fig. 5.5, the geographic rainfall factor is 1.1 for Raleigh. The average rainfall intensity is

$$i = 0.2 \text{ in./h} \times 1.1 = 0.22 \text{ in./h } (5.6 \text{ mm/h})$$

The rainfall amount in 24 h is

$$i = 0.22 \text{ in./h} \times 24 \text{ h} = 5.3 \text{ in. } (135 \text{ mm})$$

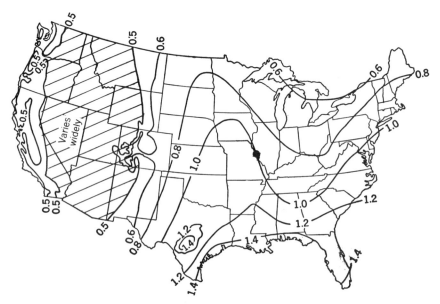

FIGURE 5.5   Geographic rainfall factor to estimate storm intensities and amounts relative to St. Louis, Missouri. (Source: Hamilton and Jepson, 1940.)

RUNOFF

When rainfall contacts the land surface or when water is released from melting snow, some water will infiltrate, some will be stored in surface depressions until the depressions are filled, and the excess will flow overland. This overland flow becomes concentrated in increasingly larger runoff channels and eventually stream channels, lakes, and oceans.

Planning for surface drainage systems, water-control structures, and water-storage reservoirs that handle natural surface runoff may require estimating peak runoff rate and/or total runoff volume. In the planning process, the manager must determine both storm (or snowmelt) properties and watershed properties. Storm properties include depth, duration, distribution, and return period of a given event. More intense storms produce greater peak runoff rates, whereas longer duration storms produce more total runoff. Watershed properties include geologic and soil hydrologic properties, land use and/or farming practice, and soil water content at the beginning of the runoff event. Soils higher in clay will have greater runoff, whereas sandy soils have less runoff. Smaller watersheds, forests and rangeland generally have lower runoff, whereas row crop farming systems will have greater runoff. Topographical features that affect runoff include area, length, slope, shape, and relationship of watershed orientation to the direction of storm travel. Shorter, steeper watersheds will have greater peak runoff rates than longer, flatter watersheds. Urban development can significantly alter runoff patterns, as it impacts on many of these runoff determining factors.

5.8 Runoff Hydrographs

A hydrograph is a graph of watershed runoff rate versus time. Fig. 5.6 is a typical hydrograph from a small watershed. The storm in Fig. 5.6 began at time zero. Before runoff could begin, the ground surface had to be wet enough that the rainfall rate exceeded the infiltration rate and surface depressions overflowed with water. For the storm in Fig. 5.6, it took about 2 hours for these processes to occur. Once runoff began, the rate increased rapidly, as seen in the rising limb of the hydrograph, because parts of the watershed more remote from the measuring point began to contribute to the runoff. When all the watershed was contributing to the runoff, the peak runoff rate of 400 cfs occurred 4.3 h from the start of the storm. After the storm was over, the flow from the watershed slowly decreased, as illustrated by the falling limb of the hydrograph. The runoff ended 10.3 h after the start of the storm. The total amount of runoff from the storm can be found by integration (graphical or mathematical) of the area under the hydrograph curve.

Hydrographs are useful in routing flood flows through reservoirs and river systems to predict maximum flood depths. The predictions of the major flooding experienced by the Upper Midwest in the summer of 1993 were determined by combining hydrographs from the various tributaries to the

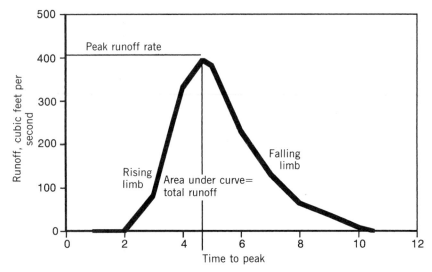

FIGURE 5.6 A typical runoff hydrograph from a small watershed.

Mississippi and Missouri Rivers. Analyses of hydrographs can tell planners much about the hydrology of a watershed, especially peak flow rate and volume of flow.

Numerous methods, including sophisticated computer prediction programs, have been developed to determine hydrograph shapes, runoff volumes, and peak runoff rates. A runoff peak rate and volume prediction method similar to that employed by the NRCS will be presented in this text, but the reader is encouraged to determine local codes and practices before applying the principles presented in this text to a local problem.

5.9 Curve Number

With the NRCS method, the effects of soil and surface vegetation, as well as management, on runoff are estimated with the curve number *(CN)*. The curve number is an approximate percentage of incident rainfall that becomes runoff. Starting in the 1930s, curve numbers were developed from numerous runoff research plots scattered throughout the United States by comparing runoff amounts with measured rainfall. Some of these plots are still providing runoff data used to develop more sophisticated runoff prediction models. Table 5.3 presents typical curve numbers for the hydrologic soil groups described in Table 5.4. An impervious area (for example, a parking lot) would have a curve number of nearly 100, whereas a sandy forest soil, with only a small fraction of the rainfall becoming runoff, could have a curve number as low as 25. Frequently, subareas of a watershed will have different curve numbers. For these heterogeneous watersheds, a weighted average may be used to find the curve number, as demonstrated in Example 5.2.

TABLE 5.3 Runoff Curve Numbers for Average Antecedent Rainfall Conditions

Land Use and Hydrologic Condition	A	B	C	D
	Runoff Curve Numbers (CN) Hydrologic Soil Groups [a]			
Fallow	77	86	91	94
Row crop				
Poor	72	81	88	91
Good	67	78	85	89
Contoured—good	65	75	82	86
Contoured and terraced—good	62	71	78	81
Small grain				
Poor	65	76	84	88
Good	63	75	83	87
Contoured and terraced—good	60	72	80	83
Meadow—continuous grass, no grazing	30	58	71	78
Pasture				
Poor	68	79	86	89
Good	39	61	74	80
Woods-grass combination (orchard)	35	56	70	77
Woodland				
Poor	45	66	77	83
Good	30	55	70	77
Farmsteads	59	74	82	86
Roofs and paved areas	—	—	—	90

Source: SCS (1990)

[a]See description in Table 5.4.

TABLE 5.4 SCS Hydrologic Soil Groups

Soil Group	Description	Final Infiltration Rate (in./h)
A	Lowest runoff potential. Includes deep sands with little silt and clay and deep permeable loess.	0.3–0.5
B	Moderately low runoff potential. Mostly sandy and loess soils less deep than A, but above average infiltration.	0.5–0.3
C	Moderately high runoff potential. Comprises shallow soils and high clay soils with below average infiltration.	0.04–0.15
D	Highest runoff potential. High clay soils with high shrink/swell potential and some shallow soils with impermeable subhorizons.	0.0–0.04

Example 5.2.   Find the curve number for the watershed in Fig. 5.7 if it contains 70 acres of row crop (poor practice) in a loam soil, and 40 acres of rolling sandy loam woodland (good practice). Both soils are in hydrologic group B.

Solution.   From Table 5.3, the curve numbers are determined to be 81 and 55. These values are multiplied by the respective areas of each crop, the products summed, and the total divided by the total area to determine the weighted average curve number in the following table.

Land Use	Area, A	Curve No, CN	A × CN
Row crop	70	81	5670
Woodland	40	55	2200
Total	110		7870

The weighted average is then found by division.

$$CN = 7870/110 = 72$$

The soil water content at the beginning of a storm describes its *antecedent hydrologic condition*. The curve numbers presented in Table 5.3 are for average antecedent hydrologic conditions. Soils that have higher than usual water content at the beginning of a storm will have greater runoff amounts than the same soils with average or low antecedent hydrologic conditions. Details on estimating the quantitative effects of antecedent water by adjusting the curve number can be found in Schwab et al. (1993) and other hydrologic references.

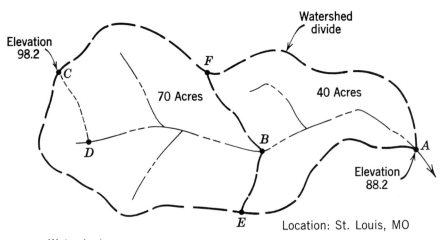

FIGURE 5.7 Watershed map for Examples 5.2 and 5.3.

5.10   Peak Runoff Rate

There are many methods for determining peak runoff rates, ranging from simple equations to complex computer models. The peak runoff depends on length, slope, and area of the watershed, as well as curve number. Greater runoffs are associated with watersheds with larger areas, shorter lengths, steeper slopes, and higher curve numbers. One of the hydrologic features of a watershed important in determining the peak runoff rate is the *time of concentration (tc)*. The time of concentration is the time that it takes for a drop of rainwater to travel from the most hydraulically remote point of a watershed to the outlet. It is approximately equal to the "time to peak" shown on the hydrograph in Fig. 5.6. On watersheds with a greater *tc* but the same area, the peak runoff rate will be lower. A simplified NRCS method of predicting the time of concentration and peak runoff is presented in Barfield et al. (1981). This method is recommended for the type of storms common in the central United States and for peak runoff rates from watersheds with relatively uniform curve numbers. The duration of all storms with this method is 24 h. The relationship between time of concentration, length of the watershed, curve number, and average slope of the watershed can be estimated from the NRCS relationship (Barfield et al., 1981 and Schwab et al., 1993).

$$tc = \frac{L^{0.8}\left(\dfrac{1000}{CN} - 9\right)^{0.7}}{1140s^{0.5}} \tag{5.1}$$

where *tc* is the time of concentration in hours, L is the watershed length in feet, CN is the curve number, and s is the average percent slope of the watershed. There are numerous relationships between the time of concentration and the peak runoff rate from a storm. One relationship between the time of concentration and peak runoff rate is (Barfield et al., 1981)

$$\log(q) = 2.51 - 0.7\log(tc) - 0.15\,(\log(tc))^2 + 0.071\,(\log(tc))^3 \tag{5.2}$$

Fig. 5.8 combines Eqs. 5.1 and 5.2 to predict the time of concentration and the peak runoff rate in a graphical form. To predict peak runoff for other regions, or for more complex watersheds, advanced references like Barfield et al. (1981) or Schwab et al. (1993) should be consulted.

Example 5.3.   Determine the time of concentration and the peak runoff rate for a 50-year return period storm if the watershed in Fig. 5.7 is near St. Louis, Missouri. The areas and curve numbers are described in Example 5.2. The maximum length of flow *(CDBA)* is 2000 ft.

Solution.   From Table 5.2, the 50-year 24-hour storm is 6.72 in. The slope of the watershed is

$$s = \frac{98.2 \text{ ft} - 88.2 \text{ ft}}{2000 \text{ ft}} \times 100\% = 0.5\%$$

The flow length of 2000 ft is found on the upper left quadrant of Fig. 5.8. From here, a line is drawn right until it intersects the 0.5 percent line. A second line is drawn

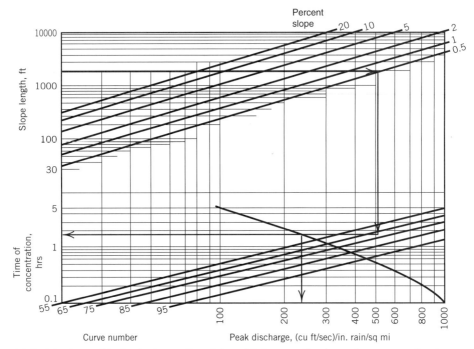

FIGURE 5.8 A graphical solution to Eqs. 5.1 and 5.2 to predict time of concentration in hours and peak runoff in cfs/sq mile/inch. This applies to natural watersheds for the central United States with 24-hour rainfall events on watersheds with small variations in curve numbers.

vertically downward from the point of intersection until it intersects the interpolated position of a curve number of 72. From this point, a line drawn to the left will give the time of concentration of *1.6 h.*

A line drawn vertically down from the intersection with the peak discharge curve gives a peak discharge of 225 cfs/in./sq mile. From this value, with 640 acres in 1 sq mi, the peak runoff rate is

$$q = 225 \text{ cfs/in./sq mile} \times 6.72 \text{ in.} \times \frac{110 \text{ acres}}{640 \text{ acres/sq mile}} = 260 \text{ cfs } (7.35 \text{ m}^3/\text{s})$$

5.11 Total Runoff Volume

The total runoff volume from one or more storms is of interest where flood-control reservoirs or other structures are to be designed. The total annual runoff amount or water yield is needed when planning reservoirs to ensure adequate runoff is available for irrigation or water supply. Frequently, long-term water-yield estimates are based on stream flow records rather than individual runoff techniques.

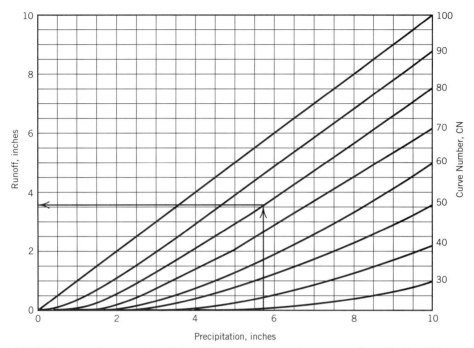

FIGURE 5.9 Runoff versus rainfall for different curve numbers as predicted by Eqs. 5.3 and 5.4.

The NRCS method estimates storm runoff using the curve number of the watershed and the relationship (SCS, 1990)

$$Q = \frac{(P - 0.2S)^2}{(P + 0.8S)} \tag{5.3}$$

where Q is the runoff volume in inches and P is the precipitation in inches. S is the precipitation surface storage in inches of water before the onset of runoff. Surface storage is estimated from the curve number by

$$S = \frac{1000}{CN} - 10 \tag{5.4}$$

A graphic solution to Eqs. 5.3 and 5.4 is presented in Fig. 5.9. If $CN = 100$, then $S = 0$, and runoff is equal to rainfall (which could occur on a smooth, impermeable paved or roofed area).

Example 5.4. Determine the runoff volume in inches for designing a storm-water detention reservoir. Assume a 25-year return period 24-h storm near St. Louis, Missouri. The soil is in hydrologic group C, with an average antecedent hydrologic condition. In the 100-acre urban watershed, 50 percent of the area is in a woods-grass mixture, and 50 percent in roads and roofed areas.

Solution. From Table 5.2 note that the 25-year 24-h storm will produce 5.76 inches.

From Table 5.3, note that the curve numbers for the two components of the watershed are 70 for the grass and 90 for the roofs, for a weighted average of

$$CN = \frac{70 \times 50 + 90 \times 50}{100} = 80$$

Entering Fig. 5.9 at 5.76 inches and going up to the position of the 80 line, and then left to the vertical axis, note that the runoff is 3.6 inches.

In addition to the depth of runoff as calculated in Example 5.4, the total volume of runoff is frequently required for hydrologic designs. It is assumed that the calculated depth of runoff comes from the entire area of the watershed. To obtain the volume of runoff, the depth is multiplied by the area of the watershed. Common units of runoff volume are acre-inches or acre-feet. Reservoir designers frequently use acre-feet in calculating storage volumes for water supply or flood routing.

Example 5.5.   For the 100-acre watershed in Example 5.4, what is the runoff volume in acre-feet for the given storm?

Solution.   From Example 5.4, the runoff depth was 3.6 in. Multiplying this depth by the area and converting inches to feet will give the runoff volume:

$$Volume = 3.6 \text{ in.} \times 100 \text{ acres} \times \frac{1 \text{ ft}}{12 \text{ in.}} = 30 \text{ ac-ft}$$

REFERENCES

Barfield, B. J., R. C. Warner, and C. T. Haan (1981) *Applied Hydrology and Sedimentology for Disturbed Areas.* Oklahoma Technical Press, Stillwater, Oklahoma.

Hamilton, C. L., and J. G. Jepson (1940) *Stock-water Developments; Wells, Springs, and Ponds,* U.S. Dept. Agr. Farmers' Bull. 1859. USDA, Washington, D.C.

Hershfield, D. M. (1961) "Rainfall Frequency Atlas of the United States," *U.S. Weather Bureau,* Tech. Paper 40, May.

Schwab, G. O., D. D. Fangmeier, W. J. Elliot, and R. K. Frevert (1993) *Soil and Water Conservation Engineering,* 4th ed. Wiley, New York.

U.S. Soil Conservation Service (SCS) (1990) *Engineering Field Manual for Conservation Practices.* USDA Soil Conservation Service, Washington, D.C.

Weiss, L. L. (1962) "A General Relation between Frequency and Duration of Precipitation," *Monthly Weather Rev.,* March:87–88.

PROBLEMS

5.1 For your present location, determine the rainfall intensity for a 10-year return period storm with durations of 60 min and 24 h, using values from Table 5.1 and Fig. 5.5. Compute the total rainfall that will fall for each of these durations.

5.2 Compare rainfall amounts obtained in Problem 5.1 to the actual records published by the Weather Service or from other local records. Compute the percentage difference for each of the storms.

5.3 Determine the peak runoff rate for a 10-year return period for a 100-acre watershed near St. Louis, Missouri. The watershed area is row crop, good condition, on a silty clay loam soil (group B). The watershed gradient is 0.5 percent and the maximum length of flow is 3540 ft.

5.4 Solve Problem 5.3 if the watershed was at your present location.

5.5 Determine the peak runoff rate at point B of the watershed shown in Fig. 5.7. The elevation of B is 91.0 ft and the distance BDC is 1200 ft. The watershed area contains 30 acres of pasture (poor condition) with a clay loam soil (group C) and 40 acres of row crop (good condition) with a sandy loam soil (group A). Assume a return period of 10 years.

5.6 Determine the peak runoff rate at point A in Fig. 5.7 if the curve numbers for areas $AEBF$ and $BECF$ are 60 and 35, respectively. The maximum length of flow for $ABDC$ is 2000 ft. Assume a return period of 10 years.

5.7 Determine the weighted curve number for a 120-acre watershed having 40 acres of woodland (good condition) on a clay loam soil (group C), and 80 acres of small grain (good condition) with a heavy clay soil (group D).

5.8 For your present location, determine the peak runoff rate for a 25-year return period storm for a 50-acre watershed in permanent woodland (good condition) on hydrologic soil group C. The slope is 8 percent, and the maximum flow length is 3000 ft.

5.9 A 25-acre watershed has a group B soil. For a 5-year return period rainfall rate of 3 in./h, compare the runoff rate from row crop, poor condition to pasture, good condition.

5.10 Determine the storm runoff volume in inches from a 50-year return period 24-h storm at your present location on a 60-acre watershed. One third of the watershed is in woodland (good condition), and two thirds in row crop (good condition). All soils are in group B.

5.11 Calculate the volume of runoff for the watershed described in Problem 5.10 in cubic feet and acre-feet.

CHAPTER 6
SOIL EROSION
BY WATER

Soil erosion is one of the most important natural resource management problems in the world. It is a primary source of sediment that pollutes streams and fills reservoirs. Total sheet and rill erosion estimates from cropland were 1.2 billion tons annually in 1992 in the United States, a decrease of 30 percent since 1982 because of improved management practices and a reduction in cropland acres due to the conservation reserve program (SCS, 1994).

Since the early 1970s, greater emphasis has been placed on erosion as a contributor to nonpoint pollution. *Nonpoint* refers to pollution from the land surface rather than from point sources like industrial wastes, feedlots, or gullies. Eroded sediment can carry nutrients, particularly phosphates, to waterways and contribute to eutrophication of lakes and streams. Adsorbed pesticides are also carried with eroded sediments, adversely affecting surface water quality.

The two major types of erosion are geological erosion and accelerated erosion, which results from human or animal activities. Geological erosion includes soil-forming as well as soil-eroding processes that generally maintain the soil in a favorable balance, usually suitable for the growth of most plants. Geological erosion has contributed to the formation of the soils and their distribution on the surface of the earth. This long-time eroding process caused many of the present topographical features, such as canyons, stream channels, valleys, and deltas. Human- or animal-induced erosion from tillage and removal of natural vegetation leads to a breakdown of soil aggregates and accelerated removal of organic and mineral particles.

Water erosion is the detachment and transport of soil from the land by water, including runoff from melted snow and ice. Types of water erosion include interrill (raindrop and sheet), rill, gully, and stream channel erosion. Water erosion is accelerated by farming, grazing, forestry, and construction activities.

The importance of soil losses from erosion is indicated by the effect of the degree of past erosion on crop yield, as shown in Fig. 6.1. Schertz et al. (1989) reported that much of the reduced yield observed on eroded soils was due to a decrease in the amount of water available to the plant on eroded soils. On some soils, these crop yield decreases can be largely overcome by higher fertilization levels. On other soils, particularly more shallow soils on sloping terrain, erosion may destroy productivity if appropriate conservation practices are not initiated (USDA, 1989). Erosion from watersheds with limited agriculture, or with con-

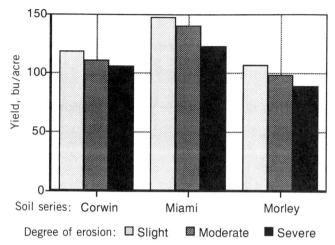

FIGURE 6.1   Effects of erosion degree and soil series on average corn yield (Schertz et al., 1989).

servation practices, is usually considerably less than from watersheds farmed intensively with no conservation practices.

6.1   Factors Affecting Erosion by Water

The major factors affecting soil erosion are climate, soil, vegetation, and topography. Of these, the vegetation, and to some extent the soil and topography, may be controlled. The climatic factors are beyond the power of humans to control.

CLIMATE   Climatic factors affecting erosion are precipitation, temperature, wind, humidity, and solar radiation. Temperature and wind are most evident through their effects on evaporation and transpiration. However, wind also changes raindrop velocities and the angle of impact. Humidity and solar radiation are somewhat less directly involved by their influence on the rate of soil water depletion.

The ability of rainfall to cause erosion is called rainfall *erosivity*. The relationship between precipitation characteristics, and runoff and soil loss is complex. In one study, the most important single measure of the erosion-producing factor of a rainstorm was a combination of rainfall energy and intensity (Wischmeier, 1959). Studies on individual erosion processes have found that interrill erosion varies with rainfall intensity to a power varying from 1.56 to 2.09 (Watson and Laflen, 1986). Rill erosion is a function of runoff rate, which depends on rainfall intensity and soil infiltration rates, and slopes.

SOIL   The capability of a soil to cause runoff, be detached, and be transported is known as soil *erodibility*. Physical properties of soil affect the infiltration capacity and the extent to which particles can be detached and transported. Soil detachment increases as the size of the soil particles or aggregates increase, and soil transport rate increases with a decrease in the particle or aggregate size. That is, clay particles are more difficult to detach than sand, but clay is more easily transported. The properties that influence erosion include soil structure, texture, organic matter, water content, clay mineralogy, and density, as well as soil chemical and biological characteristics.

VEGETATION   The major effects of vegetation in reducing erosion are (1) interception of rainfall by absorbing the energy of the raindrops and thus reducing surface sealing and runoff, (2) retardation of erosion by decreased surface velocity, (3) physical restraint of soil movement, (4) improvement of aggregation and porosity of the soil by roots and plant residue, (5) increased biological activity in the soil, and (6) transpiration, which decreases the amount of soil water, resulting in increased water storage capacity and less runoff. These vegetative influences vary with the season, crop, degree of maturity of the vegetation, soil, and climate. Residues from vegetation protect the surface from raindrop impact and improve the soil structure. Residue and tillage management practices can have a dramatic effect on soil erosion.

TOPOGRAPHY   Topographic features that influence erosion are degree of slope, shape and length of slope, and size and shape of the watershed. On steep slopes, runoff water is more erosive and can more easily transport detached sediment downslope. On longer slopes, an increased accumulation of overland flow tends to increase rill erosion. Concave slopes, with flatter slopes at the foot of the hill, are less erodible than convex slopes.

6.2   Raindrop Erosion

Raindrop or splash erosion is soil detachment and transport resulting from the impact of water drops directly on soil particles or on thin water surfaces. Raindrops break down and detach soil particles. The detached sediment can reduce the infiltration rate by sealing the soil pores, which increases runoff and sediment transported from the field. Although the impact of raindrops on shallow streams may not splash soil, it does increase turbulence, providing a greater sediment-carrying capacity.

During a storm, tremendous quantities of soil are splashed into the air, most particles more than once. The amount of soil splashed into the air is 50 to 90 times greater than the runoff losses. On bare soil, it is estimated that as much as 100 tons per acre are splashed into the air by heavy rains. The effect of a single drop as it strikes is shown in the high-speed photograph in Fig. 6.2. Splashed particles may move more than 2 ft in height and more than 5 ft laterally on level surfaces.

Factors affecting the direction and distance of soil splash are slope, wind, surface condition, and impediments to splash such as vegetative cover and mulches. On sloping land, the splash moves farther downhill than uphill, not only because the soil particles travel farther, but also because the angle of im-

FIGURE 6.2 Raindrop splash. Top: impact from a single drop. Bottom: splash pattern during a storm. (Courtesy U.S. Soil Conservation Service.)

pact causes the splash reaction to be in a downhill direction. Components of wind velocity up or down the slope have an important effect on soil movement by splash. Surface roughness and impediments to splash tend to counteract the effects of slope and winds. Contour furrows and ridges break up the slope and cause more of the soil to be splashed uphill. If raindrops fall on gravel, crop residue, or growing plants, the energy is absorbed and soil splash is reduced.

6.3   Sheet or Interrill Erosion

In the past, sheet erosion was considered the uniform removal of soil in thin layers from sloping land, resulting from sheet or overland flow. However, fundamental erosion studies indicate that this idealized form of erosion rarely occurs. During a storm, minute rilling takes place almost simultaneously with the first detachment and movement of soil particles. The constant meander and change of position of these microscopic rills obscure their presence from normal observation, hence establishing the false concept of sheet erosion. The eroding and transporting power of sheet flow is a function of the rainfall intensity, infiltration rate, field slope, and soil properties. Raindrop and sheet erosion are often combined and called interrill erosion.

6.4   Rill Erosion

Rill erosion is the detachment and transport of soil by a concentrated flow of water. Rills are eroded channels that are small enough to be removed by normal tillage operations. Rill erosion is the predominant form of erosion in most conditions. It is most serious where intense storms occur on soils with high runoff-producing characteristics and highly erodible topsoil. Rill erosion is a function of runoff rate, topography, and soil properties. Sites with longer or steeper slopes, or soils with more runoff, tend to experience more rill erosion.

6.5   Gully Erosion

Gully erosion produces channels larger than rills. These channels carry water during and immediately after rains and, as distinguished from rills, gullies cannot be obliterated by tillage. Fig. 6.3 shows an example of severe gully erosion. The amount of sediment from gully erosion is usually less than from upland areas, but the nuisance from fields divided by large gullies has been the greater problem. The best way to stop gully growth is to divert the water from the head of a gully to another waterway (Chapter 7) or to construct a concrete erosion control structure to absorb the energy of the falling water at the head of the gully (Chapter 8).

6.6   Estimating Soil Losses

Estimating soil losses or relative erosion rates for different management systems assists natural resource managers, farmers, and government agencies to evaluate existing management systems or to develop plans to decrease soil losses. In the period from 1945 until 1965, a method of estimating losses based on statistical analyses of field plot data from small plots located in many states was developed, which resulted in the Universal Soil Loss Equation (USLE). A revised version of the USLE (RUSLE) has been developed for computer applications, allowing more detailed consideration of management practices and topography for erosion prediction (Renard et al., 1991).

FIGURE 6.3 Severe gully erosion. (Courtesy U.S. Soil Conservation Service.)

Since the mid-1960s, researchers have been developing process-based erosion computer programs that estimate soil loss by considering the processes of infiltration, runoff, detachment, transport, and deposition of sediment. Numerous research programs have been developed, and computer applications are being improved for field use. The process-based Water Erosion Prediction Project (WEPP) model is replacing the USLE in the NRCS and other U.S. government agencies (Foster, 1988). The WEPP model predicts erosion from interrill and rill processes.

6.7 The Universal Soil Loss Equation

The USLE continues to be a widely accepted method of estimating sediment loss despite its simplification of the many factors involved. The average annual soil loss, as determined by Wischmeier and Smith (1978), can be estimated from the equation

$$A = RK(LS)CP \tag{6.1}$$

where A = average annual soil loss in tons/acre

   R = rainfall and runoff erosivity index for geographic location (Fig. 6.4)

   K = soil-erodibility factor (Table 6.1)

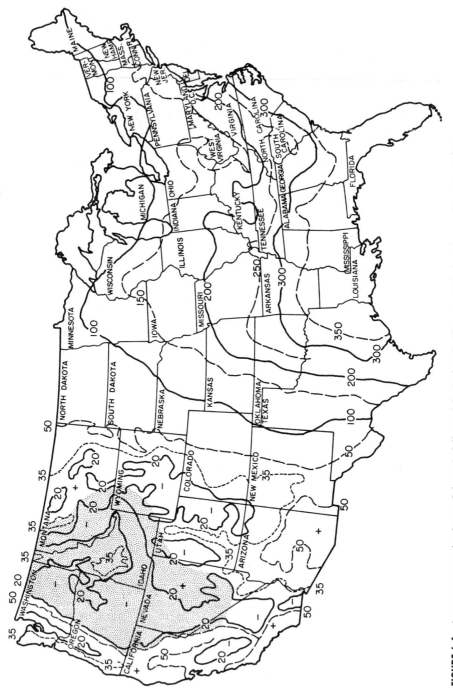

FIGURE 6.4 Average annual values of the rainfall erosivity index R. (Source: Wischmeier and Smith, 1978.)

TABLE 6.1 Soil-Erodibility Factor K by Soil Texture in Tons per Acre[a]

Textural Class	Organic Matter Content in Percent		
	0.5	2	4
Fine sand	0.16	0.14	0.10
Very fine sand	0.42	0.36	0.28
Loamy sand	0.12	0.10	0.08
Very fine loamy sand	0.44	0.38	0.30
Sandy loam	0.27	0.24	0.19
Very fine sandy loam	0.47	0.41	0.33
Silt loam	0.48	0.42	0.33
Clay loam	0.28	0.25	0.21
Silty clay loam	0.37	0.32	0.26
Silty clay	0.25	0.23	0.19

[a]Selected from USDA-EPA vol. I (1975) and are estimated averages of specific soil values. For more accurate values by soil types, use the local recommendations of the NRCS or state agencies (1 t/a = 2.24 Mg/ha).

LS = slope length and steepness factor (Fig. 6.5)
C = cover management factor (Example 6.2)
P = conservation practice factor (Chapter 7)

In developing the values for the factors of the USLE, a standard research plot was developed with a slope of 9 percent and a slope length of 72.6 ft (22 m), giving an LS of 1.0 (Fig. 6.6). The standard fallow condition had a C

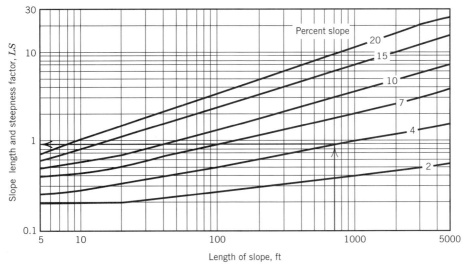

FIGURE 6.5 Slope length and steepness factor LS as a function of slope length and slope steepness. (Based on: McCool et al., 1987 and McCool et al., 1989.)

FIGURE 6.6 Typical USLE fallow erosion plot near Pullman, Washington. (Photograph by W. Elliot.)

factor of 1.0, and there were no practices, so $P = 1.0$. Fallow plots then determined the K factor, and nonfallow plots with various crops and plots with various practices were used to determine the other factors. In recent years, rainfall simulators assisted researchers to determine additional factors for many more conditions.

The average annual rainfall and runoff erosivity index R can be found from Fig. 6.4 for the continental United States. Estimates of R for single storms with 2- to 20-year return periods have also been formulated, and the values for 180 sites are given in Wischmeier and Smith (1978).

Soil erodibility K values are generally available from county soil surveys and local NRCS offices. Table 6.1 can serve as a general guide.

The topographic factor LS adjusts the predicted erosion rates to give greater erosion rates on longer and/or steeper slopes. The LS value can be determined from Fig. 6.5. The slope length is measured from the point where surface flow originates (usually the top of the ridge) to the outlet channel or a point downslope where deposition begins. The USLE assumes a uniform slope, but RUSLE can also consider nonuniform concave or convex slopes. Process-based erosion prediction programs can evaluate slopes of very complex shape and are also able to estimate downslope deposition.

Example 6.1. Determine the soil loss for the following conditions: Location, Memphis, Tennessee, $K = 0.1$ t/a, slope length = 400 ft, slope steepness = 10 percent, $C = 0.18$ (approximate for corn-corn-oats-meadow rotation with good management).

Solution. From Fig 6.4, read $R = 300$. From. Fig. 6.5, $LS = 2.4$. Assume that in the absence of any practice information, $P = 1.0$. Substituting into Eq. 6.1:

$$A = 300 \times 0.1 \times 2.4 \times 0.18 \times 1.0 = 13 \text{ t/a } (29 \text{ Mg/ha})$$

The cover-management factor C includes the effects of cover, crop sequence, productivity level, length of growing season, tillage practices, residue management, and the expected time distribution of erosive events. The effects of vegetative cover on erosion vary with the crop growth stage throughout the growing season. Table 6.2 gives a cover-management variable as a percent of the soil loss for crops to that for continuous fallow for different crop growth stages. The annual distribution of the rainfall erosivity index will vary with geographic location. The C factor for a given crop rotation is found by first multiplying the soil loss ratios for each growth stage in a crop rotation (Table 6.2) by the percent of annual erosion during each respective period (Table 6.3). These products are then summed and expressed as a decimal value to give the C factor (Example 6.2). Many state agencies have developed C factor tables for their respective climate pattern and cropping practices to aid field staff in working with the USLE. Some typical C factors for Ohio are summarized in Table 6.4.

Example 6.2. Determine the cover management factor C for a rotation with one year corn, fall tilled, followed by one year small grain and one year meadow for the site described in Example 6.1.

Solution. A weighted C factor is determined by dividing the three years of the given rotation into seasons with the same crop stage in Table 6.2.

Crop (1)	Months (2)	Crop Stage (3)	Percent C (Table 6.2) (4)	Percent Annual R (Table 6.3) (5)	Weighted C Factor (4) × (5) (6)
Corn, first year	Jan.–Mar.	F	15	21	0.032
(Line 6ᵉ)	Apr.	SB	32	12	0.038
	May	1	26	11	0.029
	Jun.	2	20	11	0.022
	Jul.–Sep.	3	13	26	0.034
Small grain,	Oct.–Feb.	F	15	32	0.048
with meadow	Mar.	SB	18	8	0.014
seeding	Apr.	1	14	12	0.017
(Line 131ᵉ)	May–Jun.	2	12	22	0.026
	Jul.–Aug.	3	4	20	0.008
	Sep.	4	22	6	0.013
Meadow	Oct.–Oct.	—	0.6	105	0.006
(Line 182)	Nov.–Dec.	F	15	14	0.021

Average Annual $C = 0.308/3 = 0.103$ Three-year total: 0.308

TABLE 6.2 Ratio of Soil Loss (Percentage) from Cropland to Corresponding Loss from Continuous Fallow—Factor C

Line [b] No.	Cover, Crop Sequence, and Management	Spring Residue (lbs/A)	F (%)	SB (%)	1 (%)	2 (%)	3[c] (%)	4[d] (%)
					Crop-Stage Period[a]			
	Corn After Corn, Grain Sorghum, Small Grain, or Cotton in Meadowless Systems							
	Moldboard Plus Conv. Till							
2	RdL, spring TP	3400	36	60	52	41	24	30
6	RdL, fall TP	GP	49	70	57	41	24	—
10	RdR, spring TP	GP	67	75	66	47	27	62
14	RdR, fall TP	GP	77	83	71	50	27	—
28	No-till plant in crop residue	3400	—	8	8	8	8	19
	Corn in Sod-Based Systems							
2[e]	RdL, spring TP 1st year after sod (2–3 t/a)	3400	11	27	23	20	13	20
6[e]	RdL, fall TP 1st year after sod (2–3 t/a)	GP	15	32	26	20	13	—
10[e]	RdR, spring TP 1st year after sod (2–3 t/a)	GP	20	34	30	24	15	40
14[e]	RdR, fall TP 1st after sod year (2–3 t/a)	GP	23	37	32	25	15	—
6[e]	RdL, fall TP 2nd year after sod (2–3 t/a)	GP	37	60	48	37	23	—
14[e]	RdR, fall TP 2nd year after sod (2–3 t/a)	GP	58	71	60	45	26	—
104	No-till plant in killed sod, 1–2 t/a yield	—		2	2	2	2	2
107	Strip till, 1–2 t/a yield, 40% cover	—		4	4	4	4	6
	Corn After Soybeans							
110	Spring TP, conv. till	GP	47	78	65	51	30	37
113	Fall TP, conv. till	GP	53	81	65	51	30	—
121	No-till, plant in crop residue	GP	—	33	29	25	18	33

(continued)

The C factor in Example 6.2 is greater than that given in Table 6.4 ($C = 0.072$) because the relationship between the erosivity distribution, R, and the C factor (Table 6.2) is not the same in Tennessee as it is in Ohio. Also, different assumptions may have been made for the values presented in Table 6.4 than were made for this example.

The conservation practice factor, P, applies to contour farming, contour strip cropping, and terracing. Conservation practices are discussed in detail in Chapter 7. In the absence of any conservation practices, $P = 1.0$.

TABLE 6.2 Ratio of Soil Loss (Percentage) from Cropland to Corresponding Loss from Continuous Fallow—Factor C *(continued)*

Soybeans After Corn								
124	Spring TP, RdL, conv. till	GP	39	64	56	41	21	—
127	Fall TP, RdL, conv. till	GP	52	73	61	41	21	—
Small Grain After Corn, Small Grain, Grain Sorghum or Cotton in Sod-Based Systems								
131[e]	Spring grain in disked residues, 50% cover, 2nd year after sod (2–3 t/a)	3400	15	18	14	12	4	22
143[e]	Winter grain after fall TP, RdL, 50% mulch 2nd year after sod (2–3 t/a)	GP	—	39	34	25	7	22
Small Grain After Summer Fallow								
149	With grain residues	1000	—	26	21	15	7	—
155	With row crop residues	1000	—	40	31	24	10	—
Established Sod or Meadow								
181	Grass or legume mix, 3–5 t/a			0.4 for entire year				
182	2–3 t/a			0.6 for entire year				
183	1 t/a			1.0 for entire year				

Source: Wischmeier and Smith (1978).

[a]Crop-stage period

  F (rough fallow), plowing to secondary tillage.

  SB (seedbed), secondary tillage for seedbed preparation until crop has 10% canopy cover.

  1 (establishment), end of SB until crop has 50% canopy cover.

  2 (development), end of 1 until 75% canopy cover.

  3 (maturing crop), end of 2 until crop harvest.

  4 (residue or stubble), harvest to plowing or new seeding.

[b]Numbers from table in Wischmeier and Smith (1978).

[c]For 90% canopy cover at maturity.

[d]For all residues left on field.

[e]Adjusted from Table 5–D for 2–3 t/a hay yield from Wischmeier and Smith (1978).

Symbols: GP, good productivity; M, grass and legume meadow; RdL, crop residues left on field; RdR, crop residues removed; TP, plowed with moldboard; t/a, tons per acre.

6.8 Erosion from Nonagricultural Sites

Natural resource managers may wish to estimate erosion from nonagricultural sites. Although the principles are the same, care needs to be exercised when applying principles developed for agriculture to other conditions. Generally, undisturbed conditions, like rangelands, forests, and meadows, have extremely low erosion rates (*C* factors in the range of 0.001 to 0.01). Human activities, however, can cause substantial increases in erosion in specific areas like access

TABLE 6.3 Monthly Distribution of the Rainfall and Runoff Erosivity Index at Selected Locations

Month	Percent of Annual Erosion at Locations					
	A	B	C	D	E	F
January	0	2	6	2	0	15
February	1	2	7	2	0	12
March	1	4	8	3	0	8
April	4	6	12	4	5	2
May	11	11	11	6	9	3
June	26	20	11	14	9	5
July	24	19	12	23	17	1
August	18	15	8	20	42	1
September	11	10	6	15	2	7
October	3	6	5	6	16	11
November	1	3	8	3	0	16
December	0	2	6	2	0	19

Source: Wischmeier and Smith (1978).

Locations

A, Area 2—northwestern Iowa, northern Nebraska, and southeastern South Dakota.
B, Area 16—northern Missouri, central Illinois, Indiana, and Ohio.
C, Area 22—Louisiana, Mississippi, western Tennessee, and eastern Arkansas.
D, Area 29—Atlantic Coastal Plains of Georgia and the Carolinas.
E, Pueblo, Colorado.
F, Portland, Oregon.

roads, parking lots, footpaths, or construction sites. Predicting erosion from such areas may be done with careful application of the USLE or other models, after considering the location of the disturbed area relative to waterways and other topographic features. The RUSLE program has given special emphasis to estimating erosion from rangelands. In forests, some attempts have been made to develop special erosion models, but they have been generally limited to specific geographic areas (Dissmeyer and Foster, 1981).

Construction sites in any setting, natural, agricultural, or urban, can be a major source of erosion and can cause considerable sedimentation in nearby streams and lakes. Builders may be required to employ erosion control practices during construction, including diversion of overland flow, sediment basins, silt fences, or other practices to reduce erosion or sedimentation. The USLE can provide relative erosion rates for different development plans to assist in determining an environmentally acceptable plan. For construction periods of less than one year, the appropriate R factor for the period of construction may be determined by multiplying the seasonal percent of erosivity found from Table 6.3 by the annual erosivity for the location from Fig. 6.4.

Example 6.3. Determine the estimated erosion on a construction site in central Illinois that will have bare soil ($C = 1.0$) from Apr 1 until Nov 1, if $LS = 1.8$, $K = 0.25$, $P = 1.0$.

TABLE 6.4 Example of Typical Cover-Management C Factors Developed by State Agencies for Ohio Climate and Vegetation Conditions

| Vegetation | Tillage | | | |
	Autumn Conventional	Spring Conventional	Spring Conservation	None
Agricultural Rotations				
Corn (Co) grown continuously	0.40	0.36	0.27	0.10
Corn and soybeans (Sb)	0.42	0.37	0.24	0.12
Two-year rotation with corn grown in year 1 and soybeans and wheat in year 2	0.30	0.28	0.24	0.10
Co oats (O) meadow (M)	0.072	0.065	0.042	0.040
Co Co O M	0.13	0.12	0.10	0.064
Co O M M	0.055	0.050	0.033	0.033
Co Co O M M	0.11	0.094	0.082	0.052
Co Sb O M M	0.13	0.12	0.082	0.052
Co Co O M M M	0.10	0.09	0.067	0.067
Double cropping wheat Sb	—	0.20	—	0.11
Permanent pasture poor condition				0.04
Good condition, grass				0.003
Good condition, broadleaf				0.01
Brush, 50% canopy, 60% ground cover				0.04
Woodland Poor Condition				0.004
Good Condition				0.001

Source: USDA Cooperative Extension Service and The Ohio State University, 1979.

Solution. From Fig. 6.4, determine that for central Illinois, annual erosivity $R = 200$. From Table 6.3, add up the monthly percentages of erosivity for Apr–Oct:

$$\text{Percent} = 6 + 11 + 20 + 19 + 15 + 10 + 6 = 87$$

The erosivity during the construction period is: 200×87 percent $= 174$. The estimated soil erosion is now found with Eq. 6.1:

$$A = 174 \times 0.25 \times 1.8 \times 1 \times 1 = 78 \text{ t/a } (174 \text{ Mg/ha})$$

Exposed subsoils have different erodibility values than topsoils, and compaction or disturbance by construction traffic can alter the erodibility. The *LS* values in Fig. 6.5 do not apply to slopes over 20 percent. The *C* factor and the *P* factor are generally assumed to be 1.0 during construction unless special surface treatments or management practices are carried out. For further information, consult Wischmeier and Smith (1978) and other references.

6.9   Sediment Delivery

Sediment delivery downstream in a watershed may be estimated from the USLE and a *sediment delivery ratio.* The USLE estimates gross sheet and rill erosion, but does not account for sediment deposited en route to the place of measurement nor for gully or channel erosion downstream. The sediment delivery ratio is defined as the ratio of sediment delivered at a location in the stream system to the gross erosion from the drainage area above that point. This ratio varies widely with size of area, steepness, density of drainage network, and many other factors. For watersheds of 6.4, 320, 3200, and 64 000 acres, ratios of 0.65, 0.33, 0.22, and 0.10, respectively, were suggested as average values by Roehl (1962). Computer models may estimate eroded sediment yield from the sediment size distribution and channel sediment transport capacity.

Example 6.4.   Assume that the 13 t/a/year erosion rate calculated for the field in Example 6.1 is typical of a 320-acre watershed draining into a small reservoir. Estimate the sediment delivery to the reservoir during a 20-year period.

Solution.   The average annual erosion rate is 13 t/a/year. The sediment delivery ratio for a 320-acre watershed was given as 0.33. The sediment yield from 20 years will be:

Sediment = 13 t/a/yr × 320 acres × 20 years × 0.33 = 27 500 tons (25,000 Mg)

6.10   Erosion from Irrigated Lands

Erosion may occur on irrigated land, especially from surface methods such as furrow or border, or from center pivot systems operating on rolling terrain. Most of the erosion occurs within the field, removing soil from near the water source and depositing it near the downstream ends of rows. Erosion is most severe on freshly tilled fields, during the initial irrigation. Finer particles removed at the upstream ends may decrease infiltration uniformity farther down the furrow. Upstream erosion can also alter the uniformity of the slope, affecting distribution of water within the furrow. Erosion cannot be eliminated, but it can be minimized by reducing flow rates where possible. The WEPP computer model is being developed to assist irrigation managers in determining the severity of erosion from different irrigation systems. Irrigation runoff is frequently high in sediment and chemicals, and may require special management (Chapter 15).

6.11   Erosion and Water Quality

Sediment is the biggest pollutant of surface water. The sediment itself can alter stream channel characteristics and adversely affect aquatic plant and animal life. Sediment can fill in spaces in gravel beds where certain fish like trout and salmon may spawn, or it may fill in deeper pools needed by certain species of fish to mature.

Eroded sediment may contain phosphorus and other nutrients, and organic matter that can release nitrogen upon decomposition. If an erosive event occurs within a short period after a pesticide application, eroded sediments may also contain significant amounts of pesticides. Eroded sediments tend to contain the finer textured fractions of the soil, which are usually higher in pollutants. The enrichment ratio is the fraction of pollutant concentration in eroded sediment to the concentration in the uneroded field. It ranges from 1.0 for coarse textured soils to 1.5 or higher for loams and clays, or simply stated: the concentration of a chemical in eroded sediment may be 1.5 times the concentration of that chemical in the field.

Example 6.5.   On a given field, atrazine is applied at a rate of 3 lbs/acre, which results in a concentration of 0.01 lbs of atrazine in a ton of topsoil. Assuming an enrichment ratio of 1.15, how many pounds of atrazine are lost from 20 acres if the erosion rate for a storm immediately after chemical application was 2 tons per acre.

Solution.   The amount of atrazine lost would be:

$$\text{Loss} = 2 \text{ t/a} \times 20 \text{ acres} \times 0.01 \text{ lbs atrazine/ton} \times 1.15$$
$$= 0.46 \text{ lbs atrazine (210 g)}$$

Eroded sediments can deposit in streams and lakes. Deposition concentrates the chemicals in runoff, leading to biological damage from pesticides or eutrophication (excessive algae growth) from nutrients. Research is ongoing to fully evaluate the effects of erosion hazards on environmental systems. As chemical analysis methods improve, so will the understanding of the fates of chemicals in the ecosystem.

6.12   Erosion Control Practices

In this chapter, the primary emphasis is on describing and estimating erosion by water from cultivated farmland and other nonpoint sources. Contour farming, terraces, and improved waterways are important methods to control erosion, but they will be discussed in later chapters. Two other major erosion control practices are available to natural resource managers: tillage practices and cropping or vegetation management.

Traditionally, tillage prepared a seedbed and reduced weed competition, but tillage has become increasingly important as a conservation tool. Modern tillage is intended to provide an adequate soil and water environment for the plant. Its role as a means of weed control has diminished with increased use of herbicides and improved timing of operations. Excessive tillage can damage soil structure, leading to surface sealing and increased runoff and erosion. The loss of structure is often associated with a reduction in organic matter, which can also reduce the resistance of soil to erosion.

Any practice that tends to increase the amount of surface residue on the field is considered a form of conservation tillage. Surface residue is extremely effective in reducing erosion. One of many similar studies showed that conservation ridge tillage reduced runoff by 2/3 and erosion from 0.5 to 6 tons per acre compared to conventional tillage (Kramer and Hjelmfelt, 1989). Other methods for reducing runoff, such as putting small dams in furrows or imprinting rangeland, can also reduce erosion.

Another tool for erosion control is vegetation management. Example 6.6 shows the major effect of meadow on reducing average annual erosion. Rotations with longer periods of meadow will have less erosion than row crop intensive rotations. Sowing winter cover crops in the southeastern United States has significantly reduced erosion during the winter. Incorporation of the residue into the soil in the spring has reduced summer erosion because of improved soil structure. Improved grazing management on rangeland and improved harvesting techniques in forests have also been effective in reducing erosion. Readers are encouraged to consult local state universities, extension services, or NRCS offices to determine local soil erosion reduction practices.

Example 6.6. A farm in central Ohio has a clay loam soil with 2 percent organic matter and a slope length of 300 feet with a steepness of 4 percent. Compute the predicted erosion rates for a conventional, fall-tilled continuous corn rotation, spring conservation-tilled continuous corn, and a spring-tilled rotation of corn, corn, oats, and meadow.

Solution. From Fig. 6.4, $R = 150$. From Table 6.1, $K = 0.25$. From Fig. 6.5, $LS = 0.7$. Assume that $P = 1.0$. The C factors are obtained from Table 6.4:

For fall conventional tillage, $C = 0.4$; $A = 150 \times 0.25 \times 0.7 \times 0.4 \times 1.0$
$$= 10.5 \text{ t/a/yr } (23.5 \text{ Mg/ha})$$
For spring conservation tillage, $C = 0.27$; $A = 150 \times 0.25 \times 0.7 \times 0.27 \times 1.0$
$$= 7.1 \text{ t/a/yr } (15.9 \text{ Mg/ha})$$
For spring tillage, crop rotation, $C = 0.12$; $A = 150 \times 0.25 \times 0.7 \times 0.12 \times 1.0$
$$= 3.2 \text{ t/a/yr } (7.2 \text{ Mg/ha})$$

Example 6.6 shows that conservation tillage can reduce soil erosion about 30 percent, and crop rotations by 70 percent or more. A combination of the two practices will reduce soil erosion even further.

REFERENCES

Dissmeyer, G. E., and G. R. Foster (1981) "Estimating the Cover-Management Factor *(C)* in the Universal Soil Loss Equation for Forest Conditions," *Jour. of Soil and Water Cons.*, 36(4):235–240.

Foster, G. R. (1988) *User Requirements, USDA-Water Erosion Prediction Project (WEPP)*, USDA-ARS. National Soil Erosion Lab., W. Lafayette, Indiana.

Kramer, L. A., and A. T. Hjelmfelt, Jr. (1989) *Watershed Erosion from Ridge-Till and Conventional-Till Corn*, Paper No. 89-2511. ASAE, St. Joseph, Michigan.

McCool, D. K., L. C. Brown, G. R. Foster, C. K. Mutchler, and L. D. Meyer (1987) "Revised Slope Steepness Factor for the Universal Soil Loss Equation," *ASAE Trans.*, 30:1387–1396.

McCool, D. K., G. R. Foster, C. K. Mutchler, and L. D. Meyer (1989) "Revised Slope Length Factor for the Universal Soil Loss Equation," *ASAE Trans.*, 32:1571–1576.

Renard, K. G., G. R. Foster, G. A. Weesies, and J. P. Porter (1991) "RUSLE Revised Universal Soil Loss Equation," *J. Soil and Water Conserv. Soc.*, 46:30–33.

Roehl, J. N. (1962) *Sediment Source Areas, Delivery Ratios and Influencing Morphological Factors.* Intern. Assoc. Scientific Hydrology, Commission of Land Erosion. Publ. No. 59.

Schertz, D. L., W. C. Moldenhauer, S. J. Livingston, G. A. Weesies, and E. A. Hintz (1989) "Effect of past soil erosion on crop productivity in Indiana," *J. Soil and Water Conserv. Soc.*, 44:604-608.

U.S. Dept. Agr. (USDA) (1989) *The Second RCA Appraisal, Soil, Water, and Related Resources on Nonfederal Land in the United States, Analysis of Condition and Trends.* U.S. Government Printing Office, Washington, D.C.

U.S. Dept. Agr. Cooperative Extension Service and The Ohio State University (1979) *Ohio Erosion Control and Sediment Pollution Abatement Guide.* Bulletin 594. The Ohio State University, Columbus, Ohio.

U.S. Dept. Agr. Research Service and Environmental Protection Agency (1975) *Control of Water Pollution from Cropland*, Vol. I (1976). U.S. Government Printing Office, Washington, D.C.

U. S. Dept. Agr. Soil Conservation Service (SCS) (1994) *1992 National Resources Inventory.* SCS, Washington, D.C.

Watson, D. A., and J. M. Laflen (1986) "Soil Strength, Slope, and Rainfall Intensity Effects on Interrill Erosion," *ASAE Trans.*, 29:98–102.

Wischmeier, W. H. (1959) "A Rainfall Erosion Index for a Universal Soil-Loss Equation," *Soil Sci. Soc. Am. Proc.*, 23:246–249.

Wischmeier, W. H. and D. D. Smith (1978) "Predicting Rainfall Erosion Losses—A Guide to Conservation Planning," USDA Hdbk. 537. Government Printing Office, Washington, D.C.

PROBLEMS

6.1 If the soil loss at Memphis, Tennessee, for a given set of conditions is 5 t/a, what is the expected soil loss at your present location if all factors are the same except the rainfall factor?

6.2 If the soil loss for a 10 percent slope and a 400-ft (120-m) length of slope is 4 t/a, what soil loss could be expected for a 200-ft slope length?

6.3 Determine the soil loss for a field at your present location if $K = 0.15$, slope length = 300 ft, slope steepness = 10 percent, and $C = 0.2$.

6.4 Which field, A or B, will produce the greatest soil loss:

Soil Loss Factor	Field A	Field B
Soil erodibility K	0.1	0.3
Slope length/steepness LS	6.0	2.0
Cover-management C	0.15	0.2
Conservation practice P	0.6	0.3

If the fields are located in southeastern Iowa, compute the average annual soil loss for field B.

6.5 Compute the average annual cover-management factor for a two-year corn-soybean rotation, conventional tillage, and crop residues left in the field. Assume fall plowing on November 1, crop stage SB in April and May, Stage 1 in June, 2 in July, and 3 in August through October. The location is northern Missouri. Compare your answer to the C factor for Ohio for a similar rotation given in Table 6.4.

6.6 If the allowable soil loss tolerance is 4 t/a, the soil is clay loam with 2 percent organic matter located in southern Indiana, $LS = 2.0$, and $P = 0.6$, compute the maximum value for the cover-management factor C.

6.7 A watershed in western Ohio was intensively farmed for almost 100 years and experienced severe erosion. The farmers all abandoned their farms, and the predominant vegetation on the watershed reverted to native woodland. When the woodland was about 80 years old, a large recreation and flood control reservoir was constructed by the NRCS on the stream draining the watershed. The engineers were surprised to discover sedimentation rates from the watershed were similar to cropland rates. (a) What do you think was the source of the sediment? (b) Describe a method to reduce sedimentation in the lake.

6.8 (a) Calculate a C factor for a farming system in Ohio that has a rotation of corn, corn, small grain, meadow, meadow, using conventional fall tillage. (b) Compare your result to the C factor given in Table 6.4.

6.9 Estimate the amount of erosion that will occur on a new pipeline excavation from August through November in southern Wisconsin. The pipeline excavation is on an 8 percent slope, 600 feet long. The soil is a silt loam with 0.5 percent organic matter.

6.10 A reservoir is constructed on a stream draining 6 square miles (1 square mile = 640 acres). Two thirds of the watershed has an estimated upland erosion rate of 4.5 tons/acre/year, and the other third a rate of 0.5 tons/acre/year. How much sedimentation will the reservoir experience during its 30-year life?

6.11 In a small watershed draining into a lake, the average concentration of phosphate in the topsoil is measured and found to be 0.2 lbs per ton of topsoil. If an erosion event contributes 300 tons of soil to a local lake, how many pounds of phosphate are added to the lake's ecosystem, assuming an enrichment ratio of 1.4?

CHAPTER 7

UPLAND WATER
EROSION CONTROL

In Chapter 6, methods of predicting soil erosion rates were discussed. The importance of vegetation management, including crop rotations and conservation tillage, were also emphasized. In addition to implementing residue management practices, water erosion can also be reduced by mechanical control practices. Mechanical control can broadly be grouped into contouring, strip cropping, and terracing (Fig. 7.1). Frequently, these methods are combined with one another and with grassed waterways (Chapter 8) to reduce runoff and soil loss.

CONTOURING AND STRIP CROPPING

7.1 Contouring

Contouring is the practice of performing field operations, such as plowing, planting, cultivating, and harvesting, approximately on the contour. Crops grown on the contour have rows that follow contour lines rather than running up and down the slope. Contouring reduces surface runoff by impounding water in small depressions, which reduces the development of rills. The practice can also conserve water in drier climates.

The relative effectiveness of contouring for controlling erosion on various slopes is shown by the Universal Soil Loss Equation (USLE) conservation practice factor P from Table 7.1. These P factors are applicable to fields relatively free from gullies and depressions (other than grassed waterways) and for slope lengths given. If ridge tillage is practiced, the storage capacity of the furrows is materially increased and the conservation practice factor is reduced to about half that shown for contouring. Contouring on steep slopes or under conditions of high rainfall intensity and soil erodibility may increase gullying because row breaks may collect large volumes of stored water and then release that water in a concentrated flow in a breakover. Breakovers cause cumulative damage as the volume of water increases with each succeeding row.

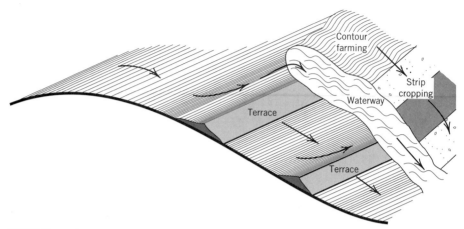

FIGURE 7.1   Relationships among waterways, terraces, contour farming and strip cropping. Arrows show the surface runoff flow direction.

TABLE 7.1   Conservation Practice Factor P for the Universal Soil Loss Equation

	Farming Up and Down Slope			P = 1.0	
	For Contour Farming			*P Factors*	
Land Slope	*Maximum Slope Length (feet)*		*Maximum Strip*		
(percent)	*Contouring*	*Strip Cropping*	*Width (feet)*	*Contour*	*Strip Crop*
1 to 2	400	800	130	0.6	0.3
3 to 5	300	600	100	0.5	0.25
6 to 8	200	400	100	0.5	0.25
9 to 12	120	240	80	0.6	0.3
13 to 16	80	160	80	0.7	0.35
17 to 20	60	120	60	0.8	0.4
21 to 25	50	100	50	Too steep	0.45

For Terraces			Use revised *LS* factor
Loss from crop			Same *P* as contouring factor
Loss from terrace with graded channel outlet			Contour *P* factor × 0.2
Loss from terrace with underground outlet			Contour *P* factor × 0.1

Source: Based on Wischmeier and Smith (1978)

Example 7.1.   In a given farming system, the slope length is 300 ft and the slope steepness is 5 percent. What are the P factors for (1) continuous row crop up and down this slope, (2) continuous row crop across the slope, and (3) strip cropping with a 4-year rotation with 2 years of row crop?

Solution.   The following values are obtained from Table 7.1 to determine the P factor for each situation.

1. Uphill and downhill farming, $P = 1.0$.
2. Contour farming, $P = 0.5$. Note that 300 ft is the maximum slope length allowed for 5 percent slope.
3. Strip cropping, $P = 0.25$.

Because of nonuniform slopes in most fields, all crop rows cannot be on the true contour. To establish row directions, a guide line (true contour) is laid out at one or more elevations in the field. On small fields of uniform slope, one guide line may be sufficient. Another guide line should be established if the slope along the row direction exceeds 1 to 2 percent in any row laid out parallel to the guide line (Fig. 7.2c). A small slope along the row is desirable to prevent runoff (from a large storm) breaking over the ridges. Any row slope should be toward a natural swale or waterway, to minimize damage caused by a breakover. Where practical, field boundaries for contour farming should be relocated on the contour or moved to eliminate odd-shaped fields that would result in short, variable rows (point rows).

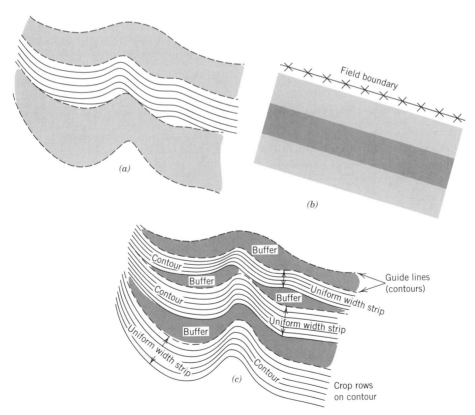

FIGURE 7.2   Three types of strip cropping: (a) contour, (b) parallel to field boundary, and (c) buffer.

7.2   Strip Cropping

Strip cropping is the practice of growing alternate strips of different crops in the same field. The three general types of strip cropping are shown in Fig. 7.2. For controlling water erosion, the strips are on the contour (Fig. 7.2a and Fig. 7.3). In contour strip cropping, layout, tillage, and planting are held closely to the contour, and the crops follow a definite rotational sequence. Rotations that provide strips of close-growing perennial grasses and legumes alternating with small grain and row crops are the most effective for controlling erosion. When strip cropping, uniform strips are placed across the general slope, parallel to field boundaries (Fig. 7.2b). When used with adequate grassed waterways, strip cropping parallel to the field boundaries may be practiced where irregular topography makes contour strip cropping impractical. Such strip cropping is a common practice in dry regions where strips are placed normal to the prevailing wind direction for wind erosion control (Fig. 7.2b and Chapter 9).

Buffer strip cropping (Fig. 7.2c) has permanent buffer strips of a grass or legume crop between contour strips of crops in the regular rotations. Buffers may be even or irregular in width, or placed on critical slope areas of the field. Their main purpose is to prevent soil detachment or to provide for areas for deposition. The recommended maximum widths for contour strip cropping are shown in Table 7.1. Strip width should be convenient for multiple-row equipment operation in mechanized farming systems.

Example 7.2.   What is the recommended strip cropping width for the farming system described in Example 7.1 with a 20-ft implement width?

Solution.   From Table 7.1, for a slope of 5 percent, maximum strip width is 100 ft. With an implement width of 20 ft, the strip width that would accommodate an even number of rounds is 80 ft (24 m) (2 rounds at 40 ft/round).

FIGURE 7.3   Contour strip cropping in southeastern Wisconsin. (Photograph by W. Elliot.)

When establishing contour lines, an Abney or tripod level and a series of surveying flags are recommended to establish a guide line, which is representative of the slope. The flags are connected by carefully plowing a contour guide line. Subsequent strips will then be established parallel to the initial guide line if the slope is uniform (Fig. 7.2a); or on a new contour, leaving buffer areas where the slope deviates more than 2 percent from the contour (Fig. 7.2c).

TERRACING

Terracing is a practice of erosion control accomplished by constructing broad earthen channels across the slope of rolling land. Since terracing requires a large additional investment and causes some inconvenience in farming, it should be considered only where other cropping and soil management practices, singly or in combination, will not provide adequate erosion control or water management. In the United States, terraces are common in the Midwest and Southeast for erosion control, and in the Great Plains for both erosion control and water conservation.

7.3 Effects of Terraces on Erosion

The primary function of terraces in humid areas is to decrease the hill slope length, thereby reducing rill erosion. Terraces can also be installed to prevent the formation of gullies by diverting water away from active gullies. Terraces allow sediment to settle from runoff, thereby improving the quality of water leaving the field. Crop rows are usually parallel to the terrace channel, so terracing usually includes contouring as a conservation practice. In drier areas, terraces retain runoff and increase the amount of water available for crop production or for recharging of shallow aquifers. Such retention of water also reduces the risk of wind erosion (Chapter 9). Terraces may also be used with surface irrigation, particularly for rice production, to ensure uniform water distribution (Chapter 16).

To estimate the effect of terracing on erosion, erosion rates for the area between terraces are first estimated by considering the effect of the terrace on the slope length and steepness factor, LS (Fig. 6.5). The P factor, which accounts for deposition in terrace channels, is determined from Table 7.1. Erosion rates for the entire hillslope, after considering deposition in the terrace channels are estimated by considering both the changes in the LS and the P factors.

Example 7.3.   For the hillslope described in the previous examples, what percent reduction in erosion will result from a terrace spacing of 100 ft compared to farming up and down the slope in (a) areas between the terraces and (b) the overall hillslope erosion, assuming continuous row crops and graded channel sod outlets?

Solution.   For a slope of 5 percent and a length of 300 ft, *LS* is 0.88 (Fig. 6.5) for uphill and downhill farming. For a slope length of 100 ft, and a slope of 5 percent, *LS* can be determined from Fig. 6.5 to be *0.63*.

The following *P* factors are obtained from Table 7.1.

	P	*LS*	*LS* × *P*	*Relative Erosion*
Up and down	1.00	0.88	0.88	100%
(a) Erosion between terraces	0.5	0.63	0.32	36%
(b) Sediment leaving terraces	0.5 × 0.2 = 0.1	0.63	0.06	7%

Thus, the predicted erosion between terraces is 36 percent of that without terracing, and the amount of sediment leaving the terraced area is only 7 percent of the same field without terraces.

7.4   Classification of Terraces

Terraces are classified by alignment, cross section, grade, and outlet (ASAE, 1994). Terrace alignment can either be nonparallel or parallel. Nonparallel terraces follow the contour of the land regardless of alignment. Some minor adjustments are frequently made to eliminate sharp turns and short rows by installing additional outlets, using variable grade, and installing vegetated turning strips. Parallel terraces eliminate point rows and simplify farming operations; they should be installed wherever possible. Parallel terraces generally require more cut and fill earthwork during construction than nonparallel systems (Fig. 7.4).

CROSS SECTIONS   Terrace cross sections include the broadbase shape with the three-segment section, the conservation bench, and the grassed backslope terrace (Fig. 7.5). The three-segment section terrace is more common in mechanized farming systems on moderate slopes (6 to 9 percent). All slopes on the three-segment section broadbase should be sufficiently flat for the operation of farm machinery. Fig. 7.6 shows a three-segment section broadbase terrace. The slope is sufficiently flat to make it difficult to discern. Lengths of all sideslopes are planned to match the width of equipment operating on those slopes.

FIGURE 7.4   Grass-backsloped parallel terraces with subsurface outlets in central Iowa. (Photograph by W. Elliot.)

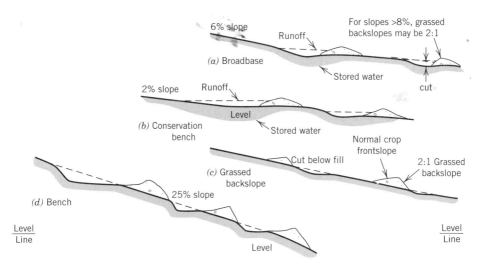

FIGURE 7.5 Comparison of terrace shapes: *(a)* three-segment section broadbase, *(b)* conservation bench, *(c)* grassed backslope pushup terrace, *(d)* bench. Note the differences between *(a)* three-segment section level terrace and *(b)* conservation bench in distribution of stored soil water.

On slopes over 2 percent, water in the channel of a three-segment section broadbase terrace is spread over a relatively small area, thus limiting the benefit to crop yield in water-deficient areas. The conservation bench cross section was designed to overcome this deficiency. The conservation bench variation incorporates a wide, flat channel bottom uphill of the embankment to provide a greater area for infiltration of runoff. A comparison of the infiltration areas between the conservation bench and the three-segment section is shown in Fig. 7.5*a* and 7.5*b*.

FIGURE 7.6 Three-segment section broadbase terrace in central Iowa. The light vegetation on the right of the picture is plants left standing around the riser from the subsurface outlet for the terrace. The field, including the terrace, had been in a crop of oats *(Avena sativa)* with undersown forage. (Photograph by W. Elliot.)

FIGURE 7.7 Bench terraces constructed by hand in central Kenya. In the foreground, a banana crop has been established. These terraces were designed to be irrigated by surface irrigation or by the sprinkler system shown in the background. (Photograph by W. Elliot.)

The grassed backslope terrace (Figs. 7.4 and 7.5c) is constructed with a 2:1 backslope that is usually seeded to permanent grass. By constructing the terrace with fill from the downhill side, the slopes between terraces are reduced, and farmability is improved. The bench terrace (Figs. 7.5d and 7.7) has improved farmability for very steep (20 to 30 percent) slopes. Bench terraces are more common in countries where labor is cheap or land is in short supply. The bench terrace provides for more efficient distribution of water under both irrigated and dry land conditions.

CHANNEL GRADES   The channel in terraces can be graded toward an outlet or can be level. Graded or channel-type terraces should remove excess water at flow velocities that minimize erosion. Terraces control erosion by reducing the slope length of overland flow and by conducting the intercepted runoff to a safe outlet at a nonerosive velocity. The reduced terrace channel velocities also allow for deposition of eroded sediment.

Level terraces are constructed to conserve water and control erosion (Fig. 7.8). In low-to-moderate rainfall regions, these terraces trap and hold rainfall for infiltration into the soil profile. Level terraces may also be suitable for permeable soils in high rainfall areas. Frequently, it is necessary to excavate soil from both sides of the embankment to achieve sufficient height to store the design runoff without the entrapped water overtopping or piping through the embankment. The channel gradient is level, and the terrace is sometimes closed at both ends to ensure maximum water retention. On slopes over 2 percent, the conservation bench cross section is recommended for level terraces (Fig. 7.5b).

FIGURE 7.8 Level terraces constructed in northern Texas to reduce erosion by water and conserve water. This photograph was taken shortly after an intense storm. Within hours, the surface water will have infiltrated. (Photograph by W. Elliot.)

OUTLETS Terraces may have blocked outlets (all water in the terrace channel infiltrates), permanently vegetated outlets (grassed waterway or a vegetated area), or surface inlets (water is removed through risers into subsurface pipes). Combinations of outlets may be employed to meet specific problems. Outlets can divert overland flow to natural drainageways, constructed channels, permanent pasture or meadow, waste land, concrete channels, subsurface drains, and stabilized gullies.

Natural drainageways, where properly vegetated, provide a desirable and economical outlet. Where these channels are unavailable, constructed waterways along field boundaries or subsurface outlets may be considered. Terrace outlets to pastureland should be staggered by increasing the length of each terrace several feet, starting with the lowest terrace (Fig. 7.9). Road ditches and

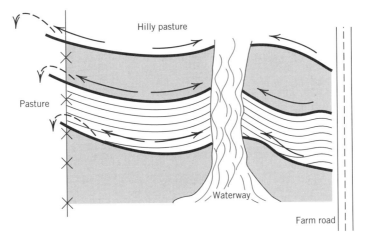

FIGURE 7.9 Typical terrace layout showing staggered outlets to pasture and outlets at both ends of long terraces (>1500 ft).

active gullies may scour or enlarge if terrace runoff is added. Connecting to road ditches may not be allowed by local or state agencies.

7.5   Planning Terrace Systems

Planning a terrace system involves specifying (1) the proper spacing and location of terraces, (2) the size a channel must have for adequate flow and/or storage capacity, and (3) a farmable cross section. For the graded terrace, runoff must be removed at nonerosive velocities in both the channel and the outlet. Soil characteristics, cropping and soil management practices, and climatic conditions are the most important considerations for planning terrace systems.

SPACING   Spacing is expressed as the vertical distance between the channels of successive terraces. For the top terrace, the spacing is the vertical distance from the top of the hill to the bottom of the terrace channel. This distance is commonly known as the vertical interval or *V.I.* The horizontal interval *H.I.*, or terrace spacing, is found by dividing *V.I.* by the slope steepness (decimal). The *H.I.* for parallel terraces is usually selected as an even number of rounds for row-crop equipment. The vertical interval is more convenient for terrace layout and construction with surveying equipment. Where soil loss data are available, spacing should be based on maximum allowable soil loss using the USLE with a contouring *P* factor and the appropriate cover-management factor.

Example 7.4.   If the soil loss was 12 tons/acre for the given contour farming system with a 5 percent slope and a 300-ft slope length, what is the maximum slope length and corresponding terrace spacing to reduce the soil loss to the terrace channel to 9 tons/acre?

Solution.   From Example 7.3, for a slope of 5 percent and a slope length of 300 feet, $LS = 0.88$.

The maximum LS to reduce loss to 9 tons/acre is $LS = 0.88 \times (9/12) = 0.66$

Determine the terrace spacing to achieve the preceding LS factor from Fig. 6.5 to be 100 ft.

Calculate the vertical interval:

$$V.I. = (5/100) \times 100 = 5.0 \text{ ft } (1.5 \text{ m})$$

In the absence of soil loss considerations, graded terrace spacing is often expressed as a function of land slope by the empirical formula (ASAE 1994)

$$V.I. = X s + Y \tag{7.1}$$

where $V.I.$ = vertical interval between corresponding points on consecutive terraces or from the top of the slope to the bottom of the first terrace in ft

   X = climatic constant (Fig. 7.10)

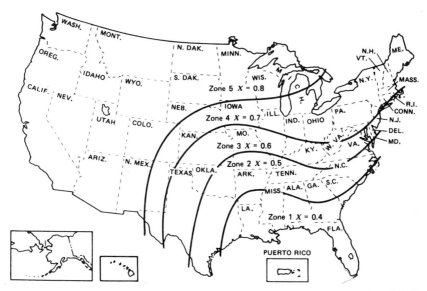

FIGURE 7.10   Geographical values of X in the terrace spacing equation. (Modified from U.S. Soil Conservation Service, 1979.)

Y = constant for soil erodibility and cover conditions during critical erosion periods ($Y = 1$, 2, 3, or 4, with the low value for highly erodible soils with no surface residue, and the high value for erosion resistant soils with conservation tillage)

s = average slope steepness above the terrace in percent

The preceding *V.I.* may be varied as much as 25 percent to allow for soil, climatic, and tillage conditions.

Example 7.5.   Compute the terrace spacing for Example 7.4 from Eq. 7.1, assuming that the soil has a moderate erodibility and that there is little cover, if the site is in central Ohio.

Solution.   From Example 7.4, $s = 5$ percent; from Fig. 7.10, for Zone 4, $X = 0.7$. For a soil with moderate erodibility and little cover, $Y = 2$.
   Substituting into Eq. 7.1:

$$V.I. = 0.7 \times 5 + 2 = 5.5 \text{ ft } (1.65 \text{ m})$$

The permissible spacing based on soil loss takes into account more of the erosion, so a vertical interval of 5 ft (1.5 m) as computed in Example 7.4 is preferred.

LEVEL TERRACES   Level terraces have a storage capacity adequate to prevent overtopping from upslope runoff, and an infiltration rate in the channel sufficiently high to prevent serious damage to crops. The spacing for level terraces is a function of channel infiltration and runoff; however, in more humid areas, where erosion control may be more critical, the slope length may limit the spacing. In addition, the depth is usually increased 3 in. (0.25 ft) for freeboard to reduce the risk of overtopping following settling, wave effects, or other unforeseen situations. The terrace channel acts as a temporary storage reservoir subjected to unequal rates of inflow and outflow. The runoff *volume* for level and conservation bench terraces should be based on a 10-year, 24-h duration storm.

Example 7.6.   If a terrace with a 70-ft spacing is to store the runoff from a 10-year, 24-h storm event (1.5 in.): (a) What is the required cross-sectional area (not including freeboard) necessary to contain the runoff and (b) if the terrace has a triangular cross section with side slopes of $8:1$, what depth is necessary, allowing for 0.25-ft freeboard?

Solution.   (a) The runoff from the 10-year, 24-h storm is 1.5 in. The runoff per foot length of terrace is

$$Q = 1.5 \text{ in.}/12 \text{ in. ft}^{-1} \times 70 \text{ ft spacing} \times 1 \text{ ft length} = 8.75 \text{ ft}^3/\text{ft length}$$

(b) To find the depth, for a triangular channel: *Volume* $= (1/2) \, bd \times 1$ ft length. With $8:1$ side slopes: $b = 2 \times 8 \times d$, so *Volume* $= (16/2) d^2$. Substituting for Q for *Volume* gives $8d^2 = 8.75 \text{ ft}^2$. Solving for d, gives $d = 1.05$ ft. Add 0.25-ft freeboard to determine a total depth of 1.3 ft (0.4 m).

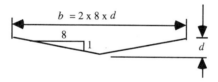

On the conservation bench terrace, the embankment is generally built up to provide a settled height of 1 ft above the level of the bench. There is no maximum length for level terraces, particularly where blocks or dams are placed in the channel about every 500 ft. These dams prevent total loss of water from the entire terrace and reduce gully damage should a break occur. The ends of the conservation bench are blocked to retain 0.5 ft of water on each bench before they overtop into a stable channel.

GRADED TERRACES   Graded terraces have a sufficient channel gradient to provide adequate drainage while removing the runoff at nonerosive velocities. The channel planning methods presented in Chapter 8 are also applicable to planning terrace channels. The minimum slope is desirable from the standpoint of soil loss. A grade of 0.4 percent is common in many regions; however, grades may range from 0.1 to 0.6 percent, depending on soil and climatic factors. Recommended minimum and maximum grades are given in Table 7.2. Generally, the steeper grades are recommended for impervious soils and short terraces.

When estimating channel capacity, a roughness coefficient of 0.06 is recom-

TABLE 7.2   Maximum and Minimum Terrace Grades

Terrace Length or Length from Upper End of Long Terraces (Feet)	Maximum Slope[a] (Percent)
100 or less	2.0
100 to 200	1.2
200 to 500	0.5
500 to 1200	0.35
1200 or more	0.3
	Minimum Slope[b] (Percent)
Soils with slow internal drainage	0.2
Soils with good internal drainage	0.0

Source: [a]Beasley (1963); [b]ASAE (1994).

mended to ensure the channel will carry the estimated runoff under the most severe channel conditions without overtopping. When estimating maximum flow velocities, a smoother roughness of 0.035 is recommended to ensure that maximum velocities are not exceeded. Recommended maximum velocities are 1.5 ft/s for erosive soils and 2 ft/s for most other soils. For graded terraces, the estimated peak runoff rate should be based on a storm return period of 10 years and a duration of 24 h (ASAE, 1994). Examples of planning channels are given in Chapter 8.

Size and shape of the field, possible outlets, rate of runoff, and channel capacity are factors that influence terrace length. The number of outlets should be a minimum, consistent with good layout and design. Graded terraces that are extremely long are undesirable. Long lengths may be reduced in some terraces by dividing the flow midway in the terrace length and draining the runoff to outlets at both ends of the terrace (Fig. 7.9). The length should be such that erosive velocities and large cross sections are not required. Terraces on permeable soils may be longer than on impermeable soils. The maximum length for graded terraces generally ranges from about 1000 to 1500 ft, depending on local conditions. The maximum applies only to that portion of the terrace that drains toward one of the outlets.

SUBSURFACE OUTLET TERRACES   A subsurface outlet terrace, shown in Fig. 7.11, may eliminate the need for grassed waterways. With a subsurface outlet, terrace alignments can be straightened at natural channels with earth fill, making it easier to have adjacent terraces parallel (Fig. 7.6). The riser pipe has an orifice plate to restrict the outflow. This restriction ensures that the subsurface drains are not overloaded, and that sediment in the runoff has time to settle in the terrace channel, improving the quality of the runoff water. The orifice is generally

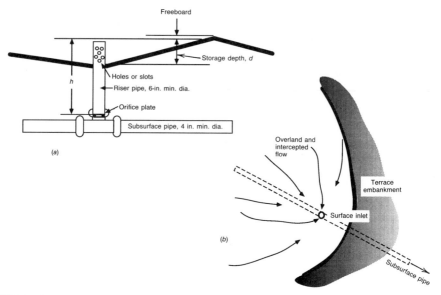

FIGURE 7.11 Subsurface outlet terrace: *(a)* details of subsurface pipe outlet, and *(b)* plan view.

planned to drain the surface runoff from a 10-year, 24-h duration storm within 48 hours. The size of the orifice is found from the orifice flow equation:

$$q = AC\sqrt{2gh} \qquad (7.2)$$

where q = design flow in ft^3/s
  A = cross-sectional area of orifice in ft^2
  C = orifice flow constant, usually assumed to be 0.6
  g = acceleration caused by gravity = 32 ft/s^2
  h = average depth of water above the orifice in ft

To provide for storage in the terrace, the top of the terrace ridge may be constructed at the same elevation along its length, even though the bottom of the channel slopes to the outlet. These terraces may be constructed with a grassed backslope or have a farmable backslope (Figs. 7.5 and 7.6).

Example 7.7.   The runoff from a 10-year, 24-h storm is 1.5 in. A terrace is 600 ft long and has a spacing of 100 ft. The average depth of water above the orifice is 4 ft. What size orifice is required to pass the estimated runoff in 48 h?

Solution.   The total runoff volume Q is

$$Q = 1.5 \text{ in.}/12 \text{ in. ft}^{-1} \times 600 \text{ ft} \times 100 \text{ ft} = 7500 \text{ ft}^3$$

The orifice capacity will need to be

$$q = 7500 \text{ ft}^3/48 \text{ hr}/3600 \text{ s/h} = 0.043 \text{ ft}^3/\text{s}$$

Solving Eq. 7.2 for the cross-sectional area A and substituting in the flow rate and the depth of water:

$$A = \frac{0.043}{0.6 \times \sqrt{2 \times 32 \times 4}} = 0.0045 \text{ ft}^2$$

From $A = \frac{3.14 \text{ dia}^2}{4}$, the diameter will be:

$$\text{dia} = 2 \sqrt{\frac{0.0045}{3.14}} \times 12 \text{ in./ft} = 0.9 \text{ in.} (23 \text{ mm})$$

CROSS SECTIONS   The terrace cross section should provide adequate runoff capacity or storage, be economical to construct with available equipment, and have broad farmable side slopes. For planning purposes, the cross section of a broadbase terrace can be considered a triangular channel as shown in Fig. 7.12a. The flow depth d is the channel depth to the top of the ridge h less a freeboard of about 0.25 ft. After smoothing, the ridge and bottom widths each will be about 3 ft, which will give a cross section that approximates the shape of a terrace after 10 years of farming (Fig. 7.12b). There are three slopes in a terrace cross section: the cutslope, the frontslope, and the backslope. For convenience, the width of each of these slopes is the same in a broadbase terrace. The relationship between the terrace dimensions in Fig. 7.12a is

$$c = f = \frac{h + sW}{2} \tag{7.3}$$

where c = depth of cut in feet
 f = height of fill in feet
 h = terrace channel flow depth plus freeboard in feet
 s = uphill and downhill slope, fraction
 W = horizontal width of the cutslope, frontslope, and backslope in feet

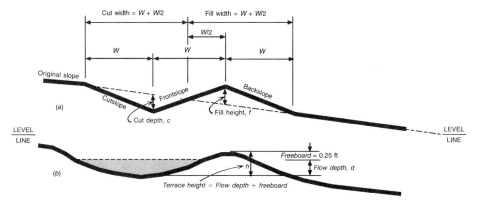

FIGURE 7.12   Terrace cross sections showing *(a)* design components and *(b)* cross section after several years of farming.

The slope of the frontslope can be found by dividing the depth plus free-board h by the width W. The cutslope and backslope slopes are approximately equal to the depth of cut or fill divided by the width W on moderate slopes. Once the depth of cut is known, the cross-sectional area of the terrace excavation or embankment can be found assuming a triangular shape:

$$A = \frac{1}{2}\, bh = \frac{1}{2}\left(W + \frac{W}{2}\right) c \tag{7.4}$$

From the cross-sectional area A, the volume of a uniformly shaped length of terrace can be calculated by multiplying the cross-sectional area by the terrace length.

Example 7.8. For a land slope of 6 percent and a farming system with six-row equipment (30-in. row width): (a) Determine the depth of cut for a flow depth of 0.8 ft; (b) Calculate the slopes of the cutslope, frontslope, and backslope; (c) Estimate the volume of earth excavated to build 100 ft of the given cross section in cubic yards.

Solution. a. The width of the slopes W will be: 6×30 in./12 in./ft $= 15$ ft

The height of the terrace h will be 0.8 ft $+ 0.25$ ft $= 1.05$ ft

The preceding values are substituted into Eq. 7.3 to determine the depth of cut (and height of fill).

$$c = \frac{1.05 + 0.06 \times 15}{2} = 1 \text{ ft}$$

b. The slope of the frontslope will be: $\dfrac{h}{W} = \dfrac{1.05}{15} = 0.07$, or $14.3:1$.

The slopes of the cutslope and backslope will be: $\dfrac{c}{W} = \dfrac{1}{15} = 0.065$ or $15:1$

c. The cross-sectional area of the cut or fill is calculated from Eq. 7.4:

$$A = \frac{1}{2}\left(W + \frac{W}{2}\right) c = \frac{1}{2}\left(15 + \frac{15}{2}\right) \times 1 = 11.25 \text{ ft}^2$$

The cut or fill volume is the cross-sectional area times the length:

$$Volume = \frac{11.25 \text{ ft}^2 \times 100 \text{ ft}}{27 \text{ ft}^3/\text{yd}^3} = 41.7 \text{ yd}^3 \ (32 \text{ m}^3)$$

7.6 Establishing Terraces

Planning for terraces can be initiated in the office with contour maps (Wittmus, 1988), but requires field surveying with a level to complete the plan similar to a reservoir (Figs. 11.3 and 11.6). Schwab et al. (1993) describe in detail the procedure for laying out terraces in the field. The steps include: (1) measure the slope and determine the vertical interval; (2) determine the desired channel grade; (3) stake out the edge of the channel of the top terrace, starting at the outlet, at about 50-ft spacing, ensuring the desired gradient; (4) adjust stakes to make terraces parallel; and (5) with a level, determine the depth of cut at each stake to ensure a uniform gradient.

Terraces are constructed with graders, scrapers (pans), and bulldozers. In some instances, farmers may use one-way plows or other implements to enhance the effects of the terraces during subsequent cultivation.

7.7   Terrace Maintenance

Proper maintenance is as important as the original planning and construction processes. A new terrace should be inspected regularly in the first few seasons to ensure there have been no failures caused by overtopping or channel erosion. Rodents should be eliminated from terraces to prevent failure from water eroding through tunnels in the embankments. Tillage operations, which are generally on the contour, should be used to enhance the effect of terraces, and should not, under any circumstances, result in reducing the terrace height.

REFERENCES

American Society of Agricultural Engineers (ASAE) (1994) "Design, Layout, Construction and Maintenance of Terrace Systems," *ASAE Standard,* S268.4.

Beasley, R. P. (1963) *A New Method of Terracing,* Missouri Agr. Expt. Sta. Bull. 699. University of Missouri, Columbia, Missouri.

Schwab, G. O., D. D. Fangmeier, W. J. Elliot, and R. K. Frevert (1993) *Soil and Water Conservation Engineering,* 4th ed. Wiley, New York.

USDA Soil Conservation Service (SCS) (1979) *Engineering Field Manual for Conservation Practices.* U.S. Government Printing Office, Washington, D.C.

Wischmeier, W. H. and D. D. Smith (1978) *Predicting Rainfall Erosion Losses—A Guide to Conservation Planning,* USDA Hdbk. 537.

Wittmus, H. (1988) "A Study of Time Required for Planning, Staking, and Designing Parallel Terrace Systems," *ASAE Trans., 30:1076–1081.*

PROBLEMS

7.1 Determine the strip width for strip cropping a field at your present location to accommodate four-row equipment if the land slope is 6 percent and the row width is 40 in. Width should correspond to that for complete rounds of the equipment.

7.2 Determine the vertical interval and the horizontal interval for graded terraces at your present location if the prevailing land slope above the terrace is 6 percent and the soil has an average intake rate with medium cover.

7.3 If the maximum soil loss to the terrace channel is 4 t/a, determine the vertical interval for terraces using the Universal Soil Loss Equation given in Chapter 6. Location is northern Missouri, $K = 0.4$ t/a, slope is 6 percent, and $C = 0.15$.

7.4 Determine the difference in elevation between the upper end and the outlet of a 1000-ft terrace if the maximum grade was selected for the entire length of the terrace.

7.5 (a) Compute the volume of water stored per foot length of terrace if the terrace height is 1 ft, the frontslope and cutslope slopes are 8:1 (12.5 percent). (b) If the terrace spacing is 100 ft, how many inches of runoff will the terrace hold?

7.6 If a subsurface outlet terrace is to store 2 in. of runoff from a 10-year return period, 24-h duration storm, determine the opening size in inches for the orifice to remove this volume in 48 h. Assume an average water depth of 2 ft above the orifice, the orifice discharge coefficient C of 0.5, and the runoff area above the terrace of 5 acres.

7.7 Determine the depth of cut, depth of channel, and cut volume per foot of terrace length for a depth of flow of 0.8 ft where the land slope is 3 percent, the freeboard is 0.25 ft, and the terrace slope widths *(W)* for the cross section are 15 ft for equipment operation.

7.8 Compute the cutslope, frontslope, and backslope slopes (horizontal:vertical) for the terrace in Problem 7.7.

7.9 A terrace has 10:1 sideslopes, a spacing of 150 ft, a length of 600 ft, and a channel gradient of 0.5 percent. The design runoff is 1 in. in 24 h. (a) Calculate the flow rate in the terrace channel in cubic feet per second. (b) Use the procedures described in Chapter 8 to determine if the velocity will be excessive when the channel is smooth (Manning's $n = 0.035$). (c) Use the procedures described in Chapter 8 to determine the depth of flow when the channel is rough ($n = 0.06$).

CHAPTER 8

CHANNEL FLOW AND EROSION CONTROL

Open channels are natural and constructed passageways or drainageways for water, with the top surface open to the atmosphere. Drainageways can be subject to erosion, which may lead to gullying. Drainageways may be grassed waterways or earth ditches. This chapter will give an overview of channel flow principles, and will then describe some typical applications of those principles. Other applications of open channels are discussed in Chapter 7 for terracing, Chapter 11 for reservoir spillways, Chapter 12 for surface drainage, Chapter 16 for irrigation canals, and Chapter 19 for measuring irrigation flows. Open channel flow principles are also widely used for other storm water conveyance systems.

OPEN CHANNELS

When planning a soil and water project that includes concentrated flows of surface water, a channel is necessary. The channel must have the capacity to carry the required flow (in cfs or m^3/s) and the stability to resist the erosive forces of the flowing water. An understanding of channel flow capacity is essential when planning or modifying any channel or other control structure that may affect surface water flow. The capacity of a channel depends on the channel cross-sectional shape and area, the channel lining or surface condition, and the gradient of the channel. Flow can also be influenced by the temperature and quality of the water, and the channel alignment. In this chapter, one common method of determining channel flow, the Manning formula, will be described. The Manning formula relates channel shape, roughness, and slope to channel velocity. The continuity equation, which relates the channel velocity and shape to the capacity, will also be presented.

8.1 Channel Cross Sections

The channel cross section is a view of the channel, generally looking upstream, from which the relative positions among the water surface, the channel sides, and the channel bottom can be determined. The cross-sectional area is the area bounded by the channel surface (when flowing full), the sides, and the bottom. Channel cross sections may be approximated as rectangular, trapezoidal, triangular, or parabolic in shape (Fig. 8.1). The dimensions for a channel cross section are shown in Fig. 8.1. For a given shape, the two most important dimensions are the channel depth of flow d and the top width t. Rectangular and trapezoidal channels have a bottom width b. With trapezoidal and triangular channels, the slopes of the channel sides are defined as $Z{:}1$, where $Z = \frac{e}{d}$ or horizontal over vertical (Fig. 8.1). From these dimensions, three other geometric properties of the channel are calculated: the cross-sectional area, the wetted perimeter, and the hydraulic radius. Formulae for the geometric properties for each of the shapes in Fig. 8.1 are given in Table 8.1. The *cross-sectional area A* is calculated from the channel shape using conventional geometric methods. The *wetted perimeter p* of a channel is the length of the channel cross-sectional surface that is wetted by the water in the channel. The *hydraulic radius R* of a channel is found by dividing the area by the wetted perimeter. The hydraulic radius reflects the relative influence of the sides and bottom of the channel on the flow within the channel. With most channels, freeboard is added to the depth of flow in the channel for a margin of safety, and to allow for unexpected turbulence or obstructions, giving a larger depth D. With the additional freeboard, a larger channel top width T is necessary to determine the extent of the area covered by a channel.

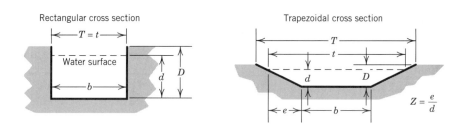

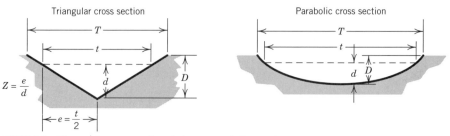

FIGURE 8.1   Channel cross-sectional shapes and dimensions.

TABLE 8.1 Relationships Among Channel Dimensions (Fig. 8.1) and
Channel Geometric Properties

Channel Shape Units	Cross-Sectional Area A ft² or m²	Wetted Perimeter p ft or m	Hydraulic Radius $R = \dfrac{A}{p}$ ft or m	Top Widths t and T ft or m
Rectangular	bd	$b + 2d$	$\dfrac{bd}{b + 2d}$	$t = b$ $T = b$
Triangular	Zd^2	$2d\sqrt{Z^2 + 1}$	$\dfrac{Zd}{2\sqrt{Z^2 + 1}}$ or $\dfrac{d}{2}$ approx.	$t = 2dZ$ $T = 2DZ$
Trapezoidal	$bd + Zd^2$	$b + 2d\sqrt{Z^2 + 1}$	$\dfrac{bd + Zd^2}{b + 2d\sqrt{Z^2 + 1}}$	$t = b + 2dZ$ $T = b + 2DZ$
Parabolic	$\dfrac{2}{3}td$	$t + \dfrac{8d^2}{3t}$	$\dfrac{t^2 d}{1.5t^2 + 4d^2}$ or $\dfrac{2d}{3}$ approx.	$t = \dfrac{A}{0.67d}$ $T = t\left(\dfrac{D}{d}\right)^{0.5}$

Example 8.1. Calculate the cross-sectional areas, wetted perimeters, hydraulic radii, and top widths for the two channel cross sections shown. The first is typical of a drainage ditch, and the second, a grassed waterway. Dimensions may be either in feet or meters.

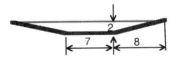

Solution. Substituting the dimensions shown into the appropriate trapezoidal equations given in Table 8.1, the following solution table is developed:

Units	b ft or m	d ft or m	Z ft or m	A ft² or m²	p ft or m	R ft or m	t ft or m
Ditch	3.5	4	1	30	14.81	2.03	11.5
Waterway	7	2	4	30	23.49	1.28	23

The preceding results demonstrate that channels with different shapes can have the same cross-sectional area. The importance of this difference in shapes will be shown in Example 8.4. The parabolic channel is similar in cross section to a natural waterway. Trapezoidal channels are frequently selected because they are more easily constructed. Trapezoidal and triangular sections tend to become parabolic under the normal actions of channel flow, deposition, and erosion.

8.2   Channel Flow Capacity

The discharge or capacity of a channel may be calculated from the fluid flow continuity equation:

$$q = Av \tag{8.1}$$

where q = channel discharge in ft³/s (m³/s)
A = channel cross-sectional area in ft² (m²)
v = average channel velocity in ft/s (m/s)

Equation 8.1 is applied to many different flow conditions, varying from open channel flow to pipe flow and air flow.

8.3   Open Channel Flow Velocity

The velocity of water flowing in an open channel depends on the channel gradient, the cross-sectional shape, and the roughness of the channel lining. One of the common relationships between these parameters and velocity is the Manning formula:

$$v = \frac{C}{n}\, R^{2/3} s^{1/2} \tag{8.2}$$

where v = velocity of flow in feet per second (meters per second)
C = Constant = 1.49 for English units (1.00 for metric units)
n = roughness coefficient (Table 8.2)
R = hydraulic radius in feet (meters)
s = channel gradient in feet per foot (meters per meter)

This equation can be solved with a scientific calculator, computer spreadsheet, numerous computer software packages, or the nomograph given in Fig. 8.2. In addition to the Manning formula, there are other similar relationships that are available to estimate the flow velocity of a channel.

Example 8.2.   Determine the average velocity of flow and the top width for a parabolic-shaped channel having a flow depth of 3.9 ft, a slope of 0.8 percent, and a roughness coefficient $n = 0.03$. The required cross-sectional area is 50 ft².

Solution.   From Table 8.1: $R = 2d/3 = 2.6$ ft. Substitute C, R, s, and n into Eq. 8.1: $v = (1.49/0.03)2.6^{2/3}0.008^{1/2} = 8.4$ ft/s (2.56 m/s) or read v from the nomograph in Fig. 8.2 to be 8.4 ft/s. From Table 8.1: $t = A/(0.67d) = 50/(0.67 \times 3.9) = 19$ ft (5.8 m).

The velocity calculated by the Manning formula is an average velocity within a channel. The velocity in actual contact with the vegetation or channel

TABLE 8.2 Roughness Coefficient n for the Manning Velocity Formula and Vegetal Retardance Class for Vegetated Waterways

Type and Description of Channel	n Values	Retardance Class
Channels, Lined		
Asphalt	0.015	
Concrete	0.012–0.018	
Concrete, rubble	0.017–0.030	
Metal, smooth	0.011–0.015	
Metal, corrugated	0.021–0.026	
Plastic	0.012–0.014	
Wood	0.011–0.015	
Channels, Vegetated		
Dense, uniform stands of green vegetation more than 30 in. high		A or B
Dense, uniform stands of green vegetation 11 to 24 in. high		B or C
Dense, uniform stands of green vegetation more than 10 in. high		
Bermuda grass	0.04–0.20	
Kudzu	0.07–0.23	
Lespedeza	0.047–0.095	
Dense, uniform stands of green vegetation 6 to 10 in. high		C or D
Dense, uniform stands of green vegetation 2 to 6 in. high		D
Dense, uniform stands of green vegetation less than 2.5 in. high		E
Bermuda grass	0.034–0.11	
Kudzu	0.045–0.16	
Lespedeza	0.023–0.05	
Earth Channels and Natural Streams		
Clean, straight, banks, full stage	0.025–0.040	
Winding, some pools and shoals, clean	0.035–0.055	
Winding, some weeds and stones	0.033–0.045	
Sluggish river reaches, weedy or with very deep pools	0.050–0.150	
Pipes		
Cast iron	0.011–0.015	
Clay or concrete (3–30 in. dia)	0.012–0.017	
Metal, corrugated	0.025	
Plastic, corrugated (3–12 in. dia)	0.015–0.02	
Smooth-wall pipe	0.009	
Steel, riveted and spiral	0.013–0.017	

Source: SCS (1979).

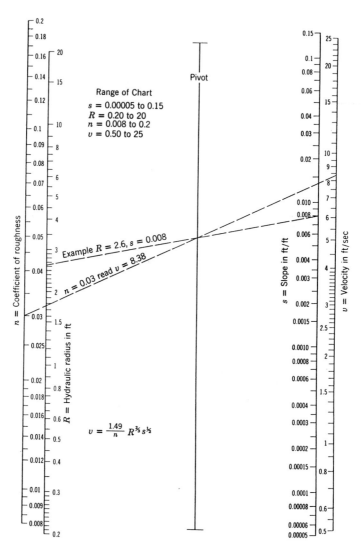

FIGURE 8.2 Nomograph for solving the Manning equation.
(Source: U.S. Soil Conservation Service, 1979.)

bed is much lower than the average, whereas the velocity of the surface of the channel is generally greater than the average velocity. A typical distribution of the velocity in a grass-lined channel is shown in Fig. 8.3. In this drawing the average velocity is about 2.5 ft/s, but the velocity in contact with the vegetation is less that 1 ft/s.

When sizing a channel to pass a given flow, Eqs. 8.1 and 8.2 are used to determine a required cross section for a given roughness, slope, and cross-sectional shape. In many cases, it is not possible to directly solve these sets of equations, so a trial-and-error procedure, as demonstrated in Example 8.3, is

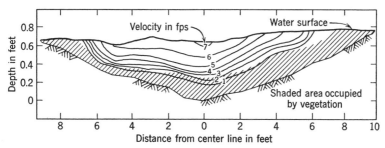

FIGURE 8.3 Velocity distribution in a grass-lined channel. (Source: Ree, 1949.)

necessary. Natural resource managers who frequently design channels will use computer solutions such as those developed by the NRCS, or will employ spreadsheet solutions to consider a range of alternative conditions to meet the desired flow capacity. Nomographs have also been developed to aid in determining channel shapes (Schwab et al., 1993).

Example 8.3. A waterway is lined with short grass and has a trapezoidal cross section. It has a bottom width of 6 ft to match the construction equipment available, channel slope of 0.5 percent, and side slopes of 10:1 to ensure ease of crossing. It is to carry a flow of 50 ft^3/s with a freeboard of 6 in. What are the total depth and width required (including freeboard) for this channel?

Solution. For the Manning formula, the slope of 0.5 percent is 0.005, the roughness coefficient n will be assumed to be 0.05, the side slope $Z = 10$.

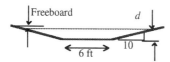

The following table is set up to allow an iterative solution. An initial depth of 1 ft is assumed to start the calculations. A, p, and R are calculated from equations presented in Table 8.1, v is calculated by Eq. 8.2, and q by Eq. 8.1. From the results of the first calculation, it is determined that the channel needs to be deeper than 1 ft, so for the second iteration, the channel capacity for a depth of 2 ft is determined. The resulting capacity is greater than that required, so a shallower depth is checked. This iterative procedure is followed until a depth is found that meets the capacity requirements. That depth is 1.4 ft.

d ft	A ft^2	p ft	R ft	v ft/s	q ft^3/s	Comment
1	16	26.10	0.61	1.52	24.3	Too shallow
2	52	46.20	1.13	2.28	118.6	Too deep
1.5	31.5	36.15	0.87	1.92	60.1	Still too deep
1.4	28	34.14	0.82	1.85	51.7	Close enough

From the preceding results, d will be 1.4 ft (0.43 m). The freeboard is added to this to determine that $D = 1.4 + 0.5 = 1.9$ ft (0.58 m). The total top width is found from the equation in Table 8.1 to be

$$T = b + 2DZ = 6 + 2 \times 1.9 \times 10 = 44 \text{ ft (13.4 m)}$$

The preceding presentation for determining channel capacity assumes that the roughness value does not change with changing depth. In practice, this assumption does not hold for vegetated channels. Figure 8.4 shows the relationship between Manning's n and the product of velocity and hydraulic radius for different retardance classes of vegetation (Table 8.2). As channel depth and/or velocity increases, channel vegetation tends to bend over and the roughness is reduced. A procedure for considering the variation in roughness is presented in Schwab et al. (1993), as well as in manuals and computer design programs developed by the NRCS and other natural resource agencies.

8.4 Stream Channel Erosion and Control

Stream channels can serve as areas of deposition for eroded sediment, or as a source of sediment through stream channel and streambank erosion. In most cases, that amount of sediment eroded from stream channels is only a small part of the channel's total sediment budget. If the source of this sediment is from the bank, and the erosion of the bank is threatening other structures, then some form of streambank protection may be necessary. Generally, a well vegetated bank has a low rate of erosion, but grazing or construction activities may expose a bank and initiate an undesirable erosion process.

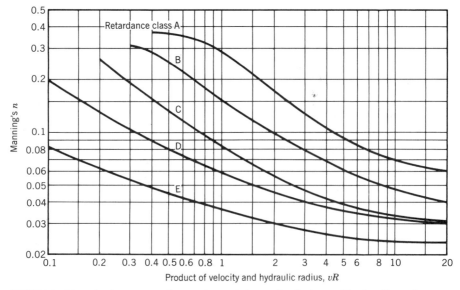

FIGURE 8.4 Manning's n versus the product of channel velocity and hydraulic radius vr in a vegetated channel. (Source: U.S. Soil Conservation Service, 1979.)

FIGURE 8.5 Box-inlet drop spillway, and riprap bank stabilization on a small river in northwestern Miss. (Photograph by W. Elliot.)

Disturbing a stream by straightening or reshaping the channel may initiate unexpected erosion from both streambanks and streambeds. Alterations of the stream may also adversely affect aquatic life in the stream as well as ecosystems in riparian areas adjacent to streams. In many cases, approval from state or federal agencies may be required prior to any streambank or streambed activities. Expert advice should be obtained before carrying out any significant channel-altering activities.

Some common streambank erosion treatments, shown in Figs. 8.5 and 8.6,

FIGURE 8.6 Demonstration of streambank erosion control methods in central Ohio. Treatments include rock riprap in foreground, an untreated section, burlap fabric, and cellular plastic as the dark treatment in the back. (Photograph by W. Elliot.)

include rock riprap, burlap fabric, and cellular plastic mats backfilled with soil. Many vegetative treatments are available including thick plantings of willow cuttings and similar shrubs that are readily established from branch cuttings. Vegetative treatments blend in more naturally with the stream environment, but mechanical treatments can be effective immediately with less risk of failure. Costs vary depending on the availability of local rock or vegetation, and the extent of the work required.

VEGETATED WATERWAYS

Vegetated or grassed waterways (Fig. 8.7) are commonly employed in soil erosion control to serve as outlets for terraces or contour furrows (Chapter 7), to reduce the risk of rill channel scour, to prevent gully erosion, to reclaim gullies, or for other surface water control systems. These waterways commonly serve as emergency spillways for reservoirs (Chapter 11) or other structures. They may be part of a surface drainage system (Chapter 12). Vegetated waterways may also be constructed across the slope to serve as diversion channels, which are similar to terraces. The function of a diversion channel is comparable to that of the top terrace. Diversion channels may divert water away from active gullies or farm buildings, protect bottomland from overland flow, or intercept and divert unwanted runoff.

8.5   Cross Sections

Waterway cross sections may be triangular, trapezoidal, or parabolic. In some cases, no earthwork is necessary to form a parabolic channel, because the natural cross section of a drainageway or meadow outlet may be adequate, and

FIGURE 8.7   Vegetated waterway in eastern Nebraska. (Photograph by W. Elliot.)

only the boundaries need to be defined. Natural drainageways have parabolic channels. Because they are easier to construct, trapezoidal or triangular cross sections are selected where the entire channel must be excavated.

Ease of maintenance and convenience in crossing a channel with machinery or vehicles should be carefully considered in the selection of a channel cross section. Side slopes of 4:1 or flatter are desirable to facilitate maintenance by mowing and provide for easy crossing. In some cases, a wider, flatter cross section may be necessary to reduce flow velocities to a nonerosive level as illustrated in Example 8.4.

Low flows in wide flat bottoms of trapezoidal channels may cause sediment deposits that result in meandering of higher flows with accompanying local damage to vegetation. Thus, a slight "V" bottom may be desirable in a trapezoidal channel bottom.

8.6 Selection of Vegetation

The vegetation selected for a given waterway is governed by (1) soil and climatic conditions, (2) duration, quantity, and velocity of runoff, (3) time required to develop a good cover and ease of establishment, (4) availability of seed or plant materials, (5) suitability for utilization as a seed or hay crop, and (6) spreading of vegetation to adjoining areas. Tolerance to herbicides such as atrazine may also be necessary in some management systems. Table 8.3 lists some of the grasses recommended for use in waterways in various regions of the United States. Additional local recommendations for species and mixes can be obtained from the NRCS.

Uniformity of cover is important, as the stability of the most sparsely covered area controls the stability of the channel. Bunch grasses do not offer good protection because they produce nonuniform flows with highly localized erosion, and their open roots do not bind the soil firmly.

8.7 Channel Velocities

The water velocity in a vegetated waterway should be slow enough to prevent scouring but sufficiently fast to minimize sedimentation. The maximum permis-

TABLE 8.3 Vegetation Recommended for Grassed Waterways

Geographical Area of U.S.	Vegetation
Northeastern	Kentucky bluegrass, tall fescue, red top
Southeastern	Kentucky bluegrass, tall fescue, brome, Reed canary, Bermuda
Upper Mississippi	Kentucky bluegrass, tall fescue, brome, Reed canary
Western Gulf	Tall fescue, Bermuda, King Ranch bluestem, native grass mixture
Southwestern	Brome, western wheatgrass, intermediate wheatgrass, tall wheatgrass
Northern Great Plains	Brome, western wheatgrass, red top, native bluestem mixture

Source: SCS (1979).

sible velocity in the channel depends on soil erodibility and vegetation condition. Generally, soils that are higher in silt content are more easily eroded, and those high in clay are more erosion resistant. A change from a steeper to a flatter slope, an increase in channel width (Example 8.4), or an increase in the resistance of the vegetation to flow will cause the velocity of water to decrease. Table 8.4 presents some typical recommendations for maximum permissible velocities. Chapter 7 presented minimum recommended velocities to limit deposition in terraces. Similar minimum velocities would apply to other channels as well.

Example 8.4. Determine the velocities for the two channel cross sections described in Example 8.1, assuming that for both cases, the slope is 4 percent, the vegetated lining is Bermuda grass less than 2.5 in. in length, and the soil is easily erodible.

Solution. From Table 8.2, note that the roughness is between 0.034 and 0.11. Select 0.07 as a midrange value, and $s = 0.04$. Substitute these values, along with the hydraulic radii from Example 8.1 into the Manning formula (Eq. 8.2) to calculate the velocities.

Ditch	$R = 2.03$ ft	$v = \dfrac{1.49}{0.07} 2.03^{2/3} 0.04^{1/2} = 6.81$ ft/s (2.08 m/s)
Waterway	$R = 1.28$ ft	$v = \dfrac{1.49}{0.07} 1.28^{2/3} 0.04^{1/2} = 5.01$ ft/s (1.53 m/s)

For the given conditions, the first cross section has a velocity in excess of the 6 ft/s maximum recommended velocity given in Table 8.4 for an easily eroded soil, whereas the second cross section has an acceptable velocity.

8.8 Drainage

Waterways are often located where there is a low flow over long periods of time. Subsurface (pipe) drain outlets, springs, or seeps flowing into a waterway may cause continually wet conditions, which may prevent the development and maintenance of good vegetal cover. Diversion of such flow, either by subsurface or surface means, is essential to the success of the waterway.

Detailed discussion of subsurface drainage of wet drainageways will be found in Chapter 13, where the interception of seepage along the sides or upper end of a waterway is discussed. Surface water entering a waterway at some point may be intercepted by a catch basin and diverted to a subsurface drain similar to a terrace subsurface outlet (Fig. 7.11). In some situations, it may be desirable to provide a concrete or asphalt trickle channel in the bottom of the waterway to carry prolonged low flow.

8.9 Establishment of Vegetated Waterways

Vegetation in waterways can be established by either seeding or sodding. Prior to either operation, the soil should be adequately tilled, fertilized, and limed to ensure a satisfactory growth. Manure or other organic matter incorporated into

TABLE 8.4 Permissible Velocities for Channels Lined with Vegetation

		Permissible Velocities[a] ft/s	
Vegetative Cover	Slope Range Percent	Easily Eroded Soils	Erosion Resistant Soils
Bermuda grass	0–5	6	8
	5–10	5	7
	over 10[b]	4	6
Kentucky bluegrass	0–5	5	7
Brome	5–10	4	6
Tall fescue	over 10[b]	3	5
Grass mixture	0–5	4	5
Wheatgrass	5–10	3	4
Reed canary grass			
Redtop, bluestem	0–5[c]	2.5	3.5
Common lespedeza[d]	0–5[e]	2.5	3.5
Sudan grass			

Source: SCS (1979).

[a]Use velocities exceeding 5 ft/s only where good covers and proper maintenance can be obtained. Easily eroded soils include those high in silts and fine sands. Clays and loams are generally more erosion resistant. (Compare relative erodibilities on Table 6.1.)

[b]Do not use on slopes steeper than 10 percent except for vegetated side slopes in combination with a stone, concrete, or highly resistant vegetative center section.

[c]Do not use on slopes steeper than 5 percent, except for vegetated side slopes in combination with a stone, concrete, or highly resistant vegetative center section.

[d]Annuals—use on mild slopes or as temporary protection until permanent covers are established.

[e]Use on slopes steeper than 5 percent is not recommended.

the waterway will increase soil erosion resistance. Sprinkler irrigation may be beneficial in some climates to aid in vegetation establishment.

Waterway seeding mixtures should include some quick-growing annuals for temporary protection, as well as a mixture of hardy perennials for permanent protection (Table 8.3). Seeding rates should be about double those used for field seeding. Local tillage and planting practices should be followed, with extra care taken to ensure a successful establishment.

In highly erosive situations, organic fiber or plastic meshes with seeds in the fabric are available. The organic fabric decays as the grass becomes established. Manufacturers' guidelines should be followed when selecting and installing these fabrics.

If establishment is difficult because of poor soils or an adverse climate, machinery that mixes seeds with peat or straw and then blows or pumps the mixture onto the waterway is available. These systems generally improve germination by making more water available to the seeds from the organic matter, and the organic matter offers some erosion protection. Other products including soil stabilizers and asphalt mulches are also available to assist in establishment or to increase channel erosion resistance.

Direct sodding may be the surest means of establishing vegetation, but it is the most expensive. Generally, sodding is used only for high-velocity channels or for critical situations, such as short, steep "sod chutes" or spillways.

8.10   Waterway Construction

Waterways are usually located in existing drainageways. Slopes are generally determined by the natural topography, and channel cross sections by methods discussed in Sections 8.5 and 8.7. Prior to construction, flags or stakes are set out along the center of the waterway, and then along either side to note the top width T (Fig. 8.1). The center depth of excavation is noted on one of the side flags.

Generally, earth-moving equipment is used for excavating the desired shape of the waterway, and farm equipment for establishing a seedbed for revegetation. If excavations are more than 1 ft deep, it may be desirable to stockpile topsoil from the site, shape the waterway, and then return the topsoil to aid in revegetation.

8.11   Waterway Maintenance

The condition of the vegetation in a waterway is dependent not only on design and construction but also on subsequent management. Waterways should not be used for vehicle or livestock traffic. The end of tillage on the shoulders should be staggered to prevent flow concentration down the edges of the watercourse. Extra care may be required to ensure that erosion associated with ends of furrows entering a waterway is minimal, and that any sign of major rill erosion is removed before gullying becomes a problem. Tillage implements should be raised when crossing the waterway.

During the first few years after seeding, vegetation should be mowed several times during the growing season to stimulate spreading and growth and to control weeds. Dense cuttings and bunches of cut grass, especially from rotary mowers, should be removed to prevent the smothering of established vegetation. An annual application of fertilizer will help maintain a dense sod. Vegetation may be harvested as a hay or seed crop without interfering with its primary purpose of erosion control.

High runoff during the period of vegetation establishment may result in small gullies. These should be filled, and the vegetation reestablished, or the channel may need to be reshaped and reseeded. If possible, risks of washout can be avoided by diverting the water to another channel during the period when grass is being established.

Accumulations of sediment in the waterway may smother vegetation and restrict channel capacity. The best method of minimizing sediment problems in waterways is by reducing erosion on the upland watershed. Rank or matted vegetal growth can cause water to back up leading to sedimentation. Selecting suitable vegetation and carrying out timely mowing will reduce such vegetation problems. Sediment may deposit at the lower end of the waterway if the slope decreases. If vegetation damage is negligible, the water may be simply allowed to spread out to allow a shallow fan of deposition to develop. If the fans are adversely affecting the waterway vegetation or the field crop, the grassed waterway may be discontinued and a narrower drainage channel established to carry the runoff at a higher velocity (Example 8.4). Extending vegetation well up the sides of the waterway and into terrace channels is recommended. Accumulated sediment may occasionally be removed or the channel reshaped to minimize damage to the vegetation and to prevent localized erosion.

GULLY CONTROL AND STRUCTURES

If overland flows are concentrated by natural topography or human structures, gullies may develop. Gullies are particularly common on soils in which a deep, highly erodible topsoil overlays a relatively impermeable subsoil. If such soils have steep gradients, gullying may be severe once the channel is disturbed. Gullies can destroy agricultural land and create problems with roads, bridges, and other structures (Figs. 6.3 and 8.8). Fields are divided by gullies making

FIGURE 8.8 Severe gully in northwestern Miss. Note the person dwarfed on the gully floor. (Photograph by W. Elliot.)

farming more difficult, and detached soil may cause sedimentation and other water quality problems downstream.

8.12 Gully Prevention and Reclamation

The hazards of gully formation can frequently be avoided by installing properly shaped channels, protected by vegetation to absorb the energy of runoff without damage to the channel, in susceptible drainageways (Example 8.4). Once a gully has formed, it may be more economical to allow the gully to revegetate with natural erosion-resistant vegetation rather than attempt to reshape the gully to the original topography with expensive earthworks. Under some conditions, trees, shrubs, and vines may be more suitable than grasses and legumes for gully control. Natural revegetation may be stimulated by diverting flow and fencing livestock out of the damaged areas. A gradual succession of plant species native to the region may become established. Planting of locally recommended species will accelerate the revegetation process. Variety in plantings will reduce the danger of destruction by disease and climatic extremes. Revegetated areas may serve as wildlife refuges.

Gully formation or growth can also be reduced or halted by reducing the flow through a problem channel. Channel flow can be reduced by diverting overland flows to more stable channels with diversion terraces, or to subsurface drains with subsurface outlet terraces (Fig. 7.11). Improved conservation practices on the watershed above gully sites can also reduce runoff (Chapter 7).

8.13 Grade Control Structures

In some situations, it may not be possible to limit the velocity in a channel by restricting the depth (Example 8.4) because of topographic considerations, steep channel gradients, or limitations on width. One additional method of limiting velocity is to reduce channel gradients by the installation of grade control structures. Such structures may also be necessary to stop further development of a gully.

Figs. 8.5 and 8.9 present three common gradient-reducing structures, the straight- and box-inlet drop spillways and the drop-inlet pipe spillway. Other structures and their designs are described in Schwab et al. (1993) and SCS (1979). The free-flow (no submergence) capacity for straight- and box-inlet drop spillways and drop-inlet pipe spillways is

$$q = CLh^{3/2} \tag{8.3}$$

where q = the discharge in ft^3/s (m^3/s)

C = weir coefficient [3.2 for most conditions (1.76 metric units)]

L = weir length in ft (m)

h = depth of flow over crest in ft (m)

The weir length L is the sum of the lengths of the three inflow sides of a box-inlet or the width of a straight inlet drop spillway. In addition, for the drop-inlet

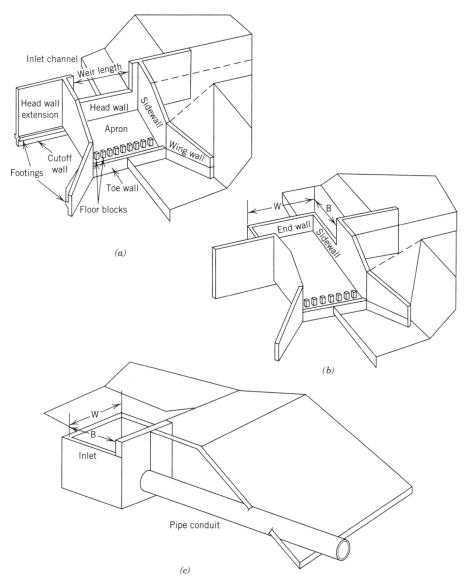

FIGURE 8.9 Structures used to reduce gradients in waterways: *(a)* straight drop spillway, *(b)* box-inlet drop spillway, and *(c)* drop-inlet pipe spillway.

pipe spillway, the capacity of the pipe conduit must also be considered, following culvert designs methods. Figure 13.6 can be used to estimate the pipe size, or methods to estimate the pipe capacity can be found in Schwab et al. (1993) and SCS (1979). The drop-inlet pipe spillway is particularly useful for gully reclamation when the change in elevation is in excess of 4 ft.

Example 8.5.   The calculated velocity in a vegetated waterway was found to be excessive, and it was determined that a straight drop spillway should be installed to reduce the channel slope by 3 ft. What width of spillway L is necessary if the flow rate is 50 ft^3/s and the depth of flow is to be 1.2 ft?

Solution.   Solve Eq. 8.3 for L, and substitute in the values given:

$$L = \frac{q}{Ch^{3/2}} = \frac{50}{3.2 \times 1.2^{3/2}} = 11.9 \text{ ft } (3.64 \text{ m})$$

REFERENCES

Ree, W. O. (1949) "Hydraulic Characteristics of Vegetation for Vegetated Waterways," *Agr. Eng.,* 30:184–187, 189.

Schwab, G. O., D. D. Fangmeier, W. J. Elliot, and R. K. Frevert (1993) *Soil and Water Conservation Engineering,* 4th ed. Wiley, New York.

U.S. Soil Conservation Service (SCS) (1979) *Engineering Field Manual for Conservation Practices.* U.S. Dept. of Commerce National Technical Information Service, Washington, D.C.

PROBLEMS

8.1 A trapezoidal irrigation canal has a bottom width of 4 ft, side slopes of 1:1, and a depth of 3 ft. The channel is earth lined, and the velocity is observed to be 2 ft/s. What is the flow rate of the channel (ft^3/s)?

8.2 Determine the top width and total depth for a trapezoidal (4:1 side slopes) grass waterway that can carry a flow of 50 ft^3/s. The channel slope is 4 percent, the bottom width is 24 ft, and a good grass stand is cut to a 4-in. height.

8.3 Determine the top width and total depth for a parabolic grass waterway for the same conditions as in Problem 8.2.

8.4 For a terrace, two roughness coefficients are considered: a maximum of 0.06 to determine the maximum depth of flow and a minimum of 0.035 to determine the maximum velocity. For a terrace that is to carry 1 ft^3/s with a triangular cross section with 8:1 side slopes and a channel gradient of 0.3 percent, what is the maximum velocity and the maximum depth of flow?

8.5 The depth of flow of a trapezoidal waterway for a reservoir spillway is 1 ft. The flow rate is 20 ft^3/s, the side slopes are 1:1, and the channel slope is 6 percent, what is the average flow velocity and required bottom width b, assuming a roughness of 0.05?

8.6 A surface drain is to have a triangular cross section with side slopes of 10:1, a gradient of 0.2 percent, and a capacity of 5 ft^3/s. Assuming a roughness of 0.08, and freeboard of 6 in., what is the width of the drain?

8.7 The runoff into a gully is estimated to be 40 ft^3/s. To halt the growth of the gully, a box-inlet drop spillway is to be installed that is 4 ft wide to match the bottom of the channel leading to the spillway. If the desired flow depth is 6 in., what breadth (in Fig. 8.9*b*) is required?

8.8 The growth of a gully is to be halted by diverting the estimated 30 ft³/s flow to a more stable waterway. The diversion is trapezoidal. The bottom width is to be 8 ft and the depth limited to 1.5 ft. The side slopes are to be 4:1. The diversion will have a minimum roughness coefficient of 0.03. The channel will be lined with Kentucky bluegrass. What is the maximum permissible slope to prevent erosion of the diversion channel on an easily eroded soil?

8.9 Contact local suppliers of erosion prevention mats and a local quarry to determine which method is cheaper to halt stream bank erosion for a site near you. Compare the costs per square foot of a 1.5-ft depth of 6-in. dia. riprap to a heavy duty erosion prevention mat.

CHAPTER 9

WIND EROSION AND CONTROL PRACTICES

Wind erosion is a worldwide problem that is particularly severe in large continental wheat-growing regions, such as the U.S. Great Plains, in dune areas along coastlines, and in intensively farmed organic soils. Approximately 6 percent of the United States is subject to severe wind erosion (Fig. 9.1). Eroded soil carries nutrients, adversely affecting productivity. Wind erosion damages land and crops (Fig. 9.2), increases road maintenance costs, and adversely affects human respiratory health. The eroded dust can carry harmful chemicals, risking the health of humans or livestock many miles from the source of the sediment. There are now EPA guidelines for limiting dust (particulate matter less than 10 microns in diameter or PM10) in the atmosphere. These limitations may restrict activities that cause wind erosion in or near cities. Wind erosion can also create visibility problems along major highways. In recent years, fatal accidents in many states have been directly attributed to limited visibility caused by eroded dust in the air. For example, in November, 1991, 104 vehicles were involved in an accident on Interstate 5 in California, resulting in at least 15 deaths and 150 injuries.

The severe dust storms of the 1930s brought national attention to wind erosion when sediment eroded from the Great Plains was deposited in Washington D.C. This deposition in the nation's capital was one of the main stimuli leading to the development of the Soil Conservation Service (name changed in 1994 to Natural Resources Conservation Service, NRCS).

9.1 Wind Erosion Principles and Processes

TYPES OF SOIL MOVEMENT   There are three types of soil movement in wind erosion: saltation, suspension, and surface creep (Fig. 9.3). Saltation is the process in which fine particles (typically 0.1 to 0.5 mm diameter) are lifted from the surface by wind turbulence and follow distinct trajectories under the influence of wind forces, air resistance, and gravity. When the returning particles impact the surface, they may rebound or become imbedded. In either case, they initiate movement of other particles to create an "avalanching" effect of additional soil movement. Most saltation occurs within 1 ft of the surface. Saltating particles tend to rebound along the soil surface until they reach an obstruction (such as a fence) that traps the particles or reduces the near-surface wind velocities to stop further activity.

Finer particles (less than 0.1 mm diameter), dislodged by saltating particles,

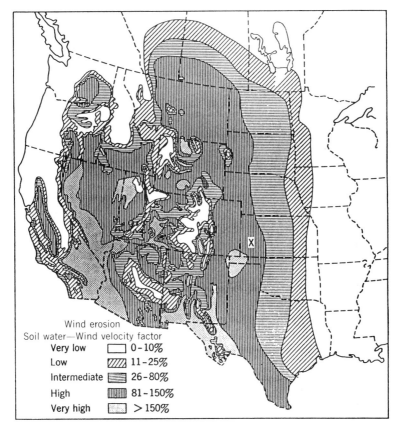

FIGURE 9.1   Areas susceptible to wind erosion in the Great Plains. The annual wind erosion climatic factor C is expressed as a percentage of that in the vicinity of Garden City, Kansas, marked by an "X". (Source: Chepil, Siddoway, and Armbrust, 1962.)

may remain suspended for an extended period. Suspended particles are often abraded by saltating particles and represent 3 to 10 percent of eroding particles. As wind velocities increase, so does the diameter of suspended particles.

Coarse and very coarse sand-size particles or aggregates (0.5 to 2 mm dia.) are set in motion by the impact of saltating particles and tend to roll or creep along the surface. Creep accounts for 7 to 25 percent of soil movement (Chepil, 1945).

MECHANICS OF WIND EROSION   To understand the mechanics of wind erosion, an analysis must be made of the nature and magnitude of the wind forces as they react on soil particles. Soil movement is initiated by wind turbulence. Winds, being variable in speed and direction, produce gusts with eddies and cross-currents that lift and transport soil. The threshold velocity is the minimum required to produce soil movement by direct action of the wind. Impact wind velocity is the minimum required to initiate movement from the impact of soil

FIGURE 9.2   Severe wind erosion in New Mexico. (Courtesy U.S. Soil Conservation Service.)

particles carried in saltation. Wind speeds of 10 mph or less, at 1-ft height, are usually considered nonerosive for mineral soils.

The quantity of soil moved is influenced by the particle size, density, gradation, wind velocity, and distance across the eroding area. The rate of soil movement increases with the distance from the windward edge of the field or eroded area. This increase is a result of the increasing amounts of moving erosive particles, which cause greater abrasion and lead to a gradual decrease in surface roughness across an exposed area. Fine particles drift and settle on the leeward side of any obstructing features or accumulate in shifting dunes.

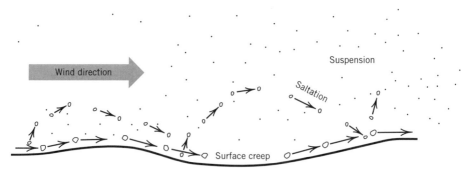

FIGURE 9.3   Types of soil movement in wind erosion: saltation, suspension, and surface creep.

Deposition of particles occurs when the gravitational force is greater than the forces holding the particles in the air. Eroding winds are turbulent, which means that their velocities generally fluctuate, with deposition occurring during periods of decreased velocity. Deposition also occurs when there is a decrease in wind velocity caused by vegetation or other physical barriers, such as ditches, vegetation, or snow fences (Fig. 9.4). Raindrops may also remove suspended sediment from the air.

9.2   Wind Erosion Factors

It is not possible to predict wind erosion by simply taking the product of erodibility parameters. Regression equations that describe the complex relationships between those parameters affecting erosion have been developed (Skidmore, 1994). A system of equations, nomographs, tables, and charts known as the Wind Erosion Equation (WEQ) were developed in the mid 1960s by Woodruff and Siddoway (1965). Since the mid 1980s, researchers have been developing a computer method for predicting wind erosion called the Wind Erosion Prediction System (WEPS) (Hagen, 1991). The WEPS program is not widely available, therefore a modified method of applying the WEQ, based on the equations presented in Skidmore (1994) will be presented to assist in understanding the important factors affecting wind erosion. The erosion rates predicted by this method are unlikely to be closer than a factor of two to any observed rates.

The Wind Erosion Equation is generally presented as a function of the following factors:

$$E = f(I, Rf, C, L, V) \tag{9.1}$$

where E = the estimated average annual soil loss in tons/acre/year
   I = the soil erodibility index in tons/acre/year
   Rf = the ridge roughness factor
   C = climate factor in percent
   L = unsheltered length of eroding field in feet
   V = vegetative cover factor

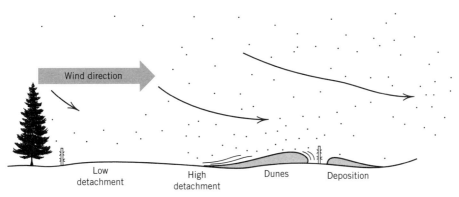

FIGURE 9.4 Variation of wind erosion within a field.

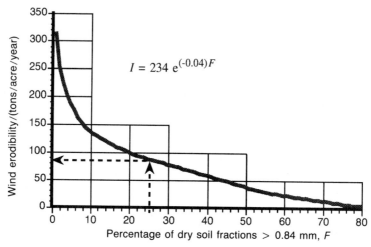

$$I = 234\ e^{(-0.04)F}$$

FIGURE 9.5   Wind erodibility I in tons/acre/year versus percent nonerodible fraction F (>0.84 mm dia by dry sieving). (Based on values given in Woodruff and Siddoway, 1965.)

ERODIBILITY INDEX I   The soil erodibility I is a function of the nonerodible soil aggregates greater than 0.84 mm dia. (Woodruff and Siddoway, 1965). Figure 9.5 shows the relationship between the nonerodible fraction of the soil (aggregates >0.84 mm dia.) and the erodibility I. The clod fraction of dry soil can vary during the season and can also be altered with changes in soil water content and organic matter. Soil erodibility can also vary with texture and organic matter. Table 9.1 gives a summary of erodibility indices for a range of different soils. Management systems that include greater amounts of surface residue can increase clod strength and decrease soil erodibility (Layton et al., 1993).

Increased wind erosion occurs on windward sides and tops of knolls (Fig.

TABLE 9.1   Wind Erodibility Indices I for Different Soil Textures

Predominant Soil Texture	*Erodibility Group*	*Erodibility Index (I) (tons/acre/yr)*
Loamy sands and sapric organic material	1	160–310 220 average
Loamy sands	2	134
Sandy loams	3	86
Clays, clay loams, and calcareous loams	4	86
Noncalcareous loams, silt loams <20% clay, and hemic organic soils	5	56
Noncalcareous and silt loams >20% clay	6	48
Silt, noncalcareous silty clay loams, and fibric organic soils	7	38
Wet or rocky soils not susceptible to erosion	8	—

Source: SCS (1988).

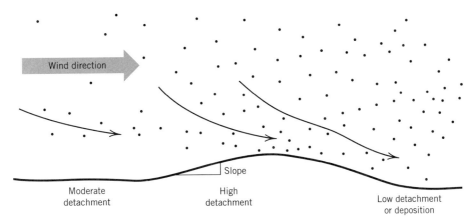

FIGURE 9.6   Effect of knolls on wind erosion.

9.6). Table 9.2 recommends adjustment factors to multiply with I to account for the knoll effect.

Example 9.1.   Calculate the erodibility index of a soil that is found to have 25 percent of the clods greater than 0.84 mm dia., in a field with numerous knolls with slopes of 3 percent.

Solution.   From Fig. 9.5, enter the bottom axis at 25 percent, draw a line up until it intersects the curve, and then go left to determine that the erodibility is 86 tons/acre/year.

From Table 9.2, note that the knoll adjustment factor for 3 percent knolls is 1.3. The adjusted erodibility factor is:

$$I = 86 \times 1.3 = 112 \text{ tons/acre/year}$$

ROUGHNESS FACTOR Rf   The roughness factor is a measure of the effect of ridges on erosion rate made by tillage and planting implements. Ridges absorb and deflect wind energy and trap moving soil particles (Fig. 9.7). To obtain the

TABLE 9.2   Knoll Erodibility Adjustment Factors

Knoll Slope in Prevailing Wind Erosion Direction (%)	Factor	
	Average	*At Knoll Crest*
3[a]	1.3[a]	1.5
4	1.6	1.9
5	1.9	2.5
6	2.3	3.2
8	3.0	4.8
10	3.6	6.8

Source: SCS (1988).

[a]Values used in Example 9.1.

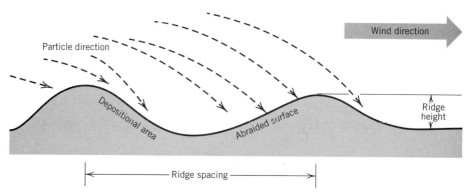

FIGURE 9.7   Effect of surface roughness and ridges on wind erosion.

roughness factor, ridge roughness Rr must first be determined. Ridge roughness can be estimated from the equation:

$$Rr = 4\ h^2/d \qquad (9.2)$$

where Rr = ridge roughness in in.
h = ridge height in in.
d = ridge spacing in in.

From the ridge roughness Rr, a roughness factor Rf can be determined
from Fig. 9.8. If there is a dominant wind direction, and the ridges are normal
to that direction, then Rf is assumed to equal 1.00 regardless of the soil roughness. In Fig. 9.8, the lowest Rf value, where the least erosion occurs, is a condition with a ridge roughness of about 3 in. Smooth surfaces offer little resistance

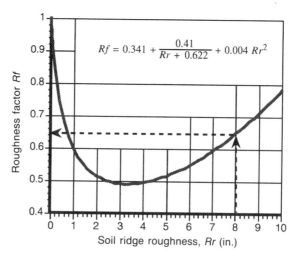

FIGURE 9.8   Relationship between the roughness factor Rf and ridge roughness Rr (in inches). (From
SCS, 1988.)

to detachment and transport of sediment by wind, but as roughness increases, growing amounts of detached sediment are deposited on the downwind side of the ridges (Fig. 9.7). Once ridge roughness exceeds about 3 in., however, the exposed ridge tops themselves become susceptible to increased erosion, because of the additional wind turbulence created under these conditions.

Example 9.2. A tillage implement leaves 4-in. high ridges at an 8-in. spacing. Calculate the ridge roughness factor.

Solution. From Eq. 9.2, calculate the ridge roughness.

$$Rr = 4 \frac{4^2}{8} = 8 \text{ in.}$$

From Fig. 9.8 for Rr = 8 in., read Rf = 0.65.

CLIMATE FACTOR C The climate factor is an index of climatic erosivity that includes the wind speed and the soil surface water content. There is ongoing research to determine the best method for predicting the climate factor from readily available weather data. It is expressed as a percent of the C factor for Garden City, Kansas. One set of C factors is presented in Fig. 9.1 for the Western U.S. C may be determined for other climates based on the average monthly rainfall and temperature, and the average annual wind velocity. Monthly rainfall and temperature can be used to determine a potential evapotranspiration index (Thornthwaite's PE Index). Figure 9.9 assists in estimating the monthly PE

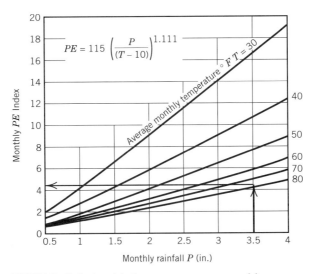

FIGURE 9.9 Relationship between average monthly temperature, rainfall, and Thornthwaite's PE Index. The determination of the PE index for June in Example 9.3 is shown. [Based on SCS (1988) equations.]

index. The monthly *PE* Indices can be summed to obtain the annual *PE* Index. The annual *PE* Index and the average annual wind velocity determine the climate *C* factor from Fig. 9.10 [based on SCS (1988)]. Other methods, which better account for climate variability and the interaction between climate and soil, are being developed to improve the estimation of a climate index (Skidmore, 1986).

Example 9.3. Calculate the climate factor *C* for the following conditions:

Solution. From the following temperatures and precipitations, the average annual wind velocity is found and the *PE* value is calculated for each month using Fig. 9.9:

Month	Avg. Temp. °F	Precip. Inches	PE Fig. 9.9	Avg. Wind Vel. mph
Jan	27[a]	0.23[b]	2.2	11.41
Feb	32	0.54	1.9	12.21
Mar	40	1.20	3.2	14.18
Apr	52	1.86	3.6	13.96
May	62	3.43	5.6	12.93
Jun	73[c]	3.51[c]	4.7[c]	13.22
Jul	79	3.54	4.2	11.95
Aug	77	2.96	3.6	11.54
Sep	68	2.00	2.7	12.24
Oct	57	1.17	1.9	11.83
Nov	41	0.78	1.9	11.61
Dec	31	0.31[b]	1.8	11.61
Annual *PE*			37.3	
Average Wind Velocity *v* mph				12.39

[a]In months less than 28°, use 28°F
[b]In months with less than 0.5 inches of precipitation, use 0.5 inches.
[c]Example in Fig. 9.9

From the total *PE* and the average wind speed, the climate factor *C* is found from Fig. 9.10:

$$C = 47 \text{ percent}$$

UNSHELTERED DISTANCE L The *L* factor represents the unsheltered distance (in feet) along the prevailing wind direction for the field or area to be evaluated (Figs. 9.11 and 9.12). This distance is from a sheltered edge of a field, parallel to the direction of the prevailing wind, to the end of the unsheltered field. Figure 9.11 shows the trigonometric relationship between the unsheltered distance and the field width when the wind is not normal to the field width. A similar trigonometric analysis is necessary for other field and wind direction relationships.

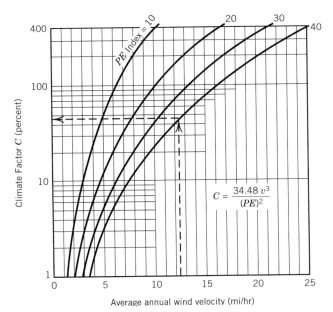

FIGURE 9.10 Relationship between Thornthwaite's *PE* Index, average annual wind velocity, and the annual wind erosion climate factor *C*. [Based on SCS (1988) equations.]

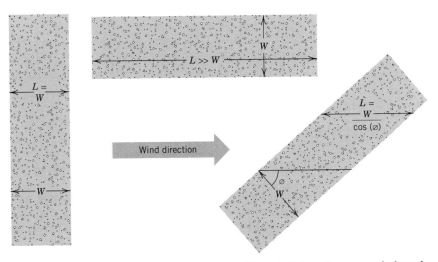

FIGURE 9.11 Effect of field orientation to prevailing wind direction on unsheltered distance *L*.

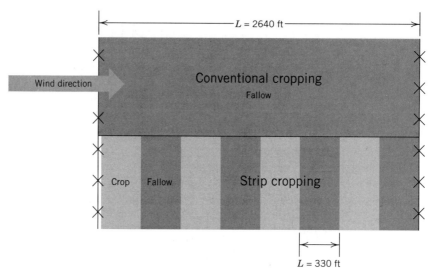

FIGURE 9.12   Effect of field strip cropping on unsheltered distance L.

Example 9.4.   A field has a width of ¼ mile in the north-south direction. If the prevailing winds are from the northwest, what is the unsheltered distance L?

Solution.   The width of the field in feet is:

$$¼ \text{ mile} \times 5280 \text{ feet/mile} = 1320 \text{ ft}$$

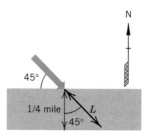

The unsheltered distance L from trigonometry is:

$$L = \frac{1320}{\cos 45^\circ} = 1867 \text{ ft}$$

VEGETATIVE COVER FACTOR V   Vegetation protects the soil surface from the erosive forces of wind and moving detached sediment. The effect of vegetative cover V in the wind erosion equation depends on the kind, amount, and orientation of vegetative material. The cover factor can be found by first determining the amount of residue R_w in pounds per acre and the residue subfactors a and b from Table 9.3, and then using Fig. 9.13 to find V.

TABLE 9.3   Crop Residue Coefficients *a* and *b* for Predicting Vegetative Cover Factor *V*

Crop	Orientation	Height (in.)	Length (in.)	Row Space (in.)	Orientation to Flow	a	b
Cotton (genus *Gossypium*)	Flat-random		10			0.067	1.17
Cotton	Standing	13		30	Normal	0.165	1.15
Forage sorghum[a] (genus *Sorghum*)	Standing	6		30	Normal	0.31	1.12
Rape (*Brassica napus*)	Flat-random		10			0.055	1.29
Rape	Standing	10		10	Normal	0.088	1.4
Silage corn (*Zea mays*)	Standing	6		30	Normal	0.20	1.14
Soybeans (*Glycine max*)	1/10 standing 9/10 flat	2.5		30		0.013	1.55
Soybeans	Flat-random		10			0.146	1.17
Sunflowers (genus *Helianthus*)	Flat-random		17			0.009	1.37
Sunflowers	Standing	17		30	Normal	0.018	1.34
Winter Wheat (genus *Triticum*)	Standing	10		10	Normal	3.85	0.97
Winter wheat	Flat-random		10			6.65	0.78

Source: Lyles and Allison (1981).

[a]Values in Example 9.5.

Example 9.5.   Determine the vegetative cover factor *V* for a field with 500 lbs per acre of standing sorghum stubble cover. The row spacing is 30 in., and the height of the stubble is 6 in.

Solution.   From Table 9.3, note that *a* is 0.31 and *b* is 1.12. Enter 500 at the top left scale in Fig. 9.13. Read the vegetative cover factor *V* = 0.33 at the lower left scale.

9.3   Estimating Wind Erosion

To calculate the estimated annual erosion, the *I*, *Rf*, *L*, *C*, and *V* factors are determined as shown in Examples 9.1 through 9.5. The estimated annual wind erosion can then be determined from Fig. 9.14, which was developed from equations presented in Skidmore (1994).

Example 9.6.   Estimate the annual erosion caused by wind for the example farming system described in Table 9.4. These values were given in the previous examples.

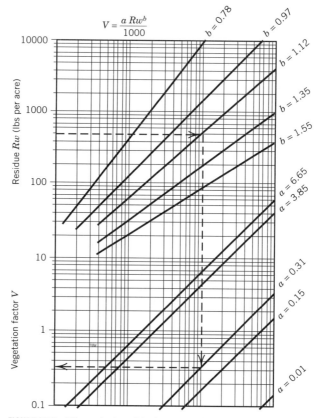

FIGURE 9.13 The relationship between the vegetative cover factor *V* and crop residue amount and type.

Solution. The soil loss is calculated as follows:

1. From Example 9.1, for 25 percent clods and 3 percent knolls, $I = 112$ tons/acre/year.
2. From Example 9.2, the ridge roughness factor $Rf = 0.65$. Calculate $I \times Rf = 112 \times 0.65 = 73$.
3. From Example 9.3, the climate factor $C = 47$.
4. From Example 9.5, the vegetative cover factor $V = 0.33$.
5. Figure 9.14 is then used to determine the average annual erosion rate.

   a. The unsheltered distance, $L = 1867$, is identified on the lower right side of Fig. 9.14.
   b. A vertical line is drawn upward until it intersects the interpolated position of $IRf = 73$.
   c. A horizontal line is drawn to the left until it intersects the interpolated position of $C = 47$.
   d. From the intersection point, a vertical line is drawn upward until it intersects the interpolated position of $IRf = 73$.

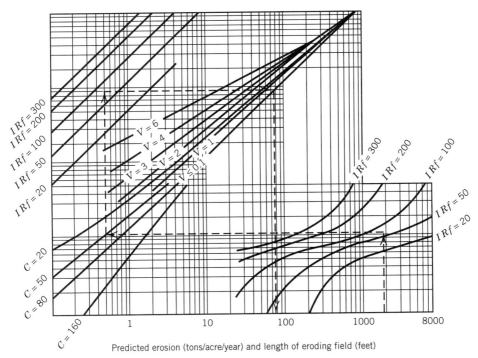

Predicted erosion (tons/acre/year) and length of eroding field (feet)

FIGURE 9.14 Relationships among length of eroding field, soil erodibility, ridge roughness, climate, vegetative cover, and average annual wind erosion.

    e. A horizontal line is drawn right from the intersection point until it intersects the interpolated position of $V = 0.33$.

    f. A vertical line is drawn downward until it intersects the bottom axis. The average annual erosion rate can then be read directly from the axis: Erosion = 77 tons/acre/yr.

TABLE 9.4 Variables Important for Predicting Wind Erosion in the Example Farming System

Parameter	Value	Unit
Clods >0.84 mm dia.	25	percent
Ridge height	4	in.
Ridge spacing	8	in.
Vegetation: standing sorghum	—	—
Vegetative cover	500	lbs/acre
Vegetative height	6	in.
Vegetative spacing	30	in.
Unsheltered distance	1867	ft
Topography: Several knolls with 3 percent slopes are present.		

If the preceding loss is unacceptable, the loss can be reduced by decreasing the erodibility index I, the roughness factor Rf, or the length of field L, or by increasing the vegetative cover factor V.

9.4 Practices to Reduce Wind Erosion

SOIL SURFACE MANAGEMENT In the absence of tillage, surface crusting caused by wetting and drying will reduce erosion for most soils. Erodibility can be decreased by increasing the amount of clods on the soil surface. To minimize wind erosion, the surface should be in a rough, cloddy condition with some plant residue exposed (to increase the vegetative cover). To obtain the maximum roughness, tillage should be carried out as soon after rain as possible. Tillage when the soil is too dry may break up a surface crust or crush clods, leading to a more erodible surface. Such tillage can also increase the rate of soil drying, which is generally undesirable in climates susceptible to wind erosion. Under severe erosion conditions, however, tillage has been widely practiced as an emergency method to reduce wind erosion by exposing less erodible, damp, clods on the surface, thus decreasing soil erodibility and increasing surface roughness.

SOIL WATER MANAGEMENT Tillage and other management practices that reduce soil water loss can reduce wind erosion. Water is lost through evapotranspiration, runoff, and deep percolation. Water conservation practices include weed control on fallow land, conservation tillage (which results in a water-saving mulch on the soil surface), water-conservation terraces (Chapter 7), or increasing surface roughness to reduce runoff. Standing residue, strip cropping (Fig. 9.12), and windbreaks (Fig. 9.15) can help to collect snow, increasing the soil water content. In strip cropping, the fallow strips accumulate water for the next season's crop, if weed growth is minimized. The climate factor is generally not

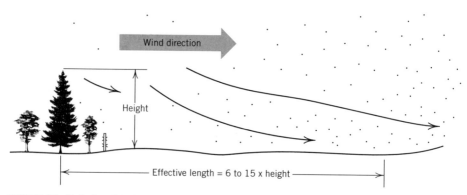

FIGURE 9.15 Relationship between height of a windbreak and length of effective erosion reduction.

altered in the WEQ, but it may be reduced if effective water conservation techniques are practiced.

ALTERING THE LENGTH OF FIELD   The length of the eroding field can be altered by strip cropping (Fig. 9.12) or by installing windbreaks perpendicular to the direction of the prevailing wind (Fig. 9.15). In the context of wind erosion, strip cropping is the practice of growing crops in strips with fallow. The strips with growing crops protect the soil beneath them while reducing the wind velocities and avalanching effects in the fallow strips. The growing strips also serve as areas of deposition during erosive events (Fig. 9.4).

Planting rows of shrubs or trees to serve as windbreaks or shelterbelts has been effective in reducing wind erosion, not only in semiarid plains, but also on sand dunes and organic soils. Wood and plastic "snow fences" have also served as windbreaks. Local recommendations for appropriate species should be followed with vegetative windbreaks. The distance protected by a windbreak may be 6 to 15 times the height of the barrier (Fig. 9.15), with effectiveness decreasing with distance. Another erosion-reducing practice is to leave rows of standing crops at regular intervals during and after harvest. The standing rows serve as natural windbreaks for the tilled areas between. Once a new crop is established, the standing rows may be removed, or left indefinitely to protect the growing crop.

Example 9.7.   A manager wants to leave rows of sorghum unharvested in the field to reduce wind erosion. The height of the sorghum is 5 ft, and the rows are perpendicular to the prevailing wind. What should be the spacing of the rows?

Solution.   Assume that the protected distance will be 10 times the height of the crop. The spacing of the sorghum rows should be: 5 ft × 10 = 50 ft.

In temperate climates, windbreaks can help to keep blowing snow in the field, increasing soil water reserves. Vegetative windbreaks do, however, compete with other crops when soil water is limited. Windbreaks can also protect livestock from wind, provide shade, serve as a source of food, forage, and/or fuel wood, and provide wildlife cover.

VEGETATION MANAGEMENT   Management of both growing crops and postharvest residue can reduce wind erosion. Closely spaced crops are more effective than row crops, and pastures may even accumulate soil that was detached from adjoining fallow or cultivated fields. In row crops, managers may alternate several rows of less wind-resistant crops, such as cotton, with one or two rows of more resistant crops, such as sorghum. The management of residue from harvested crops is also important, and leaving greater amounts of surface residue will reduce erosion risks.

The establishment and subsequent growth of vegetation is crucial to stabilize dune areas. In some areas, sand dunes have also been stabilized with surface treatments, such as spray-on adhesives and soil stabilizers.

REFERENCES

Chepil, W. S. (1945) "Dynamics of Wind Erosion. I. Nature of Movement of Soil by Wind," *Soil Sci.,* 60:305–320.

Chepil, W. S., F. H. Siddoway, and D. V. Armbrust (1962) "Climatic Factor for Estimating Wind Erodibility of Farm Fields," *J. Soil and Water Conserv.,* 17(4):162–165.

Hagen, L. J. (1991) "A Wind Erosion Prediction System to Meet User Needs," *J. Soil and Water Conserv.,* 46(2):106–111.

Layton, J. B., E. L. Skidmore, and C. A. Thompson (1993) "Winter-Associated Changes in Dry-Soil Aggregation as Influenced by Management," *Soil Sci. Soc. Am. J.,* 57:1568–1572.

Lyles, L., and B. E. Allison (1981) "Equivalent Wind-Erosion Protection from Selected Crop Residues," *ASAE Trans.,* 24(2):405–408.

Skidmore, E. L . (1986) "Wind Erosion Climatic Erosivity," *Climatic Change,* 9:195–208.

Skidmore, E. L. (1994) "Wind Erosion," Chapter 11. In Lal, R. (ed.) *Soil Erosion Research Methods,* 2nd ed. Soil and Water Conservation Society, Ankeny, Iowa.

U.S. Soil Conservation Service (SCS) (1988) *Wind Erosion,* Chapter 5. National Agronomy Manual, Washington, D.C.

Woodruff, N. P. and F. H. Siddoway (1965) "A Wind Erosion Equation," *Jour. Soil Sci. Soc. Am. Proc.,* 29:602–608.

PROBLEMS

9.1 Compare the ridge roughness factors that result from a moldboard plow (ridge height = 4 in., ridge spacing = 14 in.) and a one-way disk (ridge height = 2 in., ridge spacing = 10 in.).

9.2 Determine the climate erosivity factor C for a location near you.

9.3 Calculate the estimated annual wind erosion for a 1500-ft-wide field with a climate factor of 65 that has a soil with 30 percent nonerodible clods (>0.84 mm), several knolls with 4 percent slopes in the field, a crop of flat wheat residue estimated to be 800 lbs/acre, and a ridge roughness of 2 in.: *(a)* if the prevailing wind direction is at right angles to the field, and *(b)* if the prevailing wind direction is 30° to the field width.

9.4 A field of wheat stubble is tilled. The following table gives the roughness parameter values before and after tillage for a site with several 5 percent knolls, a climate factor of 80, a calcareous loam soil, and an unsheltered distance of 1800 ft.

Parameter	Before	After
Ridge height	2 in.	4 in.
Ridge spacing	6 in.	8 in.
Residue: wheat stubble	Standing	Flat random
Amount of residue	2400 lbs/ac	1200 lbs/ac

Calculate the percent change in predicted erosion rates due to tillage. Discuss the effect of cultivation on wind erosion.

9.5 What would be the predicted erosion if the width of the eroding field in Problem 9.3 was reduced to 300 ft by strip cropping?

9.6 Determine the spacing between 40-ft-high windbreaks, assuming an effective control is 8 times the windbreak height, and the field is oriented at 30° (Ø=30° in Fig. 9.11) to the prevailing wind.

9.7 A field is located in a region with a climate factor of 85. The field length is 1 mile (5280 ft), and the soil is a silt loam. It is free from knolls and is generally farmed with wheat leaving a stubble amount of 2000 lbs/acre after harvest. The ridge height is 3 in. and spacing is 6 in. Generally, each tillage operation buries 80 percent of the remaining residue and gives a ridge height of 4 in. and a spacing of 8 in. Determine the predicted erosion on the field before tillage, and after one, two, and three passes with the tillage implement. Suggest some methods to reduce the predicted erosion rate to under 10 tons/acre.

CHAPTER 10

WATER QUALITY
AND SUPPLY

Managers of soil and water systems must consider water quality and quantity. Water quality may be improved by the proper selection and management of the system. Conversely, water quality may deteriorate with an improper system or poor management. In addition, water sources, quantities, and delivery methods must be understood so water may be managed effectively and efficiently.

WATER QUALITY

Water quality is determined by the presence of biological, chemical, and physical contaminants. Most water pollution is the result of human activities. Biological contaminants result from human and animal wastes plus some industrial processes. Chemicals enter the water supply from industrial processes and agricultural use of fertilizers and pesticides. Physical contaminants result from erosion and disposal of man-made objects. Since all of these sources contribute to degradation of water quality, standards have been developed for drinking water by the U.S. Public Health Service (Table 10.1). These standards strive to prevent health problems by defining the quality of water considered safe for human consumption. Many local, state, and federal regulations have been instituted to prevent contamination of both surface and ground water supplies.

Sources of water pollution are recognized as either of point or nonpoint origin. Point sources include animal feedlots, chemical dump sites, storm drain and sewer outlets, acid mine outlets, industrial waste outlets, and other identifiable points of origin. Nonpoint sources include runoff from forest and agricultural land, hillside seepage, small subsurface drain outlets, and other diffuse sources. Nonpoint pollution is often more difficult to identify and to correct.

10.1 Biological Contaminants

In agriculture, biological contaminants are primarily from animal and human waste. Feedlots, dairies, and septic systems are major sources of biological pollution. Bacteria are the most common organisms, however, viruses and other microorganisms may also be present, and all can create serious health problems.

TABLE 10.1   Drinking Water Standards

A. Chemical	
1. Maximum contaminant level (mg/L)	
Arsenic	0.05
Asbestos[a]	—
Barium	2.0
Cadmium	0.0005
Chromium	0.1
Lead	0.05
Mercury	0.0002
Nitrate (as N)	10.0
Nitrite (as N)	1.0
Selenium	0.05
Alachlor	0.002
Aldicarb	0.003
Atrazine	0.003
Carbofuran	0.04
Chlordane	0.002
Endrin	0.0002
Ethylene dibromide	0.00005
Heptachlor	0.0004
Lindane	0.0002
Methoxychlor	0.04
Toxaphene	0.005
Total trihalomethanes (THMs)	0.1
2. Secondary maximum contaminant level (mg/L)	
Chloride	250
Copper	1
Fluoride	2.0
Iron	0.3
Manganese	0.05
pH	6.5 to 8.5
Sulfate	250
Total dissolved solids (TDS)	500
Zinc	5
B. Physical	
Color	15 color units
Odor	3 odor units
C. Bacteriological	
Coliform bacteria	None

Source: EPA (1990); Mancl et al. (1991)

[a]7 million fibers (longer than 10 mm)/L

Indicator organisms that occur extensively in human and animal wastes are used to assess the microbiological quality of water. The principal indicator is the coliform group of bacteria. They do not necessarily relate directly to the populations of pathogens, but provide an estimate of the extent to which the water has been recently contaminated by human or animal wastes. The fecal streptococci group is being investigated as another indicator, however, its role has been limited to supplementary testing. Biological contamination can be controlled by proper disposal of wastes, by separating septic systems, feedlots, and other contaminant sources from drinking water supplies, and by treating drinking water before consumption.

10.2 Physical Contaminants

Water temperature, color, odor, and solids concentration are physical indicators of water quality. Suspended sediment is a common physical contaminant in irrigation and runoff water; other solids include organic and dissolved solids. Most sediment occurs because of soil erosion, however, sand may be obtained during pumping from wells. Sediment must be removed from water used in microirrigation systems to prevent plugging. Sands may cause excessive wear to pump impellers and to the nozzles in sprinkler irrigation systems.

Where sediment is deposited on sandy soil, the textural composition and fertility may be improved. However, if the sediment has been derived from eroded areas, it may reduce fertility or decrease soil permeability. Sedimentation in canals or ditches may be serious, resulting in higher maintenance costs. Sediment can greatly decrease the capacity of dams and reservoirs. Mean annual sediment concentrations are highest in the west-central states, but concentration is not necessarily related to total sediment loss because runoff is lower in the western than in the eastern states.

10.3 Chemical Contaminants

Chemicals are a major source of water contamination. Some chemicals occur naturally in the water, others are introduced during water movement through geological materials, but most problems are caused by manufactured chemicals. Fertilizers and pesticides are the major contributors to water pollution by chemicals from agriculture. These chemicals may be applied to soil or foliage over large areas and hence become potential sources of nonpoint pollution. However, fertilizers and pesticides have contributed to a high quality and abundant food supply at reasonable cost for the people of the Unites States and for export.

Nitrates from fertilizers are a common chemical pollutant of water. Estimates indicate that U.S. cropland received 10.4 million tons of nitrogen in 1987 from fertilizers (U.S. Dept. of Agriculture, 1988b), and a similar amount through animal manures, crop residue, and natural sources. A maximum nitrate concentration greater than 3 mg/L was found in 20 percent of 124 000 wells analyzed over a 25-year period (Madison and Brunett, 1985). Concentrations greater than 3 mg/L usually relate to human activities, such as fertilizer applications or septic systems. The federal drinking-water standard of 10 mg/L was exceeded in 6 percent of the samples.

Pesticides include all chemical products used by farmers to control weeds, diseases, insects, and fungi. In general, insecticides are more harmful to humans than herbicides. Pesticides adsorbed onto soil are transported along with sediments in runoff water. Forty-six pesticides have been detected in ground water and confirmed to come from nonpoint sources (Williams, et al., 1988). One or more pesticides attributed to agricultural use have been detected in the ground water of 26 states. The most commonly detected pesticides are atrazine and aldicarb. Pesticide usage was about 300 000 kg in 1988 (U.S. Dept. of Agriculture, 1988a). Gianessi and Puffer (1988) reported the Cornbelt states used more than 59 percent of the U.S. total.

Eroded sediments carry attached chemical ions, such as phosphorus and potassium, which contribute to chemical pollution as well. These chemicals cause eutrophication in lakes and streams, increase the cost of treatment for domestic and industrial supplies, and adversely affect fish and other aquatic life.

Organic matter and mineralized plant and animal wastes also are significant chemical contaminants of water. Biological decomposition of organic matter in water supplies is often measured by the amount of oxygen needed to complete the decomposition process. Measures of decomposition include biochemical oxygen demand (BOD) and nitrogenous oxygen demand (NOD).

The development and implementation of practices and policies to reduce water contamination by agriculture is essential. A better understanding of the fundamental processes affecting the transport and fate of agricultural chemicals must be developed. This knowledge must be used to develop new or improved farming systems that protect, improve, or remediate the quality of water supplies. Soil and water managers have a significant role in this development process.

WATER SUPPLY

Precipitation is presently the only practical source of a continuing fresh-water supply for all agricultural, industrial, and domestic uses. Desalination of brackish or salty water can supply water for high-value uses in some locations, but precipitation will remain the dominant water source. Nearly 11 billion ac-ft of water falls on the North American continent annually. Developed water supplies in the United States use only 4 percent of the precipitation, which is only 13 percent of the residual precipitation after allowing for evaporation and transpiration from natural plants and unirrigated crops.

Thus, there is actually ample water for our needs, however, it is often not available at the time and place of need. The development of water resources involves storage and conveyance of water from the time and place of natural occurrence to the time and place of beneficial use.

In the United States and many other countries, water demand is continually increasing because of population growth and continued development of irrigated lands. In locations where the water supply is inadequate to meet the desired needs, competition occurs between agricultural and urban users. Agricultural interests have frequently developed the water supply and have established a prior water right. Urban water users can usually afford higher water costs and claim a greater economic return per unit of water. One option sometimes available is the purchase of agricultural lands for the water rights, which are then transferred to urban users. This transfer often results in prime agricultural lands being taken out of production (Chapter 1). Solutions to the water rights problem are often sought through legislative and judicial means.

10.4   Sources of Water

Water in one or more of its three physical states, solid, liquid, or gas, is present in greater or lesser quantities in or on virtually all the earth, its atmosphere, and all life forms. From the standpoint of water-resource development, water falls into the categories of surface water and ground water.

SURFACE WATER   Surface water exists in natural basins and stream channels. Where minimum flows in streams or rivers are large in relation to water demands of adjacent lands, towns, and cities, the development of surface waters is accomplished by direct withdrawal from the flow. On many streams and rivers, however, flow fluctuates widely from season to season and from year to year. Peak demands from many major rivers occur at seasons of minimum flow. As much of the annual flow as possible should be conserved for beneficial use. This situation requires the construction of reservoirs to hold the flow during seasons or years of high runoff for later release to beneficial use. Reservoirs range in size from many million acre-feet for large multipurpose reservoirs to small on-farm reservoirs with but a few acre-feet of storage.

GROUND WATER   Subsurface water available for development is normally referred to as ground water. Ground water predominantly results from precipitation that has reached the zone of saturation in the earth through infiltration and percolation. Ground water is developed for use through wells, springs, or dugout reservoirs. In many areas where ground water is an important source of water supply, it is being withdrawn much faster than it is being replenished from infiltration and percolation of precipitation. Correction of this imbalance is a major challenge facing conservationists and water managers.

WATER SUPPLY DEVELOPMENT

10.5   Surface Water

Reservoir development is basic to surface water utilization. In some situations surface waters may be recharged into ground water reservoirs, but more commonly, surface water is stored in reservoirs.

The three types of surface reservoirs are (1) dugout reservoirs fed by ground water, (2) on-stream reservoirs fed by continuous or intermittent flow of surface runoff, streams, or springs, and (3) off-stream reservoirs. The details of the construction of small reservoirs are given in Chapter 11.

DUGOUT RESERVOIRS Dugout reservoirs are limited to areas having slopes of less than 4 percent and a prevailing reliable water table within 3 ft to 4 ft of the ground surface. Design is based on the storage capacity required, depth to the water table, and the stability of the side-slope materials.

ON-STREAM RESERVOIRS The on-stream type of reservoir depends on the inflow of surface runoff for replenishment. The designed storage capacity must be based both on use requirements and on the probability of a reliable supply of runoff. Where heavy usage is expected, the design capacity of the reservoir must be adequate to supply several years' needs to assure time for recharge in the event of a sequence of several years of low runoff. Spring- or creek-fed reservoirs consist of either a scooped-out basin below a spring or a reservoir formed by a dam across a stream valley or depression below a spring. The dam may also be placed across a depression rather than a well-defined stream to catch diffuse surface water.

A spring-fed reservoir should be designed to maintain the water surface below the spring outlet. This eliminates the hazard of reversing the spring flow caused by the increased head from the reservoir. When the spring flow is adequate to meet use requirements, other surface waters should be diverted out of the reservoir to reduce sedimentation and to reduce spillway requirements.

OFF-STREAM RESERVOIR The off-stream or by-pass reservoir is constructed adjacent to a continuously flowing stream. An intake, through either a pipe or an open channel, diverts water from the stream into the reservoir. Controls on the intake permit a reduction in sedimentation, particularly if all flood water can be diverted from the reservoir. Proper location and diking are essential to protect against stream-overflow damage.

10.6 Ground Water

The classification of the earth's crust as a reservoir for water storage and movement of water is shown in Fig. 10.1. The profile of the earth is divided into two primary zones: the zone of rock fractures and the zone of rock flowage. In the rock-flowage zone, the water is in a chemically combined state and is not available. In the zone of rock fracture, interstitial water is contained in the pores of the soil or in the interstices of gravel and rock formations. This zone containing interstitial water is divided into the unsaturated zone and the zone of saturation. As considered here, ground water occurs only in the zone of saturation. However, perched water tables, shown in Fig. 10.1, are often encountered in the unsaturated zone. Capillary-fringe water exists above the ground water table and is present above perched water tables as well. The ground water table may be at the soil surface near lakes, swamps, and continuously flowing streams, but

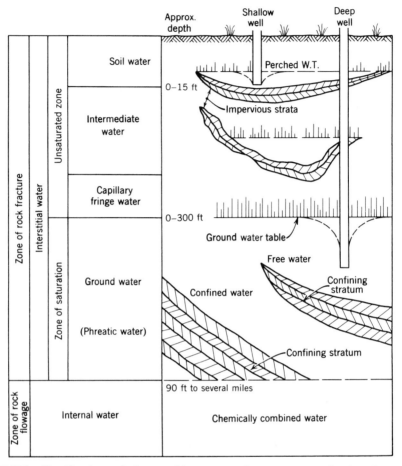

FIGURE 10.1 Classification of the earth's crust and occurrence of subsurface water. (Source: Ferris, J. G., 1959.) (Redrawn and revised with permission of J. F. Wisler.)

it may be several hundred feet deep in drier regions. Ground water is often re-ferred to as phreatic (a Greek term meaning "well") water.

Formations, from which ground water is derived in the zone of saturation, can have considerably different characteristics than the soil near the surface. The various types of deposits that furnish water supplies are shown in Fig. 10.2. Total porosity and permeability, as indicated by the size and shape of the pores, are shown in Table 10.2. Except in clay, porosity is generally a good indicator of specific yield and permeability. Usually, uncemented sand, gravel, fractured limestone, and rock formations are good water-bearing deposits.

Wells may be classed as gravity, artesian, or a combination of artesian and gravity, depending on the type of aquifer supplying the water. Gravity wells are those that penetrate the water table where water is not confined under pressure.

Gravity water may be obtained from wells, springs, and dugout reservoirs. Wells, by far the most common, are either shallow or deep, depending on the

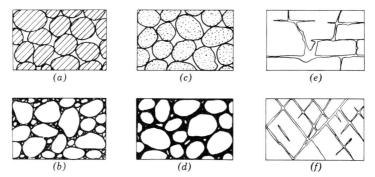

FIGURE 10.2   Several types of rock interstices and the relationship of rock texture to porosity: *(a)* well-sorted sedimentary deposit having high porosity; *(b)* poorly sorted sedimentary deposit having low porosity; *(c)* well-sorted sedimentary deposit with porous pebbles, giving the deposit as a whole a very high porosity; *(d)* well-sorted sedimentary deposit with porosity diminished by the deposition of mineral matter in the interstices; *(e)* rock rendered porous by solution; and *(f)* rock rendered porous by fracturing. (Redrawn from O. E. Meinzer, 1923.)

ground water depth. When the ground water table reaches the soil surface because the underlying strata are impervious, springs or seeps may develop. In areas where the ground water table is near the surface, dugout reservoirs or open pits are useful for access to ground water.

Artesian conditions are present when water is confined under pressure between upper and lower impervious layers (Fig. 10.3). For artesian flow, the following conditions must be present: (1) pervious stratum with an intake area, (2) impervious strata below and above the water-bearing formation, (3) inclination of the strata, and (4) source of water for recharge. When the upper impervious layer is tapped, the water will rise above the saturated zone. Thus, according to this definition, an artesian well is not necessarily a flowing well. The level to which water rises in a pipe placed in the water-bearing formation is known as the piezometric head or (in three dimensions) the piezometric surface. Where

TABLE 10.2   Approximate Characteristics of Ground Water Aquifers

Soil Material	Total Porosity[a] (percent)	Specific Yield[b] (percent)	Relative Permeability
Dense limestone or shale	5	2	1
Sandstone	15	8	700
Gravel	25	22	5000
Sand	35	25	800
Clay	45	3	1

[a]Portion of the bulk soil volume composed of voids.

[b]Portion of volume of water released from storage per unit of horizontal aquifer area per unit decline in water level.

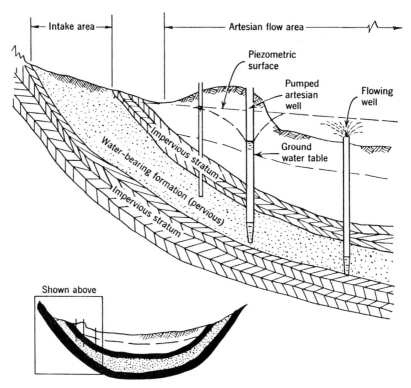

FIGURE 10.3   Diagrammatic sketch of ideal conditions for artesian flow.

the pervious stratum outcrops at the surface, artesian springs may develop. These are similar to water-table springs, except that the flow is under pressure.

WELLS

Ground water supplies are primarily obtained from wells, which are holes drilled or dug downward from the soil surface into the ground-water aquifer. A casing is installed during or after the drilling process to stabilize the hole and allow water, but not aquifer particles, to move into the hole. The lower portion of the casing is perforated or slotted and known as the well screen. The openings in the screen should be properly sized to minimize the movement of sands into the well.

10.7 Hydraulics of Wells

A cross section of a well installed in homogeneous soil overlying an impervious formation is shown in Fig. 10.4. Under static conditions, the water level will rise

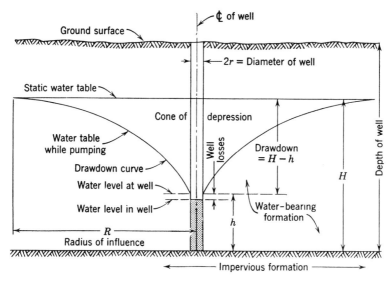

FIGURE 10.4 Cross section of a typical gravity well in a homogeneous unconfined aquifer.

to the water table. When pumping begins, the water level in the well is lowered, thus removing free water from the surrounding soil. The water table around a well assumes the general form of an inverted cone. The distance from the well to where the static water table is not noticeably lowered by drawdown is known as the radius of influence R. The water level at the edge of the well will be slightly higher than in the well because of friction losses through the perforated casing. For a given rate of pumping, the water table surrounding a well in time nearly reaches a stable condition. The radius of influence may be predicted from the physical characteristics of the aquifer. A guide for such predictions is given in Table 10.3.

For a given well there is a definite relationship between drawdown and discharge. For thick water-bearing aquifers or artesian formations, the discharge-drawdown relationship is nearly a straight line. The discharge-drawdown relationships shown in Fig. 10.5 can be obtained by pumping at various rates and

TABLE 10.3 Practical Radius of Influence of Wells

Soil Formation and Texture	Radius of Influence (ft)
Fine sand formations with some clay and silt	100–300
Fine-to-medium sand formations, fairly clean and free from clay and silt	300–600
Coarse sand and fine gravel formations free from clay and silt	600–2000
Coarse sand and gravel, no clay or silt	1000–2000

Source: Bennison, E. W. (1947). Copyright © by Wheelabrator/Johnson Screens, St. Paul, Minnnesota 55112.

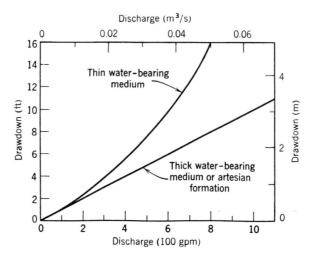

FIGURE 10.5 Relationship between drawdown and discharge of wells.

plotting the drawdown against the discharge. Test pumping a well should be continued for a considerable length of time. Short pumping tests are often misleading. Even 24-h tests are not long enough, and 30-day tests are more likely to indicate the true capacity of the well.

The rate of inflow into a gravity well (symbols illustrated in Fig. 10.4) is

$$q = \frac{\pi K(H^2 - h^2)}{\log_e (R/r)} \tag{10.1}$$

where q = rate of inflow in cfs

K = hydraulic conductivity in feet per second

H = height of static water level above the impermeable layer in feet

h = height of water level in the well above the impermeable layer in feet

R = radius of influence (measured or estimated from Table 10.3) in feet

r = radius of well casing or gravel envelope in feet

The rate of flow into a well completely penetrating a confined aquifer (symbols illustrated Fig. 10.6) is

$$q = \frac{2\pi Kd(H - h)}{\log_e (R/r)} \tag{10.2}$$

where q = rate of inflow in cfs

K = hydraulic conductivity of the aquifer in feet per second

d = thickness of the confined water-bearing stratum in feet

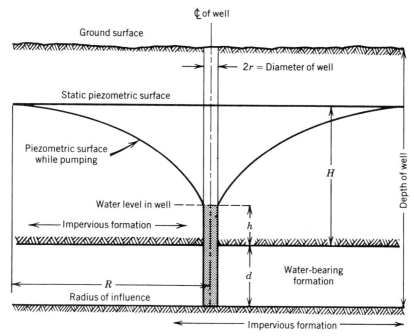

FIGURE 10.6   Cross section of a typical well in a homogeneous confined aquifer. (Source: Schwab, G. O. et al., 1993.) (Reprinted with permission of John Wiley & Sons, Inc.)

H = height of the static piezometric surface above the top of the water-bearing formation in feet

h = height of the water in the well above the top of the water-bearing formation in feet

R = radius of influence (measured or estimated from Table 10.3) in feet

r = radius of well or gravel envelope in feet

Example 10.1.   Estimate the flow rate into a well completely penetrating a confined aquifer if the well diameter is 15 in., thickness of the water-bearing stratum is 35 ft, aquifer hydraulic conductivity is 9 in./h, radius of influence is 400 ft, drawdown is 25 ft, and the piezometric surface is 50 ft above the top of the aquifer.

Solution.   Convert parameters to the required units and apply Eq. 10.2.

$$K = (9 \text{ in./h}) (1 \text{ ft}/12 \text{ in.}) (1\text{h}/3600 \text{ sec}) = 0.0002 \text{ ft/sec}$$
$$H - h = 50 \text{ ft} - 25 \text{ ft} = 25 \text{ ft}$$
$$r = (15 \text{ in.}/2)/12 \text{ in./ft} = 0.625 \text{ ft}$$
$$q = 6.28(0.0002)(35)(25)/\log_e (400/0.625) = 0.17 \text{ cfs}$$

10.8 Construction of Wells

The three general types of wells are dug, driven, and drilled. The dug well is often lined with masonry, concrete, or steel to stabilize the excavation. Because of the difficulty in digging below ground-water level, dug wells do not penetrate the ground water to a depth sufficient to produce a high yield.

Wells up to 3 in. in diameter and 60 ft in depth may be constructed by driving a well point into unconsolidated material. A well point is a section of perforated pipe pointed for driving and connected to sections of plain pipe as it is driven to the desired depth. Sometimes the penetration of well points is aided by discharging a high-velocity jet of water at the tip of the point as it is driven.

Deeper and larger-diameter wells are drilled with cable-tool (percussion) or rotary equipment. With cable tools, a heavy bit is repeatedly dropped onto material at the bottom of the well. Crushed material is removed periodically with a bailer. Cable-tool wells have been drilled to depths of 5000 ft. Deep wells are also drilled with rotary tools consisting of a bit rotated by a string of pipe. Many rotary drills use a mud slurry pumped through the drill pipe to bring cuttings to the surface as it flows up the outside of the drill pipe. Other rotary drills use compressed air for the same purpose. In unconsolidated materials, wells may be cased as drilling progresses. Casings may also be installed after drilling is completed.

Well casings are perforated where they pass through the water-yielding strata. In some situations, perforated casing may be formed in place by ripping or shooting holes through a solid casing. In most instances, better results are achieved by placing a corrosion-resistant screen at the water-bearing strata. Screens are made of brass, bronze, or special alloys to resist corrosion. Screen openings are selected to permit 50 to 70 percent of the particles in the aquifer to pass the screen. The open screen area should keep entrance velocities below 0.5 fps to minimize head loss. In aquifers of uniformly fine, unconsolidated material, a gravel pack may be placed around the screen. Figure 10.7 shows a cross section of a gravel-packed well.

After the screen is placed or the casing is perforated, a well should be developed by pumping at a high discharge rate or by surging with a plunger. These practices develop higher velocities through the screen and in the aquifer adjacent to the screen than will be developed in normal pumping from the well. This action brings fine materials into the well, where they are removed by pumping and/or bailing. As a result, the aquifer is opened for freer flow of water and is stabilized for normal operation of the well.

After a well has been developed, it should be tested to determine the well capacity and drawdown. If a pump has been used for development, the same pump can be used for testing. The pump should be run for several hours at increasing discharges and the drawdowns measured. This will produce a curve similar to Fig. 10.5, which is necessary for proper pump selection.

If the well is to be used for drinking water, it should be disinfected to kill bacteria that may have been introduced by drilling. Chlorination is a common

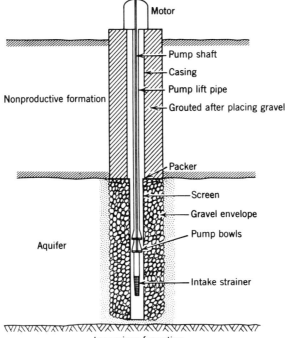

FIGURE 10.7   Cross section through a gravel-packed well. (Source: Linsley, R. K. and J. B. Franzini, 1979.) (Reprinted and revised with permission of McGraw–Hill, Inc.)

method of disinfection. A sample of water should be tested by a health department or a certified laboratory before the well is placed in service.

PUMPS

Pumps are important components of water development and management systems. Pumps add energy to water by increasing the water elevation, pressure, and/or flow rate. Typically pumps are used for sprinkler and microirrigation systems, for some drainage outlets, and to provide pressurized water supplies for livestock and domestic uses on the farm.

10.9   Types of Pumps

The most common pumps have rotating impellers or reciprocating pistons to transfer energy to the fluid. Reciprocating pumps, sometimes called piston

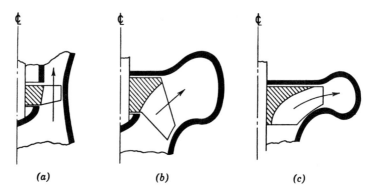

FIGURE 10.8 Flow direction through impellers with (a) axial flow, (b) mixed flow, and (c) radial flow.

pumps, are capable of developing high pressures, but their capacity is very small. A common application in water resources is with a windmill.

The flow through impeller pumps may be classified as radial, axial, or mixed (Fig. 10.8). In radial-flow pumps, the fluid moves through the impeller perpendicular to the axis of rotation of the impeller. In axial-flow pumps, the fluid moves through the pump, parallel to the axis of rotation of the impeller.

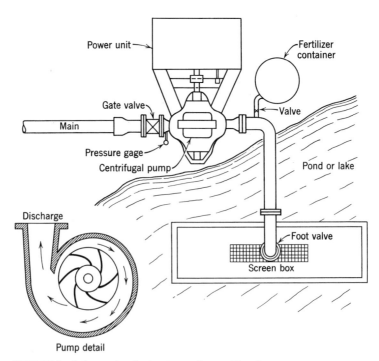

FIGURE 10.9 Schematic of a horizontal centrifugal pump.

Pumps that discharge from their impellers on angles that lie between radial and axial are mixed-flow pumps.

Impeller pumps are commonly known as centrifugal, propeller, and turbine pumps. In centrifugal pumps the flow from the impeller is radial; in propeller pumps the flow from the impeller is axial. Most turbine pumps have mixed flow. Centrifugal pumps are characterized by low discharges and high heads or pressures. Propeller pumps are characterized by high discharges and low heads. Turbine pumps have intermediate discharges and heads, however, most turbine pumps have operational characteristics closer to those of centrifugal pumps than propeller pumps. Thus, pumps may be selected from these three types for a wide range of discharge and head characteristics. Impeller pumps are also economical and simple in construction, yet they provide a smooth, steady discharge. They are small compared to their capacity, easy to operate, and capable of handling sediment and other foreign material. Example installations of a centrifugal pump, a turbine pump, and a propeller pump are illustrated in Figs. 10.9, 10.10, and 10.11, respectively.

The design of the impellers influences the efficiency, durability, and performance characteristics of the pump. Centrifugal-type impellers shown in Fig. 10.12 are classified as (a) open, (b) semienclosed, and (c) enclosed. The open-type impeller has exposed blades on all sides except where they are attached to the rotor. The semienclosed impeller has a shroud (plate) on one side; the enclosed impeller has shrouds on both sides, thus enclosing the blades completely. The open and semienclosed impellers are most suitable for pumping fluids with suspended material or trashy water. Enclosed impellers are generally not suitable where suspended sediment is carried in the water as the sediment greatly increases the wear on the impeller.

10.10   Pump Head, Flow Rate, and Performance Characteristics

Every pump has a specific relationship between head and capacity. Pump head is the energy added to the water expressed as the equivalent height of a column of water. The relationship between head and capacity changes with the size and speed of the impeller. Curves that provide these performance relationships are called pump characteristic curves. These characteristics must be known before the proper pump can be selected and are provided by the pump manufacturer as shown in Figs. 10.13, 10.14, and 10.15 for centrifugal, turbine, and propeller pumps, respectively. These curves vary in shape and magnitude, depending on the size of the pump, type and shape of impeller, and overall pump design.

The head-capacity curve shows the total head developed by the pump for different rates of discharge. At zero flow the head developed is known as the shut-off head. Head or energy losses occur in pumps because of friction and turbulence in the moving water, shock losses resulting from sudden changes in direction, leakage past the impeller, and mechanical friction, all of which reduce pump efficiency. Efficiency curves also are frequently included in the performance characteristics of pumps. For a given speed, the efficiency can be determined for any discharge. A pump should be selected that will have a high

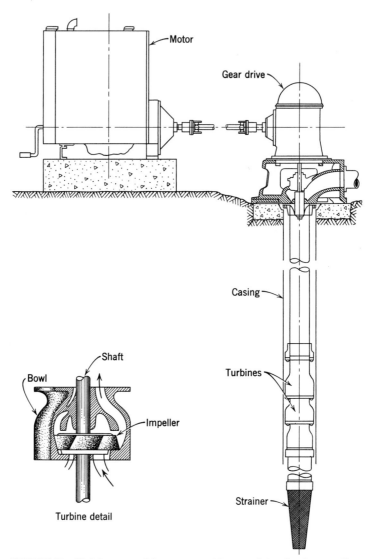

FIGURE 10.10 Multistage turbine pump with gear drive for deep wells.

efficiency for a wide range of discharges, normally in the 70 to 80 percent range. Usually the highest efficiency should occur near the normal operating head and discharge. Efficiencies decrease because of wear. Repair should be considered when the efficiency drops below an economical level.

Centrifugal pumps should operate without cavitation. Cavitation is the formation and collapse of tiny vapor bubbles in the water in the pump and is extremely destructive. Cavitation occurs when the pressure at the inlet to the impeller is too low and when the pump speed is too fast. Low inlet pressures occur when the suction lift is too high and/or when the inlet is restricted. The mini-

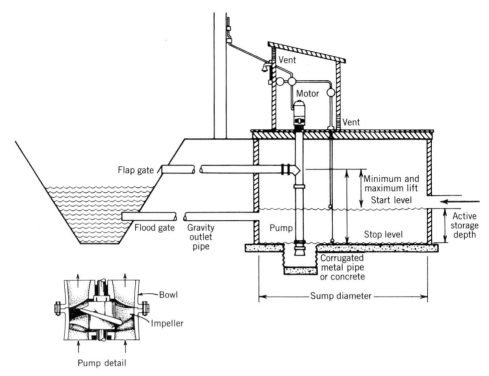

FIGURE 10.11 Schematic of a propeller pump used for a drainage outlet.

mum pressure needed at the impeller inlet is given as the required net positive suction head (NPSH) and is included in the pump characteristic curves (Figs. 10.13, 10.14, and 10.15).

10.11 Pumping Head, Rate, and Power Requirements

The total head and rate of pumping must be known before a pump and power unit can be selected. The pump capacity is the sum of all the discharges plus an allowance of about 5 percent for loss of efficiency caused by pump wear. The

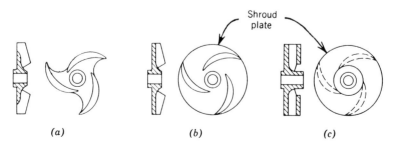

FIGURE 10.12 (a) Open, (b) semienclosed, and (c) enclosed centrifugal-type impellers.

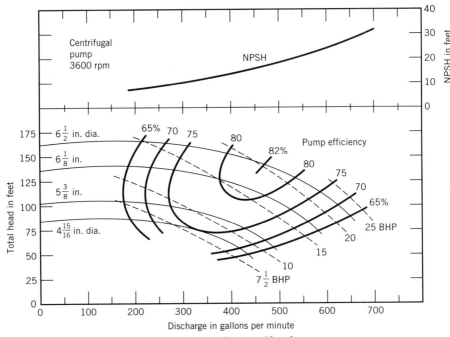

FIGURE 10.13 Performance characteristics of a centrifugal pump.

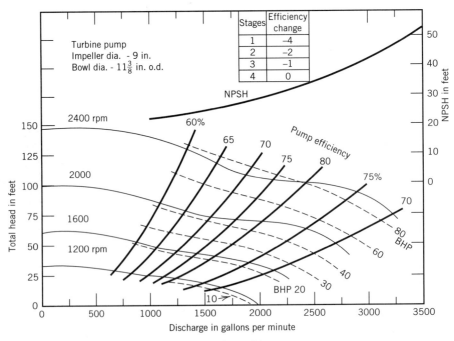

FIGURE 10.14 Performance characteristics of a turbine pump.

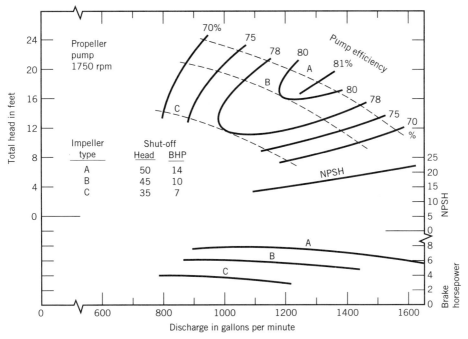

FIGURE 10.15   Performance characteristics of a propeller pump.

total head is the sum of the height to which water must be lifted, pressure head, and friction losses. It can be determined from the equation

$$H = H_L + H_p + H_f \qquad (10.3)$$

where H = total head in ft

H_L = elevation difference between the water level at the source and the pump outlet in feet

H_p = pressure head at the outlet of the pump in ft (1 psi = 2.31 ft)

H_f = friction loss in the suction line and fittings in ft

The pressure head at the outlet of the pump must be sufficient for the system to be supplied. The friction loss can be calculated based on information given in Chapter 17.

The theoretical power requirement, assuming no energy or friction losses, to lift a quantity of water is known as the water horsepower. This power can be calculated from

$$WHP = \frac{Q(cfs)\,H}{8.8} = \frac{Q(gpm)\,H}{3960} \qquad (10.4)$$

where Q = the flow rate in cubic feet per second (cfs) or gallons per minute (gpm)

H = the total head from Eq. 10.3 in ft

The input horsepower to the pump or brake horsepower of the power unit is calculated from

$$BHP = \frac{WHP}{E_p} \qquad (10.5)$$

where WHP = the water horsepower from Eq. 10.4

$\quad\;\; E_p$ = the pump efficiency obtained from the pump curves in decimal form

The actual pump efficiency must be known to determine if repair is needed. The decision to repair can be made after solving Eq. 10.5 for pump efficiency, based on measurements at the pump and power unit in the field. The actual pump discharge and total head are used to calculate *WHP*. *BHP* can be measured with a torque meter on the input shaft to the pump. It can also be estimated using the input energy to the power unit and the power unit efficiency knowing that

$$BHP = \text{input energy to power unit} \times \text{power unit efficiency} \qquad (10.6)$$

10.12 Power Units

Electric motors and internal combustion engines are commonly used for power units. Internal combustion engines may use gasoline, diesel, propane, or natural gas fuels. Large pumps are normally located at a specific site, permitting use of stationary engines or electric motors. Portable pumps are generally driven by an engine mounted with the pump on a trailer or by the power take-off from a farm tractor.

Electric motors have many advantages over internal combustion engines, such as ease of starting, low initial cost, low maintenance cost, and suitability for mounting on horizontal or vertical shafts, and can operate underwater (submersible pumps). Electric motors are not recommended for areas with frequent power outages. Direct-drive motors are preferred because gears and belts are eliminated. Special hollowshaft electric motors are commonly used on vertical turbine pumps. Watertight vertical motors are available for use with deep-well turbine pumps (also called submersible pumps). The motor is submerged in the well near the impellers, thus replacing the long shaft with an electric cable. Motors operate at peak efficiencies when the load is 75 to 125 percent of their rated load. When operating in high temperature environments or at altitudes above 3500 feet, the load on motors should be reduced to avoid overheating. Motors rated at 10 horsepower or less have efficiencies of about 88 percent, whereas larger motors have efficiencies of about 90 percent.

The major advantage of internal combustion engines is portability, since engines and their fuel supplies can be easily moved. Also, pump output can be changed simply by adjusting the throttle setting on the engine. Right-angle gear drives or belts are required for turbine pumps since they have vertical shafts, whereas engines have horizontal output shafts; however, pump speeds also can be easily changed by changing gears in a gear drive. In remote locations where electric power is not already available, it may be less expensive to use an engine than to pay the cost of installing a power line.

Internal combustion engines should not operate continuously at full load. Some engine manufacturers provide ratings based on maximum horsepower, whereas others have derated the engines for continuous operation. If the maximum horsepower of the engine is given, the horsepower rating must be greater than that determined from Eq. 10.5. Multiply the *BHP* from Eq. 10.5 by the appropriate factor from Table 10.4 to estimate the maximum horsepower rating of the unit to purchase. High air temperatures and high elevations reduce engine power and also must be considered when selecting a power unit. Engine manufacturers should be contacted to obtain specific recommendations for their engines.

Example 10.2.   The pump in Fig. 10.13 is delivering 400 gpm with a total head of 123 ft. Calculate the water horsepower, brake horsepower, and input power to an electric motor that is 90 percent efficient.

Solution.   Apply Eq. 10.4 to determine the water horsepower

$$WHP = (400 \times 123)/3960 = 12.4 \text{ hp}$$

Read a pump efficiency of 80 percent from Fig. 10.13 and apply Eq. 10.5

$$BHP = 12.4/0.8 = 15.5 \text{ hp}$$

Apply Eq. 10.6 to determine the input power to the motor

$$\text{Input Power} = BHP/\text{Motor Efficiency} = 15.5/0.9 = 17.2 \text{ hp}$$

Example 10.3   Determine the horsepower rating for a diesel engine for the pump in Example 10.2.

Solution.   From Table 10.4, the adjustment factor is 1.25
Calculate the rated horsepower from $15.5 \times 1.25 = 19.4$ or choose a 20 horsepower engine.

TABLE 10.4   Factors to Adjust Engine Horsepower Requirement for Continuous Operation

Type of Power Unit	Adjustment Factor[a]
Electric motor	1.0
Diesel engine	1.25
Gas engine, water cooled	1.45
Gas engine, air cooled	1.6

[a]Rated power is adjustment factor times continuous horsepower need of the pump.

10.13   Pump or Affinity Laws

The discharge and head of a pump can be changed by increasing or decreasing the pump speed or by decreasing the impeller diameter. It is important to recognize that changing the speed or diameter can significantly affect pump performance. If the changes are less than 10 to 20 percent, the pump performance can be estimated using the following laws.

$$\frac{Q_1}{Q_2} = \frac{rpm_1}{rpm_2} \tag{10.7}$$

$$\frac{H_1}{H_2} = \left[\frac{rpm_1}{rpm_2} \right]^2 \tag{10.8}$$

$$\frac{P_1}{P_2} = \left[\frac{rpm_1}{rpm_2} \right]^3 \tag{10.9}$$

where rpm is the pump speed in revolutions per minute, Q, H, and P are discharge, head, and power, respectively. The subscripts 1 and 2 denote initial and new values, respectively. Since the effect of pump speed and diameter are directly related, pump diameter D can be substituted for rpm in Eqs. 10.7 to 10.9. An important example of the use of these equations is to demonstrate the effect of increasing the pump speed by 10 percent. In this case, the discharge increases 10 percent, the head increases 21 percent, and the power required increases 33 percent. This example shows that the power required increases at a higher rate than the discharge and the power unit may be overloaded.

10.14   Pumping Costs

Pumping costs are the sum of fixed and variable costs. Fixed costs include initial investment, interest, taxes, and insurance. Variable costs include energy, maintenance, and repairs. Many pumping plant failures are the result of poor maintenance, such as inadequate lubrication. Poor electrical equipment or maintenance can result in lightning damage to electric motors.

Energy costs are frequently the major cost, particularly when pumping with high total heads and flow rates. When selecting a pump and power unit, the fixed and variable costs of several units should be compared to determine which unit has the lowest total cost. Energy costs for electric motors are based on both energy use and demand charges. Energy usage is obtained from a watt-hour meter and the cost is obtained by multiplying kilowatt-hours times the cost per kilowatt-hour. Since electric power is measured in kilowatts, horsepower must be multiplied by 0.746 to obtain the equivalent kilowatts. Demand charges depend on the power rating of the pump and the rate structure of the electric utility supplying the power. The demand charge is necessary to offset the fixed costs of the utility to install the power line and generating capacity needed to provide power on demand at the pump. In some irrigated areas, electric rates are lowered if the irrigator allows the utility to turn a pump off during peak electrical usage, usually during periods of high temperature. High temperatures, however, coincide with peak irrigation water demands, and this option must be carefully considered. Irrigating at night and early morning may also be an option.

Fuel costs for engines can be determined from actual fuel usage and fuel cost. To estimate fuel costs when selecting an engine, the power rating and fuel usage should be obtained from test data. Engine performance is given as horsepower-hours per gallon of fuel. Thus, when the pump power requirement and hours of operation are known, the fuel consumption and cost can be calculated.

Example 10.4. If the pump in Example 10.2 operates for 5 hours and electric energy costs $0.08/kw-h, determine the energy cost.

Solution. Convert horsepower to kilowatts $17.2 \times 0.746 = 12.8$ kw; multiply by the hours of operation to obtain kw-hs $12.8 \times 5 = 64$ kw-hs; obtain the cost by using the electrical rate.

Thus the cost is 64 kw-hs $\times$ $0.08/kw-h = $5.12.

Example 10.5. Determine the cost of operating a water-cooled, gasoline engine to operate the pump in Example 10.2 and the horsepower rating of an engine to be purchased for this pump. Gasoline costs $1.20/gal and the engine delivers 12 hp-h/gal of fuel.

Solution. The fuel required is obtained from $(15.5 \text{ hp} \times 5 \text{ h})/12 \text{ hp-h/gal.} = 75.5/12$ $= 6.5$ gal of fuel. The fuel cost is 6.5 gal $\times$ $1.20/gal = $7.80.

From Table 10.4 read the adjustment factor of 1.45, then multiply by the *BHP* or

$$15.5 \text{ hp} \times 1.45 = 22.5 \text{ hp engine.}$$

WATER CONSERVATION

10.15 Water Harvesting

The term "water harvesting" has come to be applied to any watershed modification carried out to increase and beneficially utilize surface runoff. In many arid areas, the majority of precipitation that infiltrates the soil is lost either through direct evaporation or through transpiration by natural vegetation. For example, in the Colorado River Basin less than 6 percent of the precipitation appears as streamflow. Vegetation management is an effective means of increasing streamflow, and it should receive increasing application.

The practice of water harvesting by catchments is possible when areas of concrete, sheet metal, asphalt, or treated or waterproofed soil are specifically constructed to collect and hold precipitation. Successful water harvesting requires attention not only to the collection of water, but also to the conveyance and storage of the water collected. Catchment treatment may include soil smoothing and removal of vegetation, application of salts or other chemicals that disperse the soil aggregates and greatly reduce infiltration, or application of plastic film or wax seals. Simple soil smoothing and vegetation removal can increase runoff by a factor of three.

Complete sealing will increase runoff to essentially 100 percent. Water harvesting is being applied to develop water supplies for wildlife, livestock, and occasional domestic use.

10.16   Evaporation Suppression

Reduction of evaporation from free-water surfaces can be accomplished by a reduction of the free-water surface area or by protection of the free-water surfaces. Reduction of the free-water surface is accomplished by minimizing the surface-area-to-volume ratio of reservoirs by making reservoirs deeper or more nearly square or circular. Storage of water in natural ground water reservoirs rather than in surface reservoirs also reduces evaporation losses.

Protection of free-water surfaces to reduce evaporation is uneconomical except in special situations. Methods under investigation include monomolecular films, floating membranes, and floating particles that reduce the exposure of the free-water surface. Monomolecular films are made of fatty alcohols, such as hexadecanol. The film must be applied continuously to maintain it against wind action and biological deterioration. Floating membranes of plastics or butyl rubber are practical only for small reservoirs.

10.17   Artificial Recharge

Ground-water reservoirs are supplied primarily by water percolating to them from the surface. Under natural conditions, only a small fraction of rainfall reaches the ground water. Since ground-water reservoirs provide evaporation-free storage and since surface runoff waters are often wasted, artifical recharge of ground-water reservoirs with surface runoff is a logical conservation measure.

The general methods of artificial recharge are (1) basin, (2) furrow or ditch, (3) flooding, (4) pit and shaft, or (5) injection wells. The basin method of spreading water consists of a series of small basins formed by dikes or banks. The dikes often follow contour lines, and they are so arranged that the water flows from one basin to the next. In the furrow or ditch method, the water flows along a series of parallel ditches placed close together. The flooding method consists of ponding a thin layer of water over the land surface. Pits, shafts, and injection wells, as methods of recharge, are used primarily in municipal areas and industrial centers. Regardless of the method, it is desirable to recharge only water that is relatively free of sediment. It is not unusual to use a combination of several recharge methods.

In artificial recharge basins, it is periodically necessary to remove layers of accumulated sediment and to replace them with sand. Soil conditioners, such as organic residues, dense cover crops, or chemical treatments, are effective in increasing infiltration rates. Some waters require desilting in settling basins or the use of chemical deflocculants as well as biological control treatment before they can be recharged without clogging the infiltration area or the aquifer. Before recharging, the chemical content of the water must be determined (Table 10.1). Direct injection bypasses the natural treatment provided by the soil and intervening strata, and thus contaminates the aquifer with bacteria, etc. Recharging water with harmful chemicals is also not permitted.

10.18   Control of Seepage

Water conveyance losses from canals and ditches can be greatly reduced through reduction or elimination of seepage. Concrete linings, as well as asphalt, fiberglass-reinforced asphalt, and plastic linings, are frequently placed in irrigation canals and ditches to reduce seepage. Polyphosphates and other chemical additives that tend to deflocculate the soil are successful if carefully selected in relation to soil characteristics. One group of chemical additives reduces infiltration capacity by causing soil particles to swell, making the soil hydrophobic. Swelling clays, such as bentonite, are sometimes mixed with surface soils. These swell and seal soil pores to reduce infiltration rates.

10.19   Phreatophyte Control

The term *phreatophyte* includes plants that habitually obtain their water supply from the zone of saturation or from the overlying capillary fringe. Examples are tamarisk, cottonwood, willow, and mesquite. Eighty species of phreatophytes have been identified. Phreatophytes cover approximately 17 million acres in the western United States and use an estimated 25 million acre-feet of water annually. Phreatophyte growth is largely concentrated along the lower valleys of major rivers. Consumptive use of water by phreatophytes varies with species, climate, and depth to the ground-water table. Under high water table conditions, water used by phreatophytes will approach open-pan evaporation. Over 10 ft of water use by tamarisk was measured in the Rio Grande Valley with a shallow water table.

Control of phreatophytes thus offers a great potential for water conservation. Control can be effected either by chemical or mechanical means; however, the cost of control has limited its application. To protect treated areas, replacement vegetation having relatively low water-use requirements must be established and maintained. Channelization is the most effective means of salvaging water that would otherwise be lost to phreatophytes. Through channelization and drainage, ground water can be lowered in phreatophyte-infested areas, and more of the flow can be conveyed to downstream reservoirs. The accompanying lower water table greatly reduces the consumptive use by phreatophytes; however, phreatophyte removal must be balanced against the associated loss of wildlife habitat.

WATER RIGHTS

Two basic divergent doctrines regarding the right to use water exist, namely, riparian and appropriation. They are recognized either separately or as a combination of both doctrines in different states. Both doctrines apply only to surface water in natural watercourses and to water in well-defined underground streams. A comparison of the salient features of these two doctrines is shown in Table 10.5. In the future, adjudicated water rights based on highest-value use will become increasingly important.

Water rights doctrine for percolating ground water varies greatly from state to state. The riparian doctrine for well-defined underground aquifers generally applies to the eastern states. Otherwise, either the appropriation or correlative rights doctrine applies. Under the correlative rights doctrine the landowner's use of ground water must be reasonable in consideration of the similar rights of others, and it must be correlated with the uses of others in times of shortage.

10.20   Riparian Doctrine

The riparian doctrine, which is a principle of English common law, recognizes the right of a riparian owner to make reasonable use of the stream's flow, provided the water is used on riparian land. Riparian land is contiguous to a stream or other body of surface water. The right of land ownership also includes the right of access to and use of the water, and this right is not lost by nonuse. Reasonable use of water generally implies that the landowner may use all that he or she needs for drinking, for household purposes, and for watering livestock. Where large herds of stock are watered or where irrigation is practiced, the riparian owner is not permitted to exhaust the remainder of the stream; the owner may use only his or her equitable share of the flow in relation to the needs of others similarly situated. Since few eastern states have statutory laws governing water rights, this doctrine is based mostly on court decisions.

TABLE 10.5   Comparison of Water Rights Laws[a]

Characteristic	Riparian	Doctrine of Appropriation
Acquisition of water right	By ownership of riparian land	By permit from state (state ownership)
Quantity of water	Reasonable use	Restricted to that allowed by permit
Types of use allowed	Domestic, livestock, etc., but not precisely defined	Some beneficial use required
Loss of water by nonuse	No	Yes, but continued use not always required
Location where water may be used	On riparian land, but some exceptions	Anywhere, unless specified in permit

[a]Generally applicable only for surface water in well-defined channels and for water in well-defined underground aquifers. Some state laws on ground water deviate from the above.

10.21   Doctrine of Prior Appropriation

The doctrine of prior appropriation is based on the priority of development and use, that is, the first person to develop and put water to beneficial use has the prior right to continue use. The right of appropriation is acquired mainly by filing a claim in accordance with the laws of the state. The water must be put to some beneficial use, but the appropriator has the right to all water required to satisfy his or her needs at the given time and place. This principle assumes it is better to let individuals, prior in time, take all the water, rather than distribute inadequate amounts to several owners. Appropriated water rights are not limited to riparian land and may be lost by nonuse or abandonment.

This doctrine is recognized in most of the 17 western states, although in some it is in combination with the riparian doctrine. State water rights laws differ on specific details and they may change from time to time, making generalizations that apply from state to state difficult.

REFERENCES

Bennison, E. W. (1947) *Ground Water, Its Development, Uses and Conservation.* E. E. Johnson, Inc., St. Paul, Minnesota.

Bear, J. (1979) *Hydraulics of Groundwater.* McGraw–Hill, New York.

Bouwer, H. (1978) *Groundwater Hydrology.* McGraw–Hill, New York.

Driscoll, F. G. (1986) *Groundwater and Wells.* Johnson Pump Div., St. Paul, Minnesota.

Garrity, T. A., Jr., and E. T. Nitzschke, Jr. (1967) *Water Law Atlas,* Cir. 95. State Bureau of Mines and Mineral Resources and New Mexico Institute of Mining and Technology, Socorro, New Mexico.

Gianessi, L. P., and C. M. Puffer (1988) *Use of Selected Pesticides for Agricultural Crop Production in the United States, 1982–1985.* Quality of the Environ. Div., Resources of the Future, Inc., Washington, D.C.

Linsley, R. K., and J. B. Franzini (1979) *Water-Resources Engineering,* 3rd ed. McGraw–Hill, New York.

Madison, R. J., and J. O. Brunett (1985) *Overview of Nitrate in Ground Water of the United States,* U.S. Geological Survey Water Supply Paper 2275.

Mancl, K., M. Sailus, and L. Wagenet (1991) *Private Drinking Water Supplies: Quality, Testing, and Options for Problem Waters.* NRAES-47, Northeast Regional Agricultural Engineering Service, Ithaca, New York.

Meinzer, O. E. (1923) "The Occurrence of Groundwater in The United States," *U.S. Geological Survey,* Water Supply Paper 489.

U.S. Department of Agriculture (USDA) (1988a) "ARS Strategic Groundwater Plan 1," *Pesticides* (Rep.). USDA-ARS, Washington, D.C.

U.S. Department of Agriculture (USDA) (1988b) "Agricultural Resources, Inputs, and Outlook" (Rep.). AR-9, Economic Research Service, Washington, D.C.

U.S. Environmental Protection Agency (EPA) (1990) *Federal Register 40 CFR Parts 141 to 143 National Primary and Secondary Drinking Water Regulations.* Office of Drinking Water, Washington, D.C.

Williams, W. P., P. W. Holden, D. W. Parsons, and M. N. Lorber (1988) *Pesticides and*

Ground Water Data Base, 1988 Interim Report. EPA, Office of Pesticide Programs, Washington, D.C.

PROBLEMS

10.1 Compute the flow rate into a gravity well 16 in. in diameter if the depth of the water-bearing stratum is 100 ft, the drawdown is 30 ft, aquifer hydraulic conductivity is 6 in./h, and the radius of influence is 600 ft.

10.2 Compute the flow into a well in a confined aquifer if the well diameter is 12 in., thickness of the water-bearing stratum is 50 ft, aquifer hydraulic conductivity is 1 ft/h, piezometric surface is 60 ft above the top of the aquifer, drawdown is 20 ft, and the radius of influence is 500 ft.

10.3 For the centrifugal pump shown in Fig. 10.13 and an impeller diameter of $6\frac{1}{8}$ in., determine the head, efficiency, and brake horsepower for a discharge of 400 gpm. Should this pump be recommended for these flow conditions?

10.4 For the turbine pump in Fig. 10.14 and an impeller speed of 2400 rpm, determine the recommended operating range of head and discharge.

10.5 Determine the *WHP* and *BHP* for a pump if the flow rate is 500 gpm, the total lift is 100 ft, and the pump efficiency is 75 percent.

10.6 For the propeller pump in Fig. 10.15 and impeller B, determine the head, efficiency, and power for a discharge of 1200 gpm.

10.7 An irrigator is pumping 300 gpm with a total lift of 60 ft. The pump is driven by an electric motor with an efficiency of 90 percent and the pump is 70 percent efficient. How many kilowatts are required if 1 hp = 0.746 kw. If electrical energy costs $0.10 per kw-h, how much will the energy cost for a 24-h period?

10.8 The irrigator in Problem 10.7 decides to add a sprinkler system that requires a pressure of 50 psi at the pump. Determine the energy required and cost for a 24-h period.

10.9 Using the data from Problem 10.7, estimate the cost of pumping for 24 h with a diesel engine. Assume diesel engines average 15 hp-h per gallon of fuel and fuel costs $1.00 per gallon.

10.10 A concrete-lined trapezoidal channel has a bottom width of 1 ft, side slopes of 1 horizontal to 1 vertical, channel slope of 1 ft/1000 ft, and a flow depth of 0.7 ft. Determine the mean velocity and flow rate if the roughness $n = 0.02$.

10.11 Estimate the velocity and flow rate for the channel in Problem 10.10 if weeds are allowed to grow and $n = 0.05$.

CHAPTER 11 _____
SURFACE WATER STORAGE

Surface water can be stored in streams, lakes, ponds, excavated reservoirs and pits, cisterns, above-ground reservoirs, and tanks. Terminology varies slightly in different areas. Runoff from land and roofs as well as ground water from wells and springs can be stored in these facilities. Uses for stored water include irrigation, livestock, pesticide spray water, fish production, recreation, fire protection, milkhouse sanitation, domestic purposes, etc. Water stored in the open without a tight cover and accessible to wildlife and animals must be treated for domestic use, even if it comes from a safe groundwater source.

FARM RESERVOIRS

As discussed in Chapter 10, reservoirs are constructed (1) by excavating a pit (thus forming a dugout reservoir), (2) by constructing a dam (embankment) in a natural ravine to form an on-stream reservoir, or (3) by diverting the runoff from a natural channel into an off-stream reservoir. The term *embankment,* which has a broader meaning than the word *dam,* is sometimes used interchangeably for dam. Farm reservoirs are normally built for livestock water, but they are often constructed for aesthetic and recreational purposes.

11.1 Water Storage Requirements

The water storage capacity of a reservoir depends on the water needs, evaporation from the water surface, seepage into the soil or through the dam, storage allowed for sedimentation, and the amount of water carryover from one year to the next. Water needs include the volume required for the intended uses and the desired depth and surface area to satisfy recreation or fish and wildlife requirements. Several feet of depth are needed for fish to survive the winter months. Water needs for domestic uses, livestock, spraying, irrigation, and fire protection may be estimated from Table 11.1. Evaporation from the water surface can be estimated by multiplying local pan evaporation published by the Weather Service by a factor of about 0.7. Evaporation can be reduced by selecting a site having a small surface area and deep depth. Seepage losses are

TABLE 11.1   Water Requirements for Farm Uses

Type of Use	Gallons per Day[a]	Average acre-feet per Year[b]
Household, all purposes per person	50–100	0.08
Dry cow or steer per 1000-lb weight	20–30	0.028
Milking cow per 1000-lb weight, including milkhouse and barn sanitation	35–45	0.045
Swine per 100-lb weight	3–5	0.005
Sheep per 100-lb weight	2	0.002
Chickens per 100 head	9	0.01
Turkeys per 100 head	15	0.017
Horses or mules per 1000-lb weight	12	0.013
Orchard spraying	1 gal per year of tree age per application	
Irrigation (humid region) 1–1.5 ac-ft per acre per season (arid region) 1–5 ac-ft per acre per season		
Fire protection	10 gal per min for 2 h (1200 gal)	

Source: Midwest Plan Service (1989). Reprinted with permission from *Private Water Systems Handbook, MWPS-14.* Copyright © Midwest Plan Service, Ames, Iowa 50011.

[a]Values for air temperatures of 50° F and 90° F, respectively.

[b]Acre-feet per year = 0.00112 × gallons per day.

difficult to predict because they depend on the soil and construction techniques. Sedimentation can be reduced with good vegetative cover in the watershed, especially in waterways and the area surrounding the water surface. Reservoirs with large ratios of storage volume to watershed area provide a reliable supply even in dry years. For example, if a reservoir is filled from a 100-acre watershed, the water supply is more reliable if the water storage volume is 10 ac-ft than 5 ac-ft. Thus, within reasonable limits, maximum storage capacity for a given site is desirable. Minimum storage is usually estimated from total annual needs, allowing 40 to 60 percent of the total storage for seepage, evaporation, sediment storage, and other nonusable requirements.

11.2   Site Selection

Many factors must be considered in selecting the site, and seldom will all of them be optimum at a given location. Adequate storage capacity with the least amount of earth movement is normally the most important consideration.

TOPOGRAPHY OF THE RESERVOIR SITE   The dam should be located where over 15 to 25 percent of the water will normally be at least the minimum depth indicated in Fig. 11.1. Making the dam center line perpendicular to the contour lines will result in the least amount of fill volume. Fill material is usually obtained from

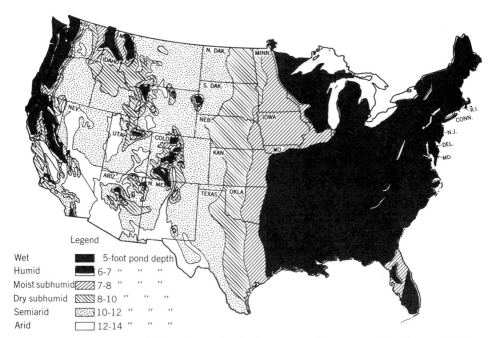

Legend

Wet	■ 5-foot pond depth
Humid	▬ 6-7 " " "
Moist subhumid	▨ 7-8 " " "
Dry subhumid	▨ 8-10 " " "
Semiarid	▨ 10-12 " " "
Arid	□ 12-14 " " "

FIGURE 11.1   Recommended minimum depths for water in farm reservoirs. (Source: SCS, 1982.)

the inundated area to give the desired depth of water. Channel slope above the dam should range from 4 to 8 percent. Steep, narrow channels will give a small surface area and volume, whereas flat, wide channels produce a large surface area and volume, with too much water at shallow depths. Shallow depths can be eliminated by steepening the side slopes along the water line or by excavation. On large reservoirs, wave damage can be reduced by locating the dam so that prevailing winds do not strike directly against the upstream face; by constructing a berm; or by using riprap. Topography is unimportant for dugout or up-ground ponds, as they can be built in flat areas.

SOIL AND UNDERLYING STRATA   Foundation material under the dam and under the water surface should be impermeable, and suitable fill soil should be available near the dam. Fill soil should generally have no more than 20 percent gravel, 20 to 50 percent sand, less than 30 percent silt, and 15 to 25 percent clay. Topsoil should be eliminated entirely in the fill. Deep peat, sand, gravel, or marl should be avoided as foundation material. Horizontal strata of sand and gravel or fractured rock at shallow depths may result in high seepage losses. Soil conditions prior to construction must be carefully evaluated because the control of seepage after construction is much more difficult.

WATERSHED AREA   The contributing watershed should be large enough to supply the desired amount of water, but not so large as to produce rapid sedimentation in the reservoir and to create an erosion problem in the flood spillway at the dam. The minimum size of watershed in acres for each acre-foot of storage capacity in the pond is shown in Fig. 11.2. (An acre-foot is a volume equal to 1 ft depth over 1 acre or 43 560 cu ft.) Watersheds larger than two to three times the minimum area in Fig. 11.2 may result in a short reservoir life or an expensive flood spillway. These problems are less serious where the watershed has low runoff-producing characteristics, such as flat topography, grass or forest cover, and permeable soils. Completely cultivated watersheds should be avoided, but well-sodded grass waterways and grass cover around the water surface will greatly reduce sediment inflow. Locations near farm buildings are desirable for fire protection, but the pollution hazard may be greater. Water from barnyards and other sources of pollution should be diverted away from the reservoir.

FLOOD SPILLWAY LOCATION   The most economical spillway is a grassed waterway around the end of the dam or across an adjacent ridge and into another natural channel. Excavated soil from the spillway can be placed in the dam. Thus, a large spillway will generally have little effect on the cost. Steep slopes along the waterway should be avoided where possible, to reduce the danger of erosion. Where good grass sod is already established, the flow may be allowed to spread out without a well-shaped channel. Concrete or permanent structures in the dam for carrying flood flows are normally not recommended because they are too costly.

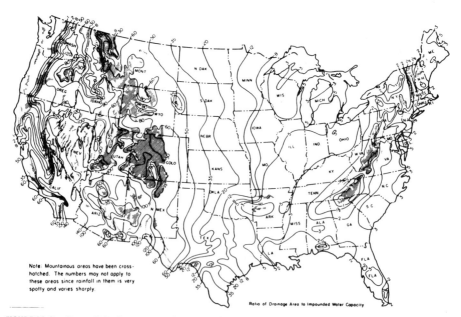

Note: Mountainous areas have been cross-hatched. The numbers may not apply to these areas since rainfall in them is very spotty and varies sharply.

Ratio of Drainage Area to Impounded Water Capacity

FIGURE 11.2   Size of drainage area in acres to impound 1 ac-ft of water storage. (Source: SCS, 1979.)

LOCATION OF WATER USE The site should be as near to the point of use as possible. For livestock in a pasture, a number of sites may be available, but for recreation or aesthetic uses, the number of choices may be more limited. Pumping from remote sites is often not practical, but where the reservoir is at a higher elevation than the point of use, gravity flow is possible.

11.3 Earth Dam Design

The design of the dam should be based on the most economical use of the available materials adjacent to the site. The most common type is a homogeneous dam with a core extending down to an impervious stratum (Fig. 11.3). When good fill material is limited, it is placed in the core or center section of the dam. Borings at the dam site should be taken at least 5 ft below the lowest level in the reservoir. Where soil or geologic conditions indicate possible seepage or foundation hazards, deeper explorations should be made. The height of the dam is determined from estimated storage requirements plus an allowance for flood storage and freeboard.

FLOOD STORAGE DEPTH The depth of water measured from the crest of the trickle spillway (normal water level) to the bottom of the flood spillway is for storage of flood water. This storage is provided so the flood spillway does not have to carry the runoff from small storms. Flood storage depth may be selected from Table 11.2 for a reservoir with a given surface area and with a given return period peak runoff rate from the watershed.

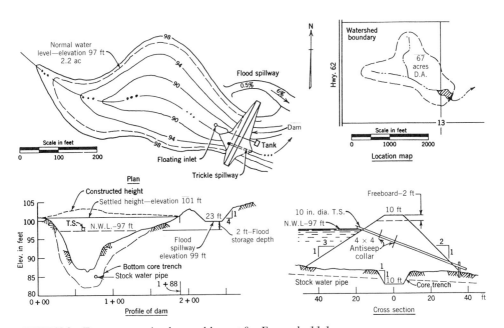

FIGURE 11.3 Farm reservoir plan and layout for Example 11.1.

TABLE 11.2 Flood Storage Depth for Farm Reservoirs (in ft)

Watershed Peak Runoff Rate (cfs)	Water Surface Area at Normal Water Level (acres)				
	0.5	1	2	3	5
15 or less	1.0	1.0	—	—	—
15 to 25	1.5	1.0	0.5	—	—
25 to 35	2.0	1.5	1.0	0.5	—
35 to 45	2.0	2.0	1.0	0.5	0.5
45 to 60	2.0	2.0	1.5	1.0	1.0
60 to 80[a]	—	2.5	2.0	1.5	1.0

[a]For runoff rates greater than 80 cfs consult the NRCS or a professional engineer.

FREEBOARD Freeboard is usually 2 ft, which is the depth from the bottom of the flood spillway to the top of the dam. It includes an allowance of 0.5 ft for frost action and a depth of flow in the flood spillway of 1.0 ft, with the remaining 0.5 ft height for wave action. Freeboard assures that the dam does not overtop. Where the dam height is over 15 ft and the length of the water surface is greater than 400 ft, the freeboard should be increased.

SIDE SLOPES For dams less than 50 ft in height with average soil, the side slopes should be not steeper than 3:1 (horizontal to vertical) on the upstream face and 2:1 on the downstream side. For coarse or uncompactible soils, the side slopes should be flatter to assure stability. Upstream side slopes should be flatter than those on the downstream side because saturated soil is less stable than the unsaturated downstream slope. For ease of mowing with machinery, the downstream slope may be reduced to 2.5:1 or flatter.

TOP WIDTH The minimum top width for dams up to 10 ft in height should be about 8 ft. The width should be increased about 0.5 ft for each additional foot of dam height. Where the top is to serve as a roadway, this minimum should be increased to 12 ft to provide a 2-ft shoulder on each side for safety.

SETTLEMENT ALLOWANCE Earth fills compacted in thin layers at optimum water content on an unyielding foundation will settle less than 1 percent. Since these conditions are usually not met, an allowance of 5 to 10 percent of the settled height should be added to the top of the fill during construction. On dams with the highest fill in the center, the top of the dam will have a slight crown sloping to either end just after construction (Fig. 11.3).

11.4 Trickle Spillway

This spillway is a pipe or other permanent structure through the dam that will maintain the normal water level at the elevation of the outlet entrance and will carry a small flow to a safe outlet below the dam. Pipe diameter may be selected from Table 11.3. The trickle spillway carries any long-duration flow, such as from a

TABLE 11.3 Trickle-Pipe Spillway Diameter (in in.)[a]

Watershed Peak Runoff Rate (cfs)	Water Surface Area at Normal Water Level (acres)				
	0.5	1	2	3	5
15 or less[b]	6	6	—	—	—
15 to 25	8	6	6	—	—
25 to 35	8	8	6	6	—
35 to 45	10	10	8	6	6
45 to 60	10	10	10	8	6
60 to 80	—	12	10	10	8

[a]For corrugated metal pipes. For smooth wall pipes, use the next smaller diameter.
[b]Trickle spillway is not essential for watersheds less than 10 acres with no ground-water flow.

drainpipe, a snowmelt, or runoff from small storms. In this way, the trickle spillway prevents long-term saturation conditions in the flood spillway that would hinder growth of stabilizing vegetation. The trickle spillway should be large enough to carry the long-duration flow, but it should not be designed to handle flood flows. The capacities of pipe spillways are given in Table 11.4. A pipe with a hood-inlet entrance or a drop-inlet pipe spillway is suitable for the trickle spillway (Fig. 11.4). The location of a typical spillway is shown in Fig. 11.3.

Any durable material (such as steel, asphalt-coated corrugated metal, or concrete) that cannot be damaged by settling, heavy loads, and the like is suitable for the trickle spillway. Where the fill material surrounding the pipe is not specially compacted, concrete or metal antiseep collars should be attached and sealed to the pipe. The number and size of the collars should be sufficient to increase the length of the creep distance at least 10 percent. The creep distance is the length along the pipe within the dam, measured from the upstream side of the dam to the point of exit on the backslope of the dam. The increase in creep distance for an antiseep collar is measured from the pipe outward to the edge of the collar (Fig. 11.4a). For example, the increase in creep distance for a 4 by

TABLE 11.4 Flow Capacity of a 100-ft Pipe Spillway in Cubic Feet per Second (cfs)[a]

Asphalt-Coated Corrugated Metal Pipe		Clay and Concrete Pipe Dia (in.)	Head H (ft) (Water Surface to Center of Outlet Pipe)			
Pipe Dia (in.)	K_c		10	20	30	40 ft
6	0.292	Min 6	1	1	1	1 cfs
8	0.199	6	2	2	3	3
10	0.148	8	3	4	6	6
12	0.116	10	5	7	9	11
15	0.086	12	10	14	17	20

[a]Calculated to the nearest cfs from the equation, $q = 8a[H/(1.6 + K_cL)]^{1/2}$, where a = cross-sectional area in square feet, H = head in ft, and K_c corresponds to a roughness coefficient of $n = 0.025$ for corrugated metal pipe.

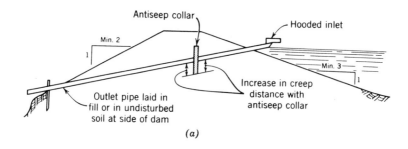

(a)

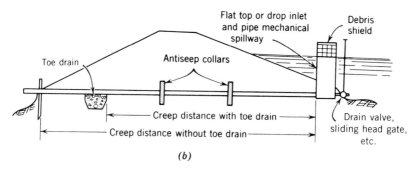

(b)

FIGURE 11.4   Types of trickle spillways and antiseep collars: (a) hooded inlet and pipe, (b) vertical riser and drainpipe combined. (Source: Schwab et al., 1993.) (Reprinted with permission of John Wiley & Sons, Inc.)

4 ft collar on a 12-in. diameter pipe would be twice the distance from the pipe or $2\,(2-0.5)=3.0$ ft. All pipes should be centered in the antiseep collar, which should be located near the middle or slightly upstream in the dam. Soil should be well tamped in layers under and around the pipe and around the collar. Antiseep collars are usually not required for dams less than 10 ft in height.

The trickle spillway may, in some cases, serve as a drain for cleaning, draining, or restocking the pond with fish. The drain valve of one such spillway, shown in Fig. 11.4b, can be operated from above the water surface. Such inlets should have a screen or debris shield to prevent clogging the pipe.

11.5   Water Pipe and Drain

A $1\frac{1}{4}$ in. diameter steel pipe through the dam is recommended for supplying water to livestock or for other uses. Two 2- by 2-ft. antiseep collars are usually sufficient because of better compaction around a small pipe. The location of a typical water pipe is shown in Fig. 11.5. Such a pipe may be used to drain the pond. All valves should be frost-proof and equipped to operate above the water surface. Water from the reservoir may be supplied to livestock by a tank and float valve. To obtain the best quality of water, a floating inlet (as shown in Fig. 11.5) should be attached to the water pipe.

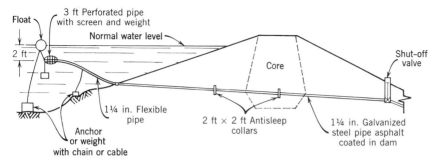

FIGURE 11.5   Floating inlet and outlet pipe for removing water from a farm reservoir.

11.6   Flood Spillway

All reservoirs filled by direct surface runoff from the watershed above should have a flood (emergency) spillway that will safely bypass floods exceeding the temporary storage capacity of the reservoir. The design capacity of the flood spillway should be the peak runoff rate for a 25-year return period storm (see Chapter 5). A return period of 50 or 100 years should be selected where potential damages are high. The flood spillway should be trapezoidal in cross section with a minimum bottom width of about 8 ft. The upper end of the spillway around the end of the dam should have a slope of not less than 0.5 percent. The width of this portion of the spillway can be computed from the weir formula. Assuming a flow depth of 1 ft.,

$$W = q/3.2 \tag{11.1}$$

where W = bottom width in feet

q = runoff rate for 25-year return period storm in cubic feet per second

The upper entrance to the flood spillway should be flared out at a 45° angle on each side, to allow smooth flow of water into the channel. The lower end of the spillway normally has a much steeper slope, depending on the topography, and should be designed as a grassed waterway as described in Chapter 8. The outlet end of the spillway should extend well below the dam or into an adjacent waterway.

11.7   Seepage Control

Seepage control can best be achieved by proper design and construction. Seepage may occur through the dam or downward into the profile where the water-covered area is underlaid by rock, sand, and gravel strata or solution channels. Knowledge of the soil conditions can best be obtained by soil evaluation at the site.

Where a limited amount of good fill soil is available, it should be placed in the center section of the dam and in the core trench, which should extend down to an impermeable layer. The core trench should have a minimum width of about 8 ft and side slopes of 1:1 or flatter (Fig. 11.3). Before the fill is placed, the topsoil or other permeable material under the dam should be removed. Maximum density of the fill should be obtained by compaction at the optimum water content. Soil is near optimum if, when compressed in the hand, it forms a ball that will not fall apart and from which free water cannot be squeezed. Soil should be carefully compacted under and around pipe or other structures through the dam. The core of good material should be as wide as possible. The next best fill material should be placed upstream in the face of the dam and the poorest in the downstream section. Steel, concrete, wood, and other materials are suitable as thin-section diaphragms in the dam, but they may be too costly for farm reservoirs.

Sealing of the reservoir area above the dam to prevent seepage may be accomplished by (1) compaction of the area to be covered by water, either with the soil in place or with a blanket of locally occurring heavy clay, (2) dispersion of clay soil with chemicals, (3) application of swelling clays, such as bentonite, and (4) lining with plastic, butyl, or asphaltic materials. Linings with sheet liners should be covered with 8 to 10 in. of stone-free soil. Livestock and large animals should be kept out of the area. Where rock, gravel, or other strata are exposed, the area should be covered with soil layers 4 to 8 in. thick and compacted with a sheepsfoot roller. The minimum depth should be 2 ft for water 8 ft deep, and it should be greater as the water depth increases. Where the existing soil is suitable, it should be ripped to a depth of 1 ft and then compacted. Where soils do not meet the criteria described in Section 11.2, bentonite or other high-swelling clays should be thoroughly mixed into the soil and then compacted. Soil should have 10 to 15 percent sand to provide the necessary soil strength. Common chemicals for dispersing the soil include salt, soda ash, sodium polyphosphates, and other salts. Effectiveness depends on the chemical and mineralogical composition of the soil. For this reason, laboratory testing is necessary to determine the rate and kind of dispersant. Thorough pulverizing and mixing of the soil and chemicals with a rotary tiller or disk is also important.

When seepage control is not possible by sealing or other measures, toe drains may be installed in the lower portion of the dam as shown in Fig. 11.4. These drains should be installed parallel to the dam to collect seepage water so it does not exit on the downstream surface and cause sloughing or failure of the dam.

11.8 Reservoir Design

The design procedure for an earth reservoir is described in the following example.

Example 11.1 Design a reservoir in northeastern Missouri to provide water for 107 steers, for irrigating a 0.5-acre garden, and for a family of 6 persons. A suitable site for the dam has a drainage area of 67 acres, and the estimated design runoff for a 25-year return period storm is 75 cfs. Seepage and evaporation losses from the reservoir are estimated as 50 percent of the storage capacity, and sediment storage allowance as 10 percent.

Solution. From Table 11.1 the annual water requirements are

107 steers (107×0.028)	3.0 ac-ft
Irrigation water (0.5×1.5)	0.75
Domestic use (6×0.08)	0.48
Total	4.23 ac-ft

Storage requirements allowing for seepage loss, evaporation, and sediment storage of 60 percent ($50 + 10$) are

$$S = 4.23 \times \frac{100}{100-60} = 10.6 \text{ ac-ft}$$

From Fig.11.2, read 6.5 acres of drainage area for each acre-foot of storage. Required watershed size is (10.6×6.5) = 69 acres. The selected site is satisfactory.

From a contour map of the reservoir area and a field investigation of the soils, a dam site was selected as shown in Fig. 11.3. By measuring the map area within each contour line, the water-level height to achieve the desired storage was determined as shown in Table 11.5. The area was measured to the center line of the dam. This procedure is sufficiently accurate, because most of the soil in the dam is usually taken from the reservoir area. (An approximate volume may be obtained by multiplying the water-surface area by 0.4 times the maximum water depth.) Storage at this site could be increased by raising the water level or by moving the dam further downstream. If these measures were unsuccessful, another site would have to be selected.

By interpolation between contours (Table 11.5) at elevations of 94 and 98 ft, 10.5 acre-ft of storage is available at an elevation of 97 ft, which is selected as the

TABLE 11.5 Volume of Reservoir-Water Storage[a]

Contour Elevation (ft)	Area within Contour Line to Center Line of Dam (acres)[b]	Average Area (acres)	Contour Interval (ft)	Volume of Storage (ac-ft)	Accumulated Storage Volume (ac-ft)
88	0.1				0
		0.4	2	0.8	
90	0.7				0.8
		1.0	4	4.0	
94	1.3				4.8
		1.9	4	7.6	
98	2.5				12.4

To elevation 97, surface = 2.2 ac and storage volume = $12.4 - (7.6/4) = 10.5$ ac-ft

[a]Computed for the reservoir in Fig. 11.3 and Example 11.1. The number of contour lines was reduced to simplify computations.

[b]Estimated from the map in Fig. 11.3.

normal water surface. For this site, the following specifications can be determined:

1. Crest elevation for trickle spillway (Table 11.5)   97.0 ft

2. Flood storage depth for 75 cfs and 2.2-ac surface area from Table 11.2 (interpolate as 1.9, round to 2.0 ft)   2.0

   Elevation of flood spillway   _____

                                                                                        99.0 ft

3. Freeboard (wave height 0.5, frost depth 0.5, and water flow depth 1.0 ft)   2.0

   Elevation of top of dam (settled height)   _____

                                                                                        101.0 ft

4. Maximum allowance for settlement at station $0 + 72$ ($10\% \times 13.7$)   1.4 ft

5. Top width of dam $[8 + 0.5 \times (13.7 - 10)]$ 10 ft

6. Select dam side slopes upstream $3:1$, downstream $2:1$

7. Trickle spillway diameter and length for 2.2-acre surface area and 75 cfs from Table 11.3   10 in.
                                                                                        60 ft (estimate from Fig. 11.3)

8. Antiseep collar on trickle spillway, required increase in creep distance $= 10$ percent $\times 60 = 6$ ft [creep distance for 4-ft collar $= (4 - 1)2 = 6.0$ ft] (outside pipe diameter assumed 1 ft)   two
                                                                                        4×4 ft

9. Livestock water pipe, diameter, and length under dam   1¼ in., 100 ft

10. Volume of fill in dam and core trench (Fig. 11.6 and Table 11.6)   1646 yd^3

11. Flood spillway width at control section (level) around end of dam from Eq. 11.1 $W = 75/3.2$ (flow depth 1.0 ft)   23 ft

12. Flood spillway width below dam for maximum slope of 6 percent, maximum velocity 6 fps, $4:1$ side slopes, trapezoidal cross section (Chapter 8)   0.6 ft depth
                                                                                        21 ft bottom width

11.9 Field Procedure

After a tentative site has been selected, a contour map of the reservoir area should be prepared (Chapter 4). Elevations normally are obtained along cross section lines taken perpendicular to the channel at intervals of 50 to 200 ft above the dam site. The watershed area may be determined from an aerial photograph on which the water-divide boundary has been located. The divide and the runoff-producing

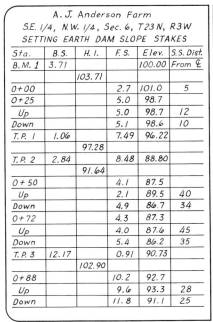

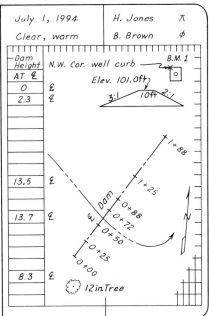

FIGURE 11.6 Field survey notes for setting slope stakes for the dam in Example 11.1. (Data for stations 1 + 25 and 1 + 88 are not shown.)

TABLE 11.6 Volume of an Earth Dam[a]

Station along Dam	Height[b] of Dam (ft)	Cross-Sectional Area (ft²)[c]	Average Cross-Sectional Area (ft²)	Length of Section (ft)	Volume of Section (ft³)
0 + 00	0	0			
0 + 25	2.3	36	18	25	450
0 + 50	13.5	590	313	25	7825
0 + 72	13.7	606	598	22	13156
0 + 88	8.3	255	430	16	6880
1 + 25	5.0	113	184	37	6808
1 + 88	0	0	57	63	3591
			Volume in dam =		38710

Volume in core trench (estimated average depth 3.3 ft and length 140 ft)

$$[(3.3 \times 3.3) + (3 \times 10)]\,140 \;=\; 5725$$

$$\text{Total earth fill volume} \;=\; 44\,435\ \text{ft}^3$$

$$\text{or } (44\,435/27) \;=\; 1646\ \text{yd}^3$$

[a] Computed for the reservoir in Fig. 11.3 and Example 11.1.

[b] Height of dam (101 − ground elevation at each station at the center line, from Fig. 11.6)

[c] Cross-sectional area = $2.5h^2 + 10h$, where h is the dam height. This equation is valid only for 3:1 and 2:1 side slopes and for 10-ft top width.

characteristics of the watershed are determined by field survey. Soil evaluation to the desired depth should be taken along the center line of the dam, in the borrow area, and at other critical points. After detailed plans have been prepared, as in Fig. 11.3, slope stakes locating the edge of the fill from the center line of the dam can be set. These stakes locate the edge of the fill for starting the dam. Distances from the center line are computed from the equation

$$d = (E_t - E)s + w/2 \tag{11.2}$$

where d = distance from center line to edge of fill in feet
  E_t = elevation of top of dam (settled height) in feet
  E = elevation of the upstream or downstream slope stake in feet
  s = side slope ratio of the dam
  w = top width of dam in feet

For the reservoir in Example 11.1, the surveying notes are shown in Fig. 11.6 and the slope-stake distances are computed in Fig. 11.7 for station $0 + 50$. Where the ground is nearly level, such as at station $0 + 25$ in Fig. 11.6, distances can be computed directly from Eq. 11.2, since elevations at the slope stakes are the same as at the center line. Where the ground is sloping (perpendicular to the dam center line), slope stakes must be located by trial and error. The procedure illustrated in Fig. 11.7 is as follows: Point A is the first trial point obtained by substituting, in Eq. 11.2, the elevation of the ground at the center line, for which $d = (101 - 87.5)3 + 5 = 45.5$ ft. The actual elevation at A is 90.5, for which $d = (101 - 90.5)3 + 5 = 36.5$ ft. This is the distance to the second trial point B. The actual elevation of B is 89.3, for which $d = (101 - 89.3)3 + 5 = 40.1$ ft. This is the location of the third point C, which has an elevation of 89.5 ft. Since this elevation is within 0.2 ft of that at point B, point C is close enough to the correct distance of 39.5 ft shown in Fig. 11.7. Usually one or two trials are sufficient. Failure to set side slopes in this manner will result in a variation of slope on the face of the dam. If the stake had been set at the

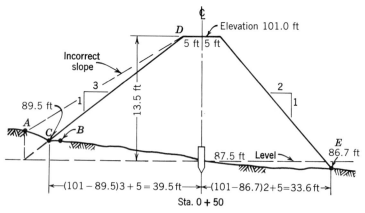

FIGURE 11.7   Setting slope stakes at station $0 + 50$ for the dam in Example 11.1.

first trial point *A* in Fig. 11.7, the side slope would have been too flat. The downstream slope stakes are set in the same manner.

The shoreline should be staked at about 50-ft intervals at the normal water level. This line will indicate where trees should be removed and where the water will be shallow. As construction proceeds, the water-supply pipe, trickle spillway, and flood spillway should be staked out.

11.10 Construction and Maintenance

All trees, stumps, and shrubs should be removed from the dam site and from the area to be inundated. Sod and topsoil should be removed and stockpiled. Before the placement of the fill, the existing soil should be thoroughly plowed and disked and should be near optimum water content. The fill material should be spread evenly in 4- to 8-in. layers over the entire dam, with a camber for drainage. The sheepsfoot roller is best suited for compacting the fill. Heavy hauling equipment should use varied travel paths to avoid overcompaction. Hand-operated pneumatic or motor tampers are best for compacting around pipes and antiseep collars. Along the shoreline, except at the dam, the slope should be increased to 2:1 to a minimum depth of 4 ft. The elimination of shallow water will minimize the growth of cattails and other water weeds.

Slopes and other areas above the waterline where the subsoil has been exposed should be covered with about 6 in. of topsoil, fertilized, and seeded with a suitable grass. The entire reservoir area should be fenced to prevent damage to the dam, spillways, and banks. Sedimentation can be reduced by protecting waterways with grass and by establishing adequate erosion-control practices on the watershed.

Normally, little maintenance is required, but the reservoir should be inspected occasionally for evidence of seepage on the downstream face of the dam, wave action, and damage by animals or humans. Weed growth and algae in the reservoir can be controlled by reducing nutrient pollution or by using suitable chemicals. Trees should not be allowed to grow near the dam.

CISTERNS AND TANKS

Cisterns and tanks are suitable water-storage structures in rural areas. These facilities provide rainwater storage where surface water is badly polluted (such as in mining areas) and where water needs are minimal (such as for a vacation home), as well as water storage where well yields are low or ground-water quality is poor. Cisterns usually store water that falls on roofs of buildings. Cisterns may be located above or below ground level, but should have closed tops. Tanks may have open or closed tops. They may be buried below ground level or elevated for gravity flow. Tank storage can also be provided as part of a pressure water system for low-yielding wells. In remote grazing areas, livestock are often watered from an open-top tank supplied by a windmill pumping from a well. In low-rainfall areas, the soil surface may be made impervious with plastic or other material and the runoff stored in a tank or a collapsible collector. This procedure is often referred to as "water harvesting" (Chapter 10).

11.11   Cisterns

Cisterns typically store water underground for domestic purposes. At best, cisterns should be considered an alternative for storage of city or community water or ground water. Further details on cisterns may be obtained from local or state health departments, state extension services, and commercial suppliers (Water Filtration Co., 1990).

11.12   Tank Water Storage, Capacity, and Location

Tanks for water storage are normally smaller than cisterns and provide much less storage than farm reservoirs. Whereas cisterns seldom, if ever, store water under pressure, tanks are often installed in pressure systems. Pressure tanks are suitable where low-yielding wells need storage to meet domestic high-flow rates for a short period of time. Most pump manufacturers provide an automatic electronic device for shutting off the pump before the water level in the well drops below the suction inlet, thus protecting the pump from damage.

Open tanks are common for livestock water in the housing area or in remote livestock pastureland. These tanks may be filled from city water, a private pressure water system, a manually operated pump, or a windmill using surface or well water. Windmills are common in remote areas where electric power is not available. Some tanks are elevated or located on high ground to provide gravity flow. In cold climates, water pipes and tanks need to be protected from freezing. Tank size will vary depending on the water-use rate and the availability and automation of the power source. Tanks provide convenient storage for hauled water. Tank location will depend largely on the source of the water and nearness to the point of use.

QUALITY OF CISTERN AND RESERVOIR WATER

All surface water and some well water need to be filtered and purified for domestic and milkhouse use. Water for milkhouse use should meet health department requirements. Except for sediment (turbidity) and bacteria, surface water is generally of good quality; but it tends to be acidic, like rainwater. Bacterial contamination is caused mainly by wildlife fecal material present on the catchment area. Heavy metals, fertilizers, pesticides, and insecticides may also accumulate in the stored water. Since these pollutants generally result from human activities, they can be controlled or reduced in the catchment area. A study in Ohio by Hill et al. (1962) showed that (1) all reservoir water was contaminated by bacteria, (2) about 50 percent was hard water (greater than 100 ppm hardness), (3) the best quality water was found just below the water surface, and (4) turbidity was lower where grassed waterways and good grass cover were present around the water area. Cistern water is usually softer and has less sediment than reservoir water, and thus is easier to treat.

REFERENCES

Hill, R. D., G. O. Schwab, G. W. Malaney, and H. H. Weiser (1962) *Quality of Water in Ohio Farm Ponds,* Res. Bull. 922. Ohio Agr. Expt. Sta., Wooster, Ohio.

Midwest Plan Service (1989) *Private Water Systems, MWPS-14,* 4th ed. Iowa State University, Ames, Iowa.

Schwab, G. O., D. D. Fangmeier, W. J. Elliot, and R. K. Frevert (1993) *Soil and Water Conservation Engineering,* 4th ed. Wiley, New York.

U.S. Bureau of Reclamation (USBR) (1987) *Design of Small Dams,* 3rd ed. Government Printing Office, Washington, D.C.

U.S. Soil Conservation Service (SCS) (1979) *Engineering Field Manual for Conservation Practices,* Lithograph. Washington, D.C.

U.S. Soil Conservation Services (SCS) (1982) *Ponds—Planning, Design, Construction,* U.S. Dept. Agr. Handb. 590. Government Printing Office, Washington, D.C.

Water Filtration Company (1990) *Alternative Water Sources for Rural Areas Using Ponds, Lakes, Springs and Rainwater Cisterns,* Brochure. Water Filtration Co., 1205 Gilman St., Marietta, Ohio.

PROBLEMS

11.1 Determine the storage requirements for a pond that is to provide sufficient water for 100 milk cows and 200 hogs, and spray water for 1000 apple trees with an average age of 10 years and with 4 applications per year. Seepage and evaporation losses are estimated to be 54 percent of the stored water.

11.2 Determine the minimum-size watershed area above an on-stream pond at your present location that will hold 11 ac-ft of water.

11.3 Determine the usable water storage for a pond at your present location that is to be filled with runoff from a 25-acre watershed. Assume seepage and evaporation loss as 60 percent. If the volume of runoff is not adequate, what measures could be taken to secure additional water? Assume that all runoff is required to fill the pond.

11.4 Determine the elevation of the normal water level to store 8.1 ac-ft of water in a pond. From a contour map of the reservoir area, the following area of each contour line to the center line of the dam was obtained: contour elevation 40, 0.2 acre; 42, 0.6; 44, 0.8; 46, 1.0; 48, 1.5; 50, 1.7. All storage below elevation 40 is reserved for sediment.

11.5 The elevation of the lowest point along a dam is 36.0 ft. The normal water level is 10 ft above this elevation, and the water surface area is 1.2 acre. The peak runoff from the watershed is 28 cfs for a 25-year return period storm. Determine the elevation of the trickle spillway, the flood spillway, and the top of the dam; the diameter for the trickle spillway; the top width of the dam; and the constructed height before and after settlement. Assume a settlement allowance of 10 percent of the settled height.

11.6 The height of fill for a dam at station $1 + 00$ along the center line is 10 ft, and at $2 + 00$ the height is 12 ft. Compute the volume of fill in cubic yards between these two stations, assuming that the top width of the dam is 10 ft and that the base of the dam is level. Side slopes for the dam are $3:1$ and $2:1$.

11.7 List the potential water sources for domestic use in your area. What storage facilities would be required?

11.8 Determine the width of the flood spillway at the level control section around the end of the dam with a maximum depth of flow of 1 ft. The runoff from a 25-year return period storm is 64 cfs. Determine the minimum bottom width of the lower end of this grassed spillway if the slope is 12 percent. Assume a maximum velocity of 7 fps. See Chapter 8.

11.9 How far from the center line of the dam should the upstream and downstream slope stakes be set if the dam is 15 ft high at the center line and the top width is 12 ft? Assume the ground is level perpendicular to the center line of the dam.

11.10 Solve Problem 11.9 if the ground slope is a uniform 10 percent toward the downstream side.

11.11 Locate and design a farm pond from a topographic map and related data supplied by your instructor.

CHAPTER 12 ———————
SURFACE DRAINAGE
AND WETLANDS
———————————————————

Surface drainage is the removal of excess water using constructed open ditches, field ditches, land grading, and related structures. In this text the application is for land that has insufficient natural slope to provide adequate drainage for good agricultural production. In humid areas many thousands of miles of main outlet ditches have been built mostly by organized drainage districts. In arid regions under irrigation, drainage ditches are necessary to remove water required for leaching undesirable salts from the soil and to dispose of excess rainfall. Outlet ditches must be large enough to carry floodwater and of sufficient depth to provide outlets for subsurface drains.

During the early development of the United States, agricultural interests prevailed and the drainage of swamps and wetlands was done without question. In more recent years, drainage of these naturally wet areas has been found to have some adverse effects on migrating wildlife and marine life as well as other environmental aspects. Wetlands have become a major political issue involving several state and federal agencies. Further discussion on the effects of wetlands on the environment and recent legislation is given in Chapter 1.

OPEN CHANNELS

Open channels include drainage ditches and irrigation canals. Drainage ditches are distinguished from surface drains mostly by steeper side slopes and they are often larger and carry greater flows. Drainage ditches provide outlets for pipe and surface drains for areas varying from small fields to several farms as well as remove surface water directly. They are often the primary purpose for having legal or organized drainage districts. Irrigation canals are designed basically the same as drainage ditches.

12.1   Channel Capacity

Channel capacity is computed from the Manning formula, which was discussed in Chapter 8. The minimum bottom width for drainage ditches is usually 4 ft and the depth and channel grade vary depending on outlet requirements. The constructed channel should provide a freeboard above the design flow depth of 20 percent of the total channel depth as shown in Fig. 12.1. For earth-lined channels, the roughness coefficient in the Manning formula will vary from 0.035 to 0.04 for medium to small ditches, respectively (Chapter 8). Most drainage ditches have trapezoidal cross sections (Fig. 12.1) with side slopes between 1.5 and 2 horizontal to 1 vertical for small to medium-size channels in clay to sandy loam soil. Spoil should be spread over the land adjacent to the channel or placed in banks having flat side slopes as shown in Fig. 12.1.

Flow velocities in open channels should be low enough to prevent scour and high enough to prevent sedimentation. Normally, velocities of 1.5 to 3 ft/s are high enough to prevent sedimentation. In drainage ditches, high velocities are not usually of as much concern as low velocities because of flat topography. To maintain the design cross section, some sedimentation at low flows may be desirable to offset scour that might occur at high flows.

Further design procedures for drainage ditches and canals are beyond the scope of this book. ASAE (1988), Schwab et al. (1993), and other references should be consulted to determine design flow requirements, water surface profiles for varied channel flow, hydraulics of channel structures, roughness coefficients, etc., especially for large channels.

12.2   Drainage Enterprises

Most states have laws regulating the organization of mutual and organized drainage enterprises, which differ from state to state. A mutual enterprise may also be known as a mutual district or drainage association. Such a district is formed by mutual agreement among a few landowners, after which operation and maintenance are turned over to the county, in a manner similar to that of

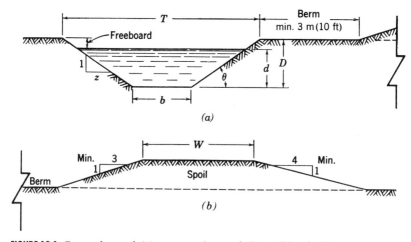

FIGURE 12.1 Open channel (a) cross section and (b) spoil bank. (Source: Schwab, G. O. et al. 1993). (Reprinted with permission of John Wiley & Sons, Inc.)

an organized enterprise. The principal advantage of the mutual district is that it can be organized quickly and inexpensively. If a large number of landowners are involved, they may have a difficult time coming to an agreement, particularly on the division of costs to each landowner. Organizing such districts often requires considerable time and legal costs. Open ditches are usually the major structures involved, but subsurface outlet drains may also be provided by the district.

An organized drainage enterprise, often called a county ditch or a drainage district, is a local unit of government established under state laws for the purpose of obtaining a satisfactory outlet for the removal of excess water. Usually bonds are issued to pay construction costs, and landowners pay taxes to cover these and maintenance costs. In the early years of land development for agriculture, much drainage was accomplished by such district enterprises, especially in the midwestern states. Most of these districts are still in existence, many of which have been reorganized several times. The key to successful operation usually depends on the collection of maintenance taxes to keep the ditches clean and functioning properly. The county engineer is usually responsible for maintenance.

SURFACE DRAINS

In this chapter, surface drainage will be limited to field-size areas and will include broad shallow surface drains that carry runoff to the point of entrance to outlet ditches. Land grading, which results in a continuous land slope toward the field ditches, is an important part of a surface drainage system. Land grading is essential for surface irrigation. It is the same as land leveling or land shaping.

Surface drainage systems must be suitable for mechanized operations on various types of topography, such as ponded areas, flat fields, and gently sloping land. Similar drains are needed to collect wastewater from surface irrigation systems. Ponded areas (potholes) are frequently found in glaciated regions, where erosion has not formed natural outlets. Flat or nearly level land with impermeable subsoils frequently requires surface drainage. Claypan, fragipan, or heavy-textured soils are examples. Flatland is defined as land with less than 2 percent slopes, with most flatland having less than 1 percent.

From an extensive survey, Gain (1964) estimated that over 100 million acres in the eastern United States and about 8 million acres in the eastern provinces of Canada would benefit from surface drainage. The location and intensity of these drainage problem areas are shown in Fig. 12.2.

12.3 Random Field Drains

Random potholes or depressional areas are common in glaciated regions and in other areas of level land. Water accumulates in these shallow depressions,

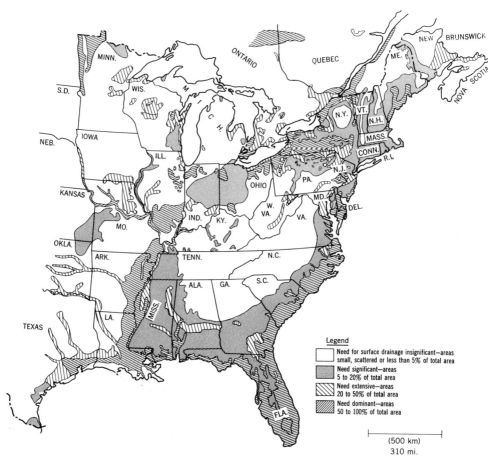

FIGURE 12.2 Lands needing surface drainage in eastern United States and Canada. (Source: Gain, 1964.)

causing crop damage. Improper tillage or deadfurrows may also result in poor surface drainage. A random field-drain system is illustrated in Fig. 12.3. The outlet for such a system may be a natural channel, a constructed ditch, or the field below, which allows the water to spread out where no distinct channel exists. Field drains connecting depressions are normally less than 3 ft in depth. Such drains should follow routes that provide minimum cuts and the least interference with farming operations.

The design of field drains is similar to the design of grass waterways, as discussed in Chapter 8. Where farming operations cross the channel, the side slopes should be flat; that is, 8:1 or greater for depths of 1 ft or less and 10:1 or greater for depths over 2 ft. Minimum side slopes of 4:1 are possible if the field is farmed parallel to the drain. The depth is determined primarily by the topography of the area, outlet conditions, and the capacity of the channel. A minimum cross-sectional area of 5 ft^2 is recommended. The grade in the channel

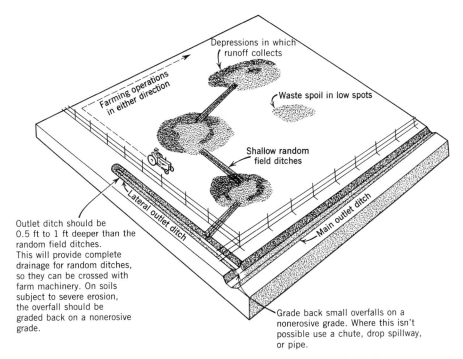

Depressions in which
runoff collects

Farming operations
in either direction

Waste spoil in low spots

Shallow random
field ditches

Lateral outlet ditch

Main outlet ditch

Outlet ditch should be
0.5 ft to 1 ft deeper than the
random field ditches.
This will provide complete
drainage for random ditches,
so they can be crossed with
farm machinery. On soils
subject to severe erosion,
the overfall should be
graded back on a nonerosive
grade.

Grade back small overfalls on a
nonerosive grade. Where this isn't
possible use a chute, drop spillway,
or pipe.

FIGURE 12.3 Random field-drain system for surface drainage. (Redrawn from Beauchamp, 1952.)

should be such that the channel will not erode or fill with sediment. Maximum velocities vary from 2.5 fps for sandy loam and 3.0 fps for silt loam to 5.0 fps for clay soil. Minimum velocities should be about 1 to 2 fps for flow depths less than 3 ft. For triangular channels, the maximum grade for sandy loam is about 0.2 percent and for clay soils it is 1.0 percent. For drainage areas in humid regions larger than 5 acres, the discharge capacity of the channel should be based on the runoff for a 10-year period storm. Because of crop damage from flooding, surface water should be removed within a period of about 24 h.

The plan, profile, and cross section of a random field ditch are shown in Fig. 12.4. The field-survey notes and a method for computing the cuts and volume of excavation are illustrated in Fig. 12.5. The detailed procedure is explained in Example 12.1. In staking out the drain in Fig. 12.4, elevations of the ground surface were taken on 50-ft stations starting from the bottom of the pothole and going over the adjacent ridge to a point providing natural drainage. An intermediate station was established at $1 + 30$, the center of another small depression. At this station, the direction of the channel was changed to minimize the depth of cuts.

Example 12.1   Determine the cuts and volume of soil to be excavated for the random field drain shown in Fig. 12.4 with elevations given in Fig. 12.5. The soil will permit a grade of 0.2 percent with a triangular channel with side slopes of 10:1. The soil is to be spread in nearby small depressions, rather than in the depression at $0 + 00$.

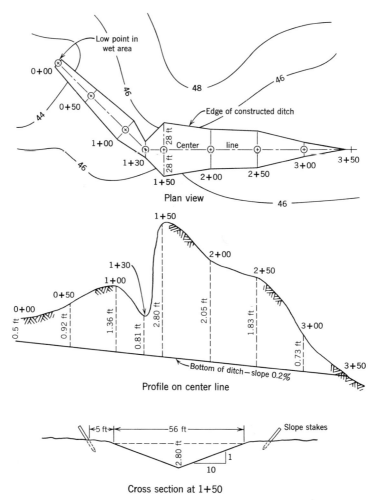

FIGURE 12.4 Plan, profile, and cross section of a random field drain.

Solution. See Fig. 12.5. At station $0 + 00$, allow a cut of 0.5 ft to provide for some sedimentation in the drain and to secure good drainage from the pothole. Compute and record the grade elevation, $44.01 - 0.5 = 43.51$ ft.

From the grade elevation at $0 + 00$, subtract the fall in 50 ft to obtain the grade elevation at $0 + 50$, which is $43.51 - (0.002 \times 50) = 43.41$ ft. Compute the grade elevations for all remaining stations.

Subtract the grade elevation from the hub elevation at each station to obtain the cut (e.g., at station $0 + 00$, $44.01 - 43.51 = 0.50$).

Compute one half the top width for each station, which is the cut times the side slope ratio ($0.5 \times 10 = 5$ ft).

Compute the cross-sectional area at each station, which is the cut times one half the top width ($0.5 \times 5 = 2.5$ ft^2)

Obtain the average cross-sectional area for each pair of adjacent stations ($(2.5 + 8.5)/2 = 5.5$ ft^2).

John Doe Farm S.E. 1/4, Sec. 6, T81N, R6W Surface Drain						May 10, 1994 Clear, cool			J. Moe R. Timm		λ φ
Sta.	B.S.	H.I.	F.S.	Hub Elev.	Grade Elev.	Cut	1/2 Top Width	X-Sect. Area	Avg. X-Sect. Area	Length (ft.)	Volume (ft.³)
B.M. 1	1.36			50.00							
		51.36									
0+00			7.35	44.01	43.51	0.50	5.0	2.5			
									5.5	50	275
0+50			7.03	44.33	43.41	0.92	9.2	8.5			
									13.5	50	675
1+00			6.69	44.67	43.31	1.36	13.6	18.5	12.5	30	375
1+30			7.30	44.06	43.25	0.81	8.1	6.5	42.5	20	850
1+50			5.35	46.01	43.21	2.80	28.0	78.4			
									60.2	50	3,010
2+00			6.20	45.16	43.11	2.05	20.5	42.0			
									37.8	50	1,890
2+50			6.52	44.84	43.01	1.83	18.3	33.5			
									19.4	50	970
3+00			7.72	43.64	42.91	0.73	7.3	5.3			
									2.7	40	108
3+50			8.71	42.65	42.81	−0.16	0	0			
B.M. 1			1.35	50.01					TOTAL − 8153		
			Error	0.01					$\frac{8153}{27}$ = 302 cu. yds.		

FIGURE 12.5 Field notes for the field drain in Fig. 12.4.

Compute the volume of excavation between each pair of adjacent stations, which is the length of the section times the average cross-sectional area ($50 \times 5.5 = 275$ ft³). Note that near the last station, the grade line would intersect the ground surface at about $3 + 40$ rather than $3 + 50$. Add the volumes of each segment to obtain the total excavation.

Surface water in large depressions may also be removed through pipe drains. With pipe outlets, the water may flow by gravity or it may be raised by a pump to a higher level. Most pumps for this purpose are of the propeller type (Chapter 10), which discharge several hundred gallons per minute with lifts from 5 to 10 ft. The design of such pumping installations is beyond the scope of this text.

As shown in Fig. 12.6, a surface inlet, sometimes called an open inlet, is an intake structure for the removal of surface water from potholes, road ditches, depressions, and farmsteads. These inlets should be placed at the lowest point along fence rows or in land that is in permanent vegetation. Where the inlet is in a cultivated field, the area immediately around the intake should be kept in grass. To prevent the entrance of trash, a fence may be constructed around it. Galvanized metal, brick, monolithic concrete, or plastic are satisfactory inlet materials. Manholes with sediment basins are sometimes used as surface inlets. At the surface of the ground a concrete collar should extend around the intake on the riser to prevent growth of vegetation and to hold it in place. On top of the riser a beehive cover or other suitable grate is necessary to prevent trash from entering the pipe drain.

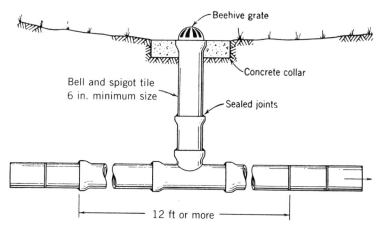

FIGURE 12.6 Surface inlet into a subsurface drain for removing surface water.

Where the surface inlet is connected to a main drain, it is good practice to offset the surface inlet several feet from the main. Such construction may prevent failure of the system if the surface inlet structure should become damaged.

Where the quantity of water to be removed is small (as in local depressions), blind inlets, also called "French drains," may be installed over the pipe drain. These are constructed by backfilling the trench with various gradations of materials, such as gravel or coarse sand, corncobs, straw, and similar substances. Although such inlets may not be permanently effective, they are economical to install and do not interfere with farming operations. As the voids in the backfill of the blind inlet become filled, its effectiveness is reduced. Since the soil surface has a tendency to seal, a narrow strip of dense sod-forming grass or other vegetation directly over the pipe drain will greatly increase the flow through the inlet. Where cultivation over the pipe can be avoided, flow may be increased by extending the coarse material to the soil surface.

12.4 Bedding

Bedding is a method of surface drainage consisting of narrow-width plow lands in which the deadfurrows run parallel to the prevailing land slope (Fig. 12.7). The area between two adjacent deadfurrows is known as a bed. Bedding is most practicable on flat slopes of less than 1.5 percent where the soils are slowly permeable and subsurface drainage is not economical. Studies in southern Iowa showed that level land gave slightly better yields than bedding.

The design and layout of a bedding system involves the proper spacing of deadfurrows, depth of bed, and grade in the channel. The width of bed depends on the land slope, drainage characteristics of the soil, and the cropping system. Bed widths recommended by Beauchamp (1952) for the Corn Belt region vary from 23 to 37 ft for very slow internal drainage, from 44 to 51 ft for slow internal drainage, and from 58 to 93 ft for fair internal

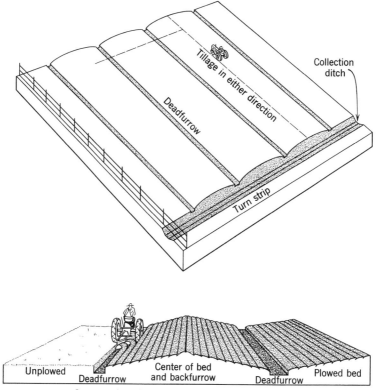

FIGURE 12.7 Bedding method of surface drainage for flat land.

drainage. The depth of bed depends on the soil characteristics and tillage practices. In the bedded area the direction of farming may be parallel or normal to the deadfurrows. Tillage practices parallel to the beds have a tendency to retard water movements to the deadfurrows. Plowing is always parallel to the deadfurrows.

12.5 Parallel Field Drain System

Parallel field drains are similar to bedding except that the channels are spaced farther apart and have a greater capacity than the deadfurrows. This system is well adapted to flat, poorly drained soils with numerous small depressions that must be filled by land grading.

The design and layout are similar to those for bedding except that drains need not be equally spaced and the water may move in only one direction. The layout of such a field system is shown in Fig. 12.8. The size of the drain may be varied, depending on grade, soil, and drainage area. The depth of the drain should be a minimum of 0.75 ft and have a minimum cross-sec-

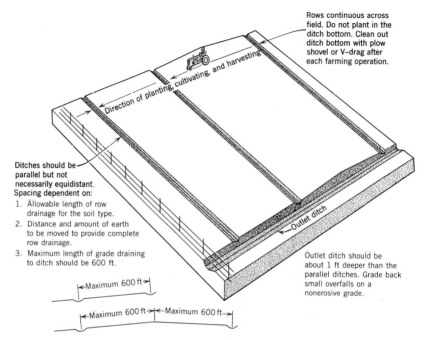

FIGURE 12.8 Parallel field-drain system for surface drainage.

tional area of 5 ft². The side slopes should be 8:1 or flatter to facilitate farm machinery traffic. As in bedding, plowing operations must be parallel to the channels; but planting, cultivating, and harvesting are normally perpendicular to them. The rows should have a continuous slope to the drains. The maximum length for rows with a continuous slope in one direction is 600 ft, allowing a maximum spacing of 1200 ft where the rows drain in both directions. In very flat land with little or no slope, some of the excavated soil may be used to provide the necessary grade. However, the length and grade of the rows should be limited to prevent damage by erosion. On highly erosive soils that are slowly permeable, the slope length should be reduced to 300 ft or less.

The cross section for field drains may be V-shaped, trapezoidal, or parabolic. The W-drain shown in Fig. 12.9 is essentially two parallel single ditches with a narrow spacing. All of the spoil is placed between the channels, making the cross section similar to that of a road. The advantages of the W-drain are (1) it allows better row drainage because spoil does not have to be spread, (2) it may be used as a turn row, (3) it may serve as a field road, (4) it can be constructed and maintained with ordinary farm equipment, and (5) it may be seeded to grass or row crops. The disadvantages of the W-drain are (1) the spoil is not available for filling depressions, and (2) the drains occupy a large area. The minimum width between the pairs of drains varies from 15 to 50 ft. The

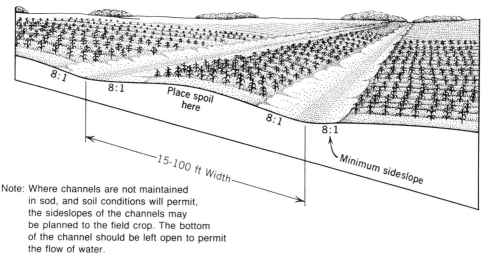

Note: Where channels are not maintained
in sod, and soil conditions will permit,
the sideslopes of the channels may
be planned to the field crop. The bottom
of the channel should be left open to permit
the flow of water.

FIGURE 12.9 W-drain for surface drainage.

W-drain should be positioned so that runoff comes from both directions into the drain.

12.6 Parallel Lateral Ditch System

This system is similar to the parallel field system except that the channels are deeper. The minimum depth for these ditches is 2 ft, with side slopes steeper than 4:1. These channels are not designed for crossing with machinery. Such a system of ditches is suitable for controlling the water table and for subirrigation. In peat and muck soils, the side slopes may be vertical. Such ditches may be installed in peat and muck to obtain initial subsidence prior to subsurface drain installation. Any ditch will provide subsurface drainage as effectively as a pipe of the same depth.

The layout of a ditch system is similar to that for the drains in Fig. 12.8. As in other methods of drainage on flat land, the surface must be smoothed and shaped so that water will move to the ditches. Farming operations must be parallel to the channels.

Parallel lateral ditches are common in the low lands of The Netherlands, West Germany, and England. Many of these lands are below sea level and because of their high water table are used mostly for pasture. Water may be in the channels for a large part of the year. Drainage water is removed entirely by pumping.

For water table control or subirrigation, the soil must be several feet deep and permeable, especially horizontally, and be underlaid with an impermeable material to prevent deep seepage. During dry seasons, the water level is held at the proper depth by suitably located control structures. In wet seasons the controls are removed so the ditches may provide surface drainage. In organic soils the water level is maintained from 1 to 4 ft below the surface to provide water for plant growth, to control subsidence, and to reduce fire and wind-erosion hazards.

12.7   Surface Drain Construction

Prior to an instrument survey or construction, heavy vegetation and residue should be removed or plowed under. For the construction of field drains with shallow cuts to 1-ft depth, moldboard and disk plows, small scrapers, blade graders, and other light equipment are suitable. For depths up to 2.5 ft, such equipment as motor graders, scrapers, and heavy terracing machines are more appropriate. Bulldozers equipped with push- or pull-back blades and carryall scrapers are useful for deeper cuts.

Setting grade stakes along the center line or along an offset line outside the area to be excavated are two methods of establishing the grade line. After the cuts are made to nearly the grade line, they should be checked with a surveying instrument or a laser.

12.8   Surface Drain Maintenance

Plowing parallel to field drains is usually adequate to maintain the channel depth and shape. Deadfurrows placed in the channels with backfurrows on the ridges between drains will maintain and accentuate the drainage system. After the first few years, it may be advisable to shift the positions of the deadfurrows and backfurrows to avoid developing too deep a drain or too high a ridge. Methods of terrace and grass waterway maintenance also apply to field drains. Land-smoothing operations may be required for several years or even every year, since fill soil has a tendency to settle. Small ridges across the slope or small depressions resulting from improper tillage can usually be removed by land smoothing.

LAND GRADING

Land grading produces a plane land surface with a continuous slope. It is the same as land leveling. A field properly graded for surface drainage may also be suitable for surface irrigation. Land grading is normally a necessary operation in conjunction with the surface drainage systems previously described.

In grading the land, the high areas are cut down and the spoil is moved into the low sections. Reduced plant growth may occur on fill areas, although the exposure of subsoil in the cuts is usually more serious. Precision in establishing the desired elevation is extremely important, because the slopes are usually a few tenths of a percent. A specified design slope and accuracy of grading are more important for irrigation than for drainage. A variable slope for drainage is not objectionable, provided it is continuous without reverse grade.

12.9   Earthwork Volumes

Earthwork volumes are computed from cuts and fills taken from a field survey with ground elevations taken to the nearest 0.1 ft on a 100-ft square grid for horizontal control. Elevations are taken at other critical points, such as highs and lows between grid stakes and in the outlet ditch.

The elevations are then recorded on a field map, as illustrated in Fig. 12.10. For ease of calculating earthwork volumes, the grid points along the bottom and left sides of the field were established 50 ft from these base lines. By plotting the ground profiles in the direction of the desired drainage as shown in Fig. 12.11, cuts and fills can be determined. A desired slope for each line is established by trial and error, which will provide an approximate balance between cuts and fills as well as reduce haul distances to reasonable limits. Cuts and fills are estimated graphically from the original ground surface and the design profile. If cross-slope drainage is desired, profiles may also be plotted at right angles to the original lines, that is, along lines 1, 2, etc. in Fig. 12.10. The profiles shown in Fig. 12.11 are normally plotted directly on the map, which has been drawn on graph paper with 10 lines per inch. An approximate method of determining earthwork volume is to assume that the cut or fill at a grid point represents the average for 50 ft from the point in four directions (see Example 12.2). More accurate methods are described in Schwab et al. (1993).

Experience has shown that, in land grading or leveling, the cut-fill ratio should be greater than 1. Compaction from equipment in the cut area that reduces the volume as well as compaction in the fill area that increases the fill volume needed are believed to be the principal reasons for this effect. On level ground between stakes, the operator has an optical illusion of a dip in the middle, and therefore in filling, crowning often occurs. The ratio of cut to fill volume for various soil conditions is shown in Table 12.1. These ratios should be adjusted where local conditions dictate. Land-grading costs are normally based on the volume of the cuts.

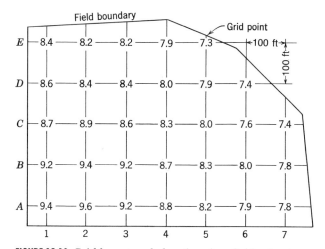

FIGURE 12.10 Grid layout and elevations in a field prior to land-grading operations.

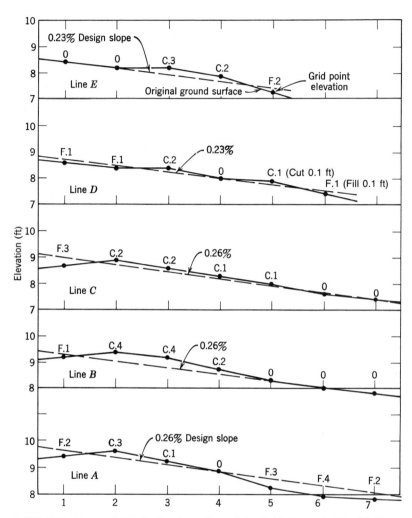

FIGURE 12.11 Cuts and fills for land grading of the field shown in Fig. 12.10.

Example 12.2. Determine the volume of cuts and fills and the cut-fill ratio for the field in Fig. 12.10.

Solution. Using the cuts and fills from the profiles plotted in Fig. 12.11 and assuming that each value represents a full grid square, the volume of cuts is (start with line *A*, cut at *A*2 is 0.3, *A*3 is 0.1, and on line *B*, *B*2 is 0.4, etc.)

$$V_c = (0.3 + 0.1 + 0.4 + 0.4 + 0.2 + 0.2 + 0.2 + 0.1 + 0.1 + 0.2 + 0.1 \\ + 0.3 + 0.2) \times (100 \times 100) \\ = 28\,000 \text{ ft}^3 \ (1037 \text{ yd}^3 \text{ or } 793 \text{ m}^3)$$

TABLE 12.1 Suggested Cut-Fill Ratios for Land Grading

Soil Conditions	Ratio: Volume of Cut to Volume of Fill
Organic soils	
Cuts less than 0.4 ft	2.0
Cuts greater than 0.4 ft	1.7
Clay loam soil	1.2 to 1.45
Clay soils	1.3 to 1.4
Medium-textured soils	1.2 to 1.25
Sandy soils, compacted	1.1 to 1.5
Bottomland soil in Mississippi and Arkansas	1.5

Source: Coote and Zwerman (1970) and other references.

The volume of fill is (start with line *A*, *A*1 is a fill of 0.2, *A*5 is 0.3, *A*6 is 0.4, *A*7 is 0.2, and on line *B*, *B*1 is 0.1, etc.

$$V_f = (0.2 + 0.3 + 0.4 + 0.2 + 0.1 + 0.3 + 0.1 + 0.1 + 0.1 + 0.2)$$
$$\times (100 \times 100)$$
$$= 20\ 000\ \text{ft}^3\ (741\ \text{yd}^3\ \text{or}\ 566\ \text{m}^3)$$

The cut-fill ratio is

$$C/F\ \text{ratio} = 28\ 000/20\ 000 = 1.4$$

This ratio may be increased by lowering one or more of the design profiles or by changing the slope to increase cuts and decrease fills. The opposite changes will decrease the ratio.

A more accurate method of computing earthwork volume for land grading is the four-point method. The volume of cuts or fills is computed for each grid square separately, and they are added together to obtain the total. By this method, the volume of cuts for each grid square is

$$V_c = \frac{L \times L(SC)^2}{4(SC + SF)} \tag{12.1}$$

and the volume of fills is

$$V_f = \frac{L \times L(SF)^2}{4(SC + SF)} \tag{12.2}$$

where L = length of one side of the grid in feet
      SC = sum of the cuts on a grid square in feet
      SF = sum of the fills on a grid square in feet

The preceding equations apply to a rectangular grid, provided the appropriate lengths are known as well as the cuts or fills for the four corners. If a grid has all cuts or all fills, the volume is simply the product of the average cut or fill and the grid area. The volume in cubic yards can be obtained by dividing each equation by 27 (cu ft/cu yd). Although the total cut and fill volume calculations for a field would be time-consuming for hand computation, a solution can be obtained on a computer.

Example 12.3   Determine the cut and fill volume by the four-point method for the 100-ft grid square $A1$, $A2$, $B1$, $B2$ in Fig. 12.10 using the cuts and fills for this grid square shown in Fig. 12.11.

Solution.   Read from Fig 12.11, $F.2$, $C.3$, $F.1$, and $C.4$. $SC = 0.4 + 0.3 = 0.7$, $SF - 0.1 + 0.2 = 0.3$, and $SC + SF = 0.7 + 0.3 = 1.0$.

Substituting in Eqs. 12.1 and 12.2,

$$V_c = \frac{100 \times 100(0.7)^2}{4 \times 1.0} = 1225 \text{ cu ft } (34.7 \text{ m}^3)$$

$$V_f = \frac{100 \times 100(0.3)^2}{4 \times 1.0} = 225 \text{ cu ft } (6.4 \text{ m}^3)$$

To obtain the volume of cuts and fills for the entire field in Fig. 12.10, the sum of the full grid squares as well as the partial areas around the boundary is required. This procedure is described in detail in Schwab et al. (1993).

12.10   Laser Surveying

Elevations of the land surface may be obtained with the scraper receiving unit and laser transmitter shown in Fig. 12.12. The receiving unit mounted on the scraper references the scraper blade height from the laser plane beam. With the scraper blade just touching the field surface and using the survey mode of the laser system, the operator drives the equipment back and forth across the field along lines for which the elevations are desired. The elevations may be hand-recorded at the desired points or automatically recorded for direct use by a computer. Some computer programs will determine the plane of best fit, a cut-fill map, earthwork vol-

FIGURE 12.12 Laser transmitter and scraper with laser receiver for grade control. (Courtesy Spectra-Physics, Laserplane, Inc., Dayton, Ohio.)

umes, and three-dimensional and profile views of a field. With the laser system, grid points do not have to be staked or preserved as with the operator-controlled system.

12.11   Land-Grading Operations

In land-grading operations, the major earthwork is accomplished with heavy scrapers or pans equipped with laser grade control systems for accurate depth control. Grid points may be marked in the field with the cut or fill as noted in Fig. 12.13. The general direction of earth movement is noted by the long arrows. Small islands left by avoiding the grid stakes and other minor irregularities on the surface are removed by land smoothing with a land plane or with laser-controlled equipment.

12.12   Land Smoothing

Land smoothing is the practice of removing small surface irregularities on the land surface after land grading with scrapers and other heavy equipment. A land plane or land leveler is required for land smoothing. They are made with blades up to 15 ft wide and wheel bases up to 90 ft. The blade on the land plane is mounted midway on the frame and is adjustable vertically so that the depth of cut and amount of soil carried can be regulated. With laser grade-control systems, land-grading equipment may be suitable for land smoothing as weil as grading. The purpose of land smoothing is to make a uniform plane surface either for surface drainage or for the uniform distribution of irrigation water. The smoothing operation is ordinarily done in the field after grading without detailed surveys or plans.

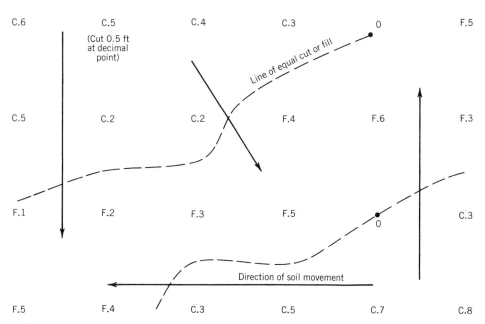

FIGURE 12.13 Cut and fill map showing directions for earth movement. (Redrawn from Anderson et al., 1980.)

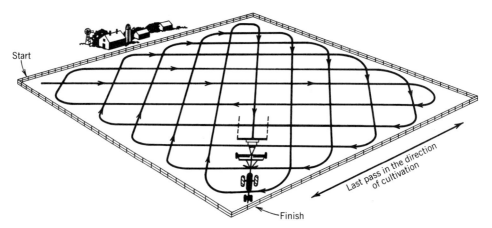

FIGURE 12.14 Procedure for land-smoothing operations. (Source: Drablos and Moe, 1984.)

A smoothing operation consists of a minimum of three passes with the leveler. The first two passes are made on opposite diagonals as shown in Fig. 12.14, and the last path is made in the direction of cultivation. The soil should be relatively free of crop residue or trash to allow a good distribution of the soil in the depressions. Because fill areas continue to settle and farming with large equipment may produce surface irregularities, land smoothing may be required annually or every few years. Scheduling the operation when soil and crop conditions are suitable is often a problem.

WETLANDS

New state and federal laws in response to public concern dictate that wetlands are an important land use and should be preserved. Some of this legislation is discussed in Chapter 1. Since about 1990, the policy and goal of the federal government is that there shall be "no net loss" of wetlands. These laws are interpreted and enforced by several federal agencies, including the Army Corps of Engineers, the Environmental Protection Agency, and the Department of Agriculture. A variety of groups have an interest in wetlands, each with differing views regarding valid uses of these areas. These groups include farmers; wildlife, conservation, and recreation organizations; developers; financial institutions; land-use planners; local and state governments; and others. During the present period of regulatory change and competing interests, much controversy exists. What constitutes a wetland and how the land is acquired and/or controlled are two important questions.

12.13 Definition of Wetlands

Wetlands are defined as areas that have a predominance of hydric soils, and that are inundated or saturated by surface or ground water at frequency and duration sufficient to support, and under normal circumstances do support a prevalence of hydrophytic vegetation typically adapted for life in saturated soil conditions, except lands in Alaska identified as having a high potential for agricultural development and a predominance of permafrost soils. This definition is given in the Food Security Act of 1985 and in the U.S. Army Corps of Engineers (1987) Wetland Delineation Manual. All federal agencies have adopted this definition. Hydric soils have anaerobic conditions and are listed by soil type in USDA (1991). These soils are usually dark in color with mottling. Other indicators include chemical deposits associated with the absence of oxygen. Evidence of flooding includes fine sediment deposits on the soil surface or on plants. On the soil surface, rafted debris may often be present at the margin of flooding. Aerial photography and satellite imagery are excellent techniques for identifying wetlands, both in extent and time. Wetlands generally occur on land in close proximity to lakes, streams, ponds, swamps, and bodies of water. In glaciated areas, potholes or low spots often indicate wetlands. Wetlands usually have ponded water for at least 1 to 2 weeks during the growing season, and many are flooded the entire year. These and other methods of identifying and delineating wetlands are described by Lyon (1993), FICWD (1989), and USACE (1987).

12.14   Benefits of Wetlands

The benefits of land for food production, human habitation, industrial production, transportation, forests, recreation, and similar uses are more evident and publicly recognized than those for wetlands. These land-use areas often include wetlands, such as water-storage reservoirs, borrow pits, ponds, lakes, etc. Some of the benefits from natural and constructed wetlands include: (1) provides life support and habitat diversity for flora and fauna, (2) provides services to human society, such as aesthetics, hunting, fishing, and recreation, (3) reduces the loss of sediment, nutrients, pesticides, toxic chemicals, etc. from point and non-point sources, thereby improving water quality downstream, (4) reduces flood peak flows downstream, (5) provides areas for the disposal of waste water if appropriate requirements and water quality standards are met, (6) has the potential for recharging ground-water supplies, and (6) enhances the treatment of acid mine drainage water.

When wetlands are large, they have the potential to store large quantities of flood water and may prolong long-duration stream flow similar to a spring or seepage from ground water. Past experience with flood control reservoirs has shown that their large-storage capacity is effective in reducing flood peak flows immediately downstream. Because wetlands are not normally provided with outlet control structures, reduction of peak flood flows may not be as much as with flood control reservoirs. Reduction in flood peak flows from wetlands is subject to speculation and disagreement among hydrologists. Studies in Ohio have shown that subsurface drainage has the potential to reduce flood peaks and the number of floods compared to poorly drained land. In Iowa, flood flows over a 20-year period on the Des Moines and Iowa Rivers were negligibly affected by the drainage of wetlands.

12.15 Wetland Development

In agriculture, development of wetlands has become a questionable and controversial issue because landowners traditionally have drained wetlands to increase crop production. Under the "Swampbuster" act, some farming of wetlands has been sustained. The U.S. Natural Resources Conservation Service, which has the responsibility for wetlands on farmland, has divided agricultural wetlands into (1) prior converted cropland, (2) farmed wetland, (3) farmed wetland pasture, and (4) wetland seasonally cropped. Where subsurface drains have been installed prior to current wetlands legislation, farming is generally allowed to continue as is maintenance to the existing drainage system. Some uncertainty exists regarding the farming of small depressions or wet spots in a large field and improvements to the existing drainage system. Farmland, purchased prior to the passage of wetland legislation, may have been overvalued because of its potential value for agriculture. Thus, the argument could be made that by reclassifying an area as wetland, such land is being devalued without due compensation.

Jurisdictional wetlands have also attracted the attention to the public, other than in agriculture. Wetlands subject to state or federal government regulation (jurisdiction) may be referred to as jurisdictional wetlands. All landowners and public agencies must obtain wetland assessments before property is developed for any other use, including residential, roads, utilities, industrial land, recreational sites, shopping malls, etc. Despite the current level of enforcement by regulatory agencies, illegal landfills still occur. Under the current policy, some wetlands may be allowed to be modified if they are constructed or set aside at another location, nearby, or on the developed area. Current laws do not restrict the construction of artificial wetlands or the improvement of existing wetlands. Thus, wetlands may be either natural or constructed. This distinction is important because regulations for constructed wetlands are sometimes less rigid than those for natural wetlands. The real issue is one of conflicting land use that should be resolved on a scientific basis of need rather than political decisions. With the growing pressure of a greater population with time, wetland policy will become of increasing importance.

REFERENCES

American Society of Agricultural Engineers (ASAE) (1986) *Design and Construction of Surface Drainage Systems on Farms in Humid Areas,* EP302.3. ASAE, St. Joseph, Michigan.

ASAE (1988) *Agricultural Drainage Outlets—Open Channels,* EP407. ASAE, St. Joseph, Michigan.

Anderson, C. L., A. D. Halderman, H. A. Paul, and E. Rapp (1980) "Land Shaping Requirements," Chapter 8. In M. E. Jensen (ed.), *Design and Operation of Farm Irrigation Systems,* ASAE Monograph. ASAE, St. Joseph, Michigan.

Beauchamp, K. H. (1952) *Surface Drainage of Tight Soils in the Midwest,* Agr. Eng. 33:208–212.

Coote, D. R., and P. J. Zwerman (1970) *Surface Drainage of Flat Lands in the Eastern United States,* Cornell University Ext. Bull. 1224.

Drablos, C. J. W., and R. C. Moe (1984) *Illinois Drainage Guide,* University of Illinois Ext. Cir. 1226.

Federal Interagency Committee for Wetlands Delineation (FICWD) (1989) *Federal Manual for Identifying and Delineating Jurisdictional Wetlands.* Government Printing Office, Washington, D.C.

Gain, E. W. (1964) "Nature and Scope of Surface Drainage in Eastern United States and Canada," *ASAE Trans.* 7(2):167–169.

Lyon, J. G. (1993) *Practical Handbook for Wetland Identification and Delineation.* Lewis Publishers, Boca Raton, Florida.

Schwab, G. O., D. D. Fangmeier, W. J. Elliot, and R. K. Frevert (1993) *Soil and Water Conservation Engineering,* 4th ed. Wiley, New York.

U.S. Army Corps of Engineers (USACE) (1987) *Wetland Delineation Manual,* Tech. Report Y-87-1. Dept. Army, Washington, D.C.

U.S. Department of Agriculture (USDA) (1991) *Hydric Soils of the United States.* Misc. Publ., Washington, D.C.

PROBLEMS

12.1 Determine the volume of excavation in cubic yards for a surface ditch with $10:1$ side slopes if the cuts at successive 50-ft stations are 0.5, 2, 4, 3, and 0 ft. Use the average end area between stations to compute the volume.

12.2 Compute the volume of soil to be excavated in cubic yards for a random field ditch with $8:1$ side slopes if the cuts at consecutive 50-ft stations are 0.5, 1.0, 2.4, 1.8, 0.8, and 0 ft.

12.3 If the depth of cut at station $0+00$ in fig. 12.5 is reduced to 0.1 ft and the grade is increased to 0.3 percent, compute the cut at each station and the total volume of excavation.

12.4 From the following survey notes, determine the depth of cut at each station for a surface ditch that is to remove water from a pothole. The slope of the drain from $0+00$ to $1+00$ is 0.2 percent. Select a uniform slope from $1+00$ to $2+00$ so that the cut at $2+00$ is zero. Compute the slope.

Sta.	B.S.	F.S.	Elev.	Cut
BM 1	4.6		50.0 ft	
$0+00$		6.2		0.2 (bottom of pothole)
$0+50$		5.5		
$1+00$		3.0		
$1+50$		4.6		
$2+00$		7.0		0

12.5 Raise the design profile on line *C* in Fig. 12.11 by 0.1 ft. Then compute the cut-fill ratio as in Example 12.2. Compute the new volume of cut and volume of fill. If the contractor charges $1.50 per cubic yard of cut, what is the cost for grading the field?

12.6 Plot the ground elevations in Fig. 12.10 along the vertical lines, that is, on lines 1, 2, etc. By selecting nearly uniform design slopes, determine cuts and fills so as to give a cut-fill ratio of about 1.3. Compute the volume of cuts and fills.

12.7 Determine cuts and fills for a field assigned by the instructor for a given cut-fill ratio. Compute the volume of cuts and fills in cubic yards.

12.8 Compute the cut and fill volume in cubic yards for a 100×100-ft grid square if two cuts are 0.2 and 0.5 ft and two fills are 0.4 and 0.2 ft.

12.9 Compute the volume of cuts, volume of fills, and the cut-fill ratio for the data in Fig. 12.13 using the four-point method. Assume the grid is 100×100 ft.

CHAPTER 13 _____
SUBSURFACE DRAINAGE
_____

Plants need air as well as water in their root zones. Excess water that is free to move to subsurface drains generally retards plant growth, because it fills soil voids and restricts aeration. For most cultivated crops, adequate surface drains are needed to remove excess ponded water for flat or undulating topography. Subsurface drains are required to remove excess water for soils with poor internal drainage and a high water table. Soils that do not have an impermeable layer below the root zone but have adequate internal drainage do not need pipe drains. For maximum productivity of most crops, both surface and subsurface (pipe) drainage are essential, regardless of whether drainage is man-made or natural. Surface drainage systems may give a greater return per invested dollar than a higher-cost subsurface system.

Subsurface drainage is needed for many other purposes than agriculture, such as drainage of roads, embankments, athletic fields, buildings, city lots, parks, recreational areas, etc. The same principles can be applied as described for agriculture in this chapter.

13.1 Benefits of Subsurface Drainage

Subsurface drainage increases crop yields by (1) removing gravitational or free water, (2) increasing the volume of soil from which roots can obtain nutrients, (3) increasing the movement and quantity of air in the soil, (4) providing conditions that permit the soil to warm up faster in the spring, (5) increasing the bacterial activity in the soil, which improves soil structure and makes nutrients more readily available, (6) reducing soil erosion, since a well-drained soil has more capacity to hold rainfall, resulting in less runoff, and (7) removing toxic substances, such as sodium and other soluble salts that in high concentrations retard plant growth. Other benefits that may not necessarily result in increased yields include a reduction in time and labor for tillage and harvesting operations. With a crop such as corn, a delay in planting date will decrease yields. Planting in wet soils is likely to decrease plant stands. A delay in harvesting time will increase field losses as well as grain damage.

As illustrated in Fig. 13.1, subsurface drains enhance deep root development by lowering the water table, especially during the spring. A plant with a deep root system can withstand droughts better than shallow-rooted plants,

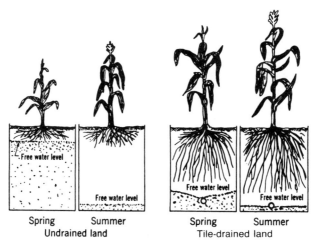

FIGURE 13.1   Root development of crops grown on drained and undrained land. (Redrawn from Manson and Rost, 1951.)

because larger quantities of water and plant nutrients are available when needed. A large root system is also desirable later in the season when the plant is larger and transpires more water.

13.2 Environmental Impacts of Drainage

As food requirements expanded during the past century, much land was drained for agriculture, thus reducing the wetland area by a large percentage. This loss of natural ecosystems has reduced the habitat for birds and wildlife as well as water storage areas that accumulate sediment and nutrients. In recent years, legislation and economic forces have encouraged the preservation of wetlands and have essentially stopped conversion of wetland to farmland (Chapter 1). Federal legislation and regulations (1994) support the concept that there should be "not net loss" of wetlands (Chapter 12). Methods of evaluating and assessing the environmental impacts of drainage are discussed by Ritzema and Braun (1994).

Increasing the degree and extent of drainage in humid areas has both negative and positive impacts on hydrology and water quality. Subsurface drains lower the water table, thus increasing the pore space, which allows for greater infiltration and storage of water in the soil profile. In Ohio, studies showed that peak runoff rates (flooding) from small plots were reduced 7 percent by subsurface drains, and the number of floods was reduced by 46 percent. Similar results were obtained by computer modeling of small watersheds in North Carolina. These results may not be consistent for all soils and hydrologic conditions. Subsurface drains reduce eroded sediment by reducing surface runoff rates and volumes. Some pollutants may be reduced, whereas others may be increased. Subsurface drainage generally reduces the loss of phosphorus, organic nitrogen, potassium, and pesticides, whereas the loss of nitrate-nitrogen and soluble salts is usually increased. Drainage water may have a lower or higher acidity than rainwater; for example, in an Ohio study pH was increased from 6 in rainwater to 7.1 from pipe drains. In the eastern Coastal Plains of the United States, shallow ground water may have a pH as low as 4.

Subsurface drainage is required for many irrigated arid lands to prevent the rise of the water table, waterlogging, and salinity build-up in the soil (Chapter 15). Natural or artificial subsurface drainage of excess soil water is the only practical way to remove salts. Sodium as well as trace elements, such as selenium, boron, molybdenum, and arsenic, have potentially serious impacts on the environment, including the death of aquatic organisms and waterfowl (Chapter 15). Storage of drainage water in a closed reservoir where chemicals may be concentrated by evaporation may cause serious problems. Discharging saline drainage water directly to the ocean (as has been done in Pakistan with a 175-mile ditch) is a possible solution.

Pollutant loads from drainage systems can be minimized by proper selection of the water management system, ranging from controlled drainage and subirrigation to cultural and structural measures. In North Carolina (Skaggs et al., 1994), controlled drainage and subirrigation (Chapter 15) reduced nitrate-nitrogen and phosphorus losses by 45 and 35 percent, respectively. Nitrate losses can also be reduced by using less fertilizer, by applying slow-release fertilizers, and by timing applications to meet the needs of the plant. No-tillage or minimum-tillage cropping systems will greatly reduce sediment loss and (to some extent) the loss of nitrates and other fertilizers, but herbicide losses may not be reduced. Although significant advances have been made in recent years, much has yet to be learned about the complex mechanisms of the movement of pollutants from drained soils.

13.3 Movement of Water into Subsurface (Pipe) Drains

Permeability and hydraulic conductivity both refer to the movement of water through the soil. Permeability is a more general term, whereas hydraulic conductivity specifically defines the rate of flow as in Eq. 13.1. Mathematically, it is directly related to the soil pore space and inversely related to water viscosity. The two terms are often used interchangeably. Hydraulic conductivity can be measured from soil cores in the laboratory or directly in the field, although both methods are time consuming and do not easily produce consistent results.

A uniformly permeable soil overlaying an impermeable layer below the drains is shown in Fig. 13.2. After several hours of drainage following rainfall or irrigation, the water table will be drawn down nearly to the drain directly over the pipe and be nearest the surface midway between parallel drains. This drawdown shape develops because the water travels a shorter distance, and the velocity is greater near the drain than at the midway point. Since flow is caused by gravity, the drains must be below the water table.

The slope of the ground-water surface at any point (x, y) is proportional to

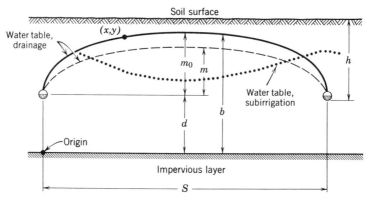

FIGURE 13.2 Water table between parallel subsurface drains from drainage and subirrigation. (Source: Schwab, G. O. et al., 1993.) (Reprinted with permission of John Wiley & Sons, Inc.)

the horizontal velocity of flow to the drain. With time, the water table will drop from one level to a lower one, as shown by the dashed line. When the drain is running full and the pipe is completely porous, the path of the water near the drain follows a radial direction toward the center of the drain. With perforations (as in plastic tubing) or circumferential joints (as in clay or concrete tile), the flow is restricted and converges to the opening. This restriction is not serious if the area of the openings is greater than about 1 percent of the outside area of the pipe.

The principles of water flow determine the appropriate spacing for parallel drains (Figs. 13.3b and 13.3c). Many theoretical and computer drain spacing solutions have been developed, which can be justified because the cost of subsurface drainage is directly related to the spacing.

Soil permeability and spacing of the drains have the greatest effect on the rate of fall or rise of the water table. Since water must move greater distances horizontally than vertically to reach the drain, the horizontal permeability is the more important. The permeability of most soils decreases with depth. This change in permeability affects the shape of the flow lines and the rate of rise or fall of the water table.

13.4 Controlled Drainage and Subirrigation Systems

In humid areas of the United States, the trend in design practice is to develop a total water management system. Such systems have been working for years in peat and muck soils with high permeability and with an impervious layer below the drains or with a naturally high water table. This practice is known as controlled drainage. Several water-control structures similar to those shown in Fig. 13.4 may be required in series, especially where the land has a gentle slope or is rolling. Commercial units are also available. Valves may be controlled automatically and even from remote locations, but most are manually operated. When the water table is lower than desired (as during a dry season), water must be

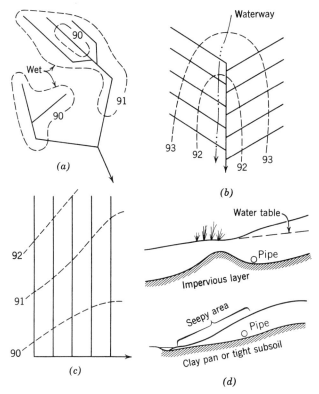

FIGURE 13.3 Common types of subsurface drainage systems: *(a)* random, *(b)* herringbone, *(c)* gridiron, *(d)* interceptor.

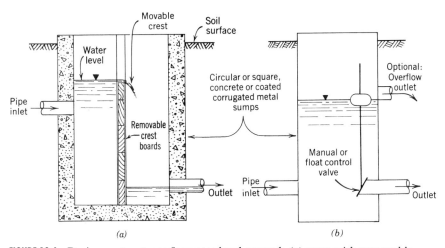

FIGURE 13.4 Drainage structures for water level control: *(a)* sump with removable crestboards (also called stop logs or flash boards), *(b)* sump with manual or float valve controls for pipe drains. (Source: Schwab, G. O. et al., 1993.) (Reprinted with permission of John Wiley & Sons, Inc.)

pumped into the existing drains. If pumping is required, the practice is referred to as subirrigation.

With controlled drainage and subirrigation, water is supplied to the soil through the drains, the flow is reversed, and the resulting water table is higher directly above the drain than midway (as shown in Fig. 13.2). In both the drainage and subirrigation mode, the most critical area is midway between the drains. Studies have shown that the spacing for a subirrigation system should be as much as 30 percent less than for subsurface drainage alone. This difference in spacing requirement indicates that existing drainage systems (designed for drainage) may not be the most efficient for controlled drainage or subirrigation.

With controlled drainage, the water table tends to be flat between the drains under static conditions. During periods of rainfall, the water table height, m, would tend to be higher midway between the drains because some water would be removed by the drains. If in the subirrigation mode, irrigation continues during rainfall, the water table would rise uniformly between the drains, but maintain a higher level over the drains.

13.5 Subsurface Drainage Systems

Random, herringbone, and gridiron systems of layout are shown in Fig. 13.3. The topography to which each of these systems is adapted is indicated by the dashed contour lines. The random system is suitable where the field is not to be completely drained. It is flexible in layout, permitting the location of drains where they are most needed. Considerable cost savings are possible with proper location and layout.

In most large installations, several of the layout patterns may be found. The drains are normally located so that the land surface along the pipes has some slope toward the outlet. Interceptor drains shown in Fig. 13.3d are laid out nearly on the contour and are placed to drain as much of the seepy area below as possible. Extensive soil borings may be required to locate the impermeable layer above which the ground water is flowing.

Where grass waterways or channels are wet from seepage, pipe drains should be placed to one side of the center line of the channel as shown in Fig. 13.3b. Such a location will more likely intercept the seepage and avoid erosion along the trench from runoff, especially before the backfill has settled. Where both sides of the channels are wet, a main drain along both sides may be necessary to intercept the seepage. Locating the impermeable layer and the source of seepage is important for the proper location of the drains. Herringbone and gridiron patterns (Figs. 13.3b and c) are installed where the entire area is to be drained. The drains are located to conform to the topography of the land.

13.6 Computer Modeling with DRAINMOD

A water management simulation computer model, DRAINMOD, has been developed by Skaggs (1980) in North Carolina. It is a complex model involving more than 20 parameters. When it is properly calibrated, it will predict crop yields for any combination of parameters desired. DRAINMOD is based on a water balance in the soil profile. It uses climatological data on an hour-to-hour basis to predict the response of the water table (and the soil water regime above it) to various combinations of surface and subsurface water management practices. By simulating the performance of alternative systems over several years of record, an optimum system can be selected on the basis of probability. The model is composed of a number of separate components including methods to evaluate infiltration, subsurface drainage, surface drainage, evapotranspiration, subirrigation, and soil water distribution.

DRAINMOD can predict trafficability in the spring for tillage and planting operations, the time available for harvesting, the hydraulic loading capacities for land application of wastewater, and the plant response to wet and dry stress during the growing season. Plant response to drainage, adjusted for different stages of plant growth, is evaluated by the height and the duration of the water table, called SEW (sum of excess water) values.

The model has been validated using measured crop yields for many soils. The computer program is operational on mainframe computers, state Natural Resources Conservation Service computers, and personal computers. The details are beyond the scope of this book (Skaggs, 1980).

13.7 Outlets

The failure of many drainage systems is caused by faulty outlets. As shown in Fig. 13.5, corrugated metal or similar pipe with a flap gate to prevent entry of small animals is recommended. If there is a danger of flood water backing up into the drain, a flood gate may be installed in place of the flap gate.

Where the depth of the outlet ditch is inadequate, an automatically controlled pump with a small sump for storage, as shown in Fig. 10.11, can be installed. Before considering such an installation, deepening of the outlet ditch should be investigated. Pump outlets require operation and maintenance costs that must be compared to those for the ditch excavation and maintenance. Where outlet ditches must pass through the land of others, right-of-way costs and legal problems often necessitate the construction of pump outlets.

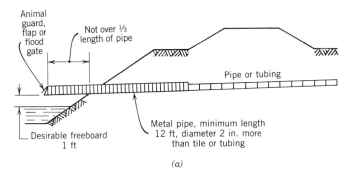

(a)

FIGURE 13.5 Gravity pipe drain outlet into an open ditch.

13.8 Depth and Spacing

Because of variable soil and climatic factors, it is difficult to make specific recommendations for the proper depth and spacing of drains. The depth and spacing of drains apply only to laterals and have nothing to do with the depth of the mains or submains. Depths for mains are governed by outlet conditions and topography of the drainage area.

Deep drains are more effective than shallow ones in soils that do not have a claypan or tight subsoil at a depth less than the drain. Two feet of soil over the top of the drain is a minimum to prevent damage from farming activities. Unless there is a tight subsoil, drains should be placed at optimum depth to provide minimum drainage cost. This depth is always measured to the bottom of the drain.

There is a definite relationship between depth and spacing of drains. For soils with a uniform profile, the deeper the drains—the wider the spacing and vice versa. Average pipe drain depth and spacing for soils of different texture are given in Table 13.1. Local field practices and recommendations of the state experiment stations should be followed. Most states have drainage guides that recommend depth and spacings by soil types and crops to be grown.

Depth and spacing for humid areas are based on the rate of drop of the water table. In arid irrigated land, however, the maintenance of a low water table (to prevent the upward movement of toxic salts) and the removal of excess salts are the criteria for depth and spacing recommendations. Because soils in irrigated regions are generally more permeable than in humid areas, spacings of drains are wider, as shown in Table 13.1. The greater depth of drains is necessary to provide a lower water table than is usually required in humid regions. Mathematical equations have been developed for computing drain spacings for either the falling water table case or the constant water table (steady state) criteria.

One of many equations for estimating drain spacing is known as the steady state ellipse equation (first developed in 1872), which is

$$S = [4K(b^2 - d^2)/i]^{0.5} \qquad (13.1)$$

TABLE 13.1 Average Depth and Spacing of Pipe Drains for Field Crops

Soil	Hydraulic Conductivity		Spacing (ft)	Depth (ft)
	Class	iph		
Clay	Very slow	0.05	30–50	3.0–3.5
Clay loam	Slow	0.05–0.2	40–70	3.0–3.5
Average loam	Moderately slow	0.2–0.8	60–100	3.5–4.0
Fine sandy loam	Moderate	0.8–2.5	100–120	4.0–4.5
Sandy loam	Moderately rapid	2.5–5.0	100–200	4.0–5.0
Peat and muck	Rapid	5.0–10.0	100–300	4.0–5.0
Irrigated soils	Variable		150–600	5.0–8.0

where S = drain spacing in ft

K = soil hydraulic conductivity in ft/day

b = depth from midway water table height to the impermeable layer (Fig. 13.2) in ft

d = depth from the center of the drain to the impermeable layer (Fig. 13.2) in ft

i = drainage or irrigation rate in ft/day

This approximate mathematical equation is suitable only where the spacing is large (compared with the depth to the impermeable layer) and the water table is constant. The flow resistance resulting from the convergence of flow lines near the drains is ignored, but it can be estimated by substituting an equivalent depth for d in the equation. The equivalent depth may be obtained from Schwab et al. (1993) and other references. Briefly, the procedure for applying Eq. 13.1 is to (1) determine the drainage or irrigation rate and the desired midway water table height, (2) estimate or measure the depth to the impermeable layer and the average hydraulic conductivity for the soil, (3) select a suitable drain depth, and (4) compute the spacing (by trial and error if necessary to obtain the equivalent depth). The ellipse equation has been applied more widely in the irrigated West than in humid areas. One of the most difficult factors to evaluate in any spacing equation or computer solution is the soil hydraulic conductivity, which has been, and will continue to be, the subject of much research.

Drain spacings can also be determined by using the simulation model DRAINMOD, previously described. For specialty or truck crops that produce a high income per acre, such as tobacco, fruit, and vegetables, closer spacings can generally be justified because drainage requirements and crop value are higher than for field crops. Spacings for field crops (Table 13.1) should be reduced as much as 25 to 50 percent for specialty crops. Pipe drainage is seldom justified for forage crops. For all crops in humid regions, pipe drainage is most beneficial during the spring months.

13.9 Grades

Maximum grades are limiting only where drains are designed for near maximum capacity or where they are embedded in unstable soil. Pipe embedded in fine sand or other unstable material may become undermined and settle out of alignment unless special care is taken to provide joints that fit snugly. Under extreme conditions it may be necessary to install bell and spigot tile, tongue and groove concrete tile, metal pipe, or nonperforated corrugated plastic tubing. On mains, steep grades up to 2 or 3 percent are not objectionable, provided the capacity at all points close to the outlet is equal to or greater than the pipe above.

A desirable minimum working grade is 0.2 percent. Where sufficient slope is not available, the grade may be reduced to 0.1 percent where fine sand and silt are not present. The minimum velocity at full flow where fine sand and silt are present should be 1.5 fps as computed from the Manning equation (Fig. 13.6 or Chapter 8).

13.10   Drainage Rates

Pipe drains are designed to remove water at a given rate, called the *drainage coefficient,* which is the depth of water in inches to be removed in a 24-h period from the drainage area. In humid regions, the drainage coefficient depends largely on the rainfall, but varies with the soil, crop, and degree of surface drainage. Recommended drainage coefficients are shown in Table 13.2.

In irrigated areas the discharge from drains may vary from about 10 to 50 percent of the water applied. Since not all of the area is irrigated at the same time, the entire drained area is not used to calculate drain flow. The drainage area is estimated from the area irrigated. Because of a wide range in flow, the drainage coefficient should be based on local recommendations.

13.11   Drainage Area

The drainage area is that actually drained by the pipe. Where surface water is to be removed by the drains, the watershed area is the drainage area (even though it may not be entirely pipe-drained). For a single drain, as in a random system, the width drained is the approximate pipe spacing for the soil.

13.12   Size of Drains

The size of drains depends on the pipe, the drainage area, the drainage coefficient, and the grade in the drain. The slope of the soil surface and the depth of the outlet can be determined by profile leveling. From these data the grade in the drain can be selected. Three inches is the minimum recommended size; however, 4-in. and 6-in. drains are considered minimum sizes in many areas. The minimum size for perforated tubing or pipes may be reduced because mis-

TABLE 13.2   Drainage Coefficients for Pipe Drains in Humid Regions

Crops and Degrees of Surface Drainage	Drainage Coefficient in ipd	
	Mineral Soil[a]	Organic Soil
Field Crops Normal[b]	$\frac{3}{8} - \frac{1}{2}$	$\frac{1}{2} - \frac{3}{4}$
With surface inlets	$\frac{1}{2} - 1$	$1 - 1\frac{1}{2}$
Truck Crops Normal[b]	$\frac{1}{2} - \frac{3}{4}$	$\frac{3}{4} - 1\frac{1}{2}$
With surface inlets	$1 - 1\frac{1}{2}$	$2-4$

Source: ASAE (1993).

[a]These values may vary, depending on special soil and crop conditions. Where available, local recommendations should be followed.

[b]Adequate surface drainage must be provided.

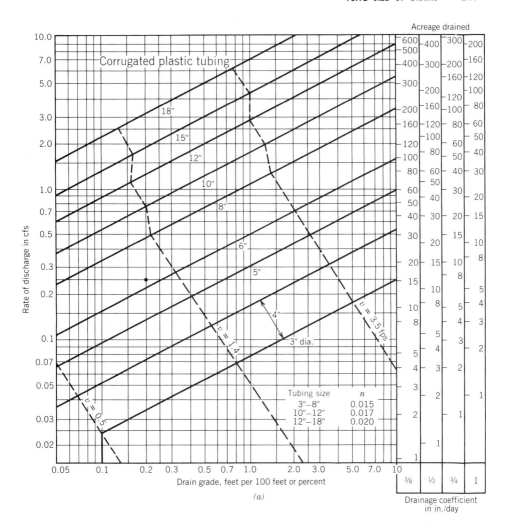

FIGURE 13.6 Pipe-size nomographs: *(a)* corrugated plastic tubing. (Source: Drablos and Moe, 1984.)

alignment at cracks or joints is not a problem. Six-inch pipe mains are often the minimum diameter recommended, as the greater capacity will reduce the duration of back flooding in the laterals when peak flows exceed the design rate.

Pipe size can be determined from Fig. 13.6, which is based on the Manning velocity equation discussed in Chapter 8. For example, if the drainage area is 15 acres, the drainage coefficient (D.C.) is $\frac{3}{8}$ in., and the slope is 0.2 percent, the required corrugated plastic tubing size is 8 in. diameter, as noted by the dot on Figure 13.6a. This dot is located at the intersection of the vertical line through 0.2 percent grade and the horizontal line through 15 acres in the $\frac{3}{8}$-in. column. Read the pipe size from the first upper diagonal line above the intersection. The discharge rate for the 15 acres at $\frac{3}{8}$ in. is about 0.24 cfs, as can

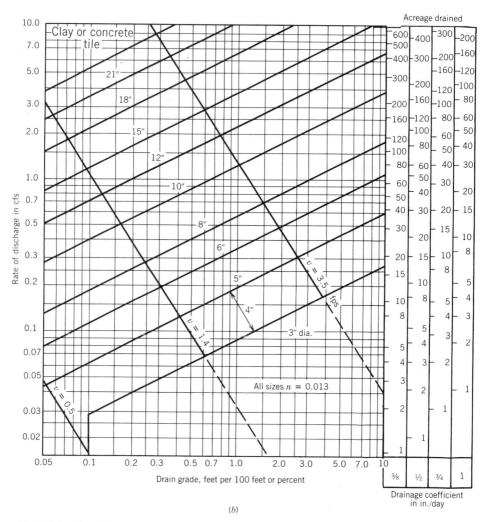

FIGURE 13.6 Pipe-size nomographs: (b) clay or concrete tile. (Source: Drablos and Moe, 1984.)

be read from the left-hand scale in Fig. 13.6a. The 8-in pipe will run only partly full, since its maximum capacity at full flow is about 0.48 cfs. This flow can be read on the left-hand scale at the intersection of the diagonal line above the 8-in. size and the 0.2 percent vertical line.

The pipe diameter may be computed directly from the Manning equation, from which

$$d = 16 \, (qn/s^{\frac{1}{2}})^{\frac{3}{8}}$$ (13.2)

where d = pipe diameter in inches
  q = required rate of discharge in cubic feet per second
  n = roughness coefficient (see Fig. 13.6)
  s = grade or slope of the drain in feet per foot

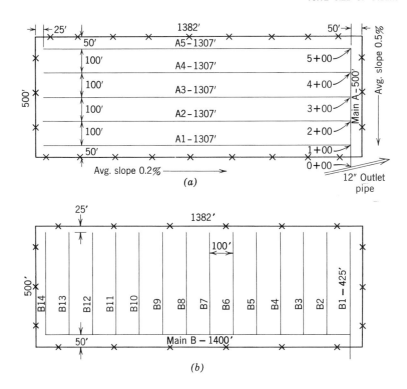

FIGURE 13.7 A comparison of main and lateral layouts for subsurface drainage systems.

| Plan | Number of Junctions | Length of Main (ft) | Length of Pipe (ft) | | | Total Length (ft) | Area Double Drained (acres) |
			8 in.	6 in.	4 in.		
(a)	5	500	200	200	6635	7035	0.5
(b)	14	1400	500	400	6450	7350	1.5

The rate of discharge may be read directly from the left-hand side of Fig. 13.6 or may be computed by multiplying the acreage by the D.C. and by the conversion factor 0.042. In the example in the preceding paragraph, $q = 15 \times 0.375 \times 0.042 = 0.24$ cfs. Substituting in Eq. 13.2 would give 6.2 in. diameter required for the drain. Since this size is greater than 6 in., the next largest available size is selected, which in the preceding example is 8-in. diameter.

Example 13.1   Determine the depth, spacing, drainage coefficient, and system to provide uniform drainage for a field of average loam soil in Georgia in which field crops are to be grown. No surface inlets are required. Size and shape of the field, general slope, and outlet are shown in Fig. 13.7*a*.

Solution.   From Table 13.1, for average loam soil, select a spacing of 100 ft and a depth of 4 ft (check with local recommendations). From Table 13.2, for normal surface drainage and for field crops, the drainage coefficient for determining the tubing size is $\frac{1}{2}$ in. Because the field is long and has adequate slope toward one corner, the gridiron system (Fig. 13.7*a*) with long laterals is most suitable. Long laterals reduce the amount of large pipe in the main and the number of junctions. For efficiency, the main and lateral are placed 50 ft from fence lines. The upper ends of the lateral lines should extend to within $\frac{1}{4}$ of the spacing from fence lines or 25 ft.

Example 13.2 For the system described in Example 13.1, determine the amount of each size pipe required. After staking and surveying was completed, the slopes were as follows: *A*1 and *A*2, 0.15 percent; *A*3, *A*4, and *A*5, 0.2 percent; and Main *A* 0.25 percent. Assume minimum size as 4-in.

Solution. The area drained by each lateral is $(1307 \times 100)/43\,560 = 3.0$ acres. Since the main is short and is partly drained by the laterals, the area drained by the main may be neglected. Determine the tubing size from Fig. 13.6*a* ($\frac{1}{2}$ in. D.C.) for each change in area drained or change in slope as in the following table.

Line	Area drained (acres)	Slope (percent)	Tubing size (in.)	Length (ft)
*A*5	3.0	0.2	4	1307
*A*4	3.0	0.2	4	1307
*A*3	3.0	0.2	4	1307
*A*2	3.0	0.15	4	1307
*A*1	3.0	0.15	4	1307
Main *A*				
4 + 00 to 5 + 00	3.0	0.25	4	100
2 + 00 to 4 + 00	9.0 max	0.25	6	200
0 + 00 to 2 + 00	15.0 max	0.25	8	200

The length of 4-in. tubing required is 6635 ft; 6-in., 200 ft; and 8-in., 200 ft. Since the amount of 4- and 6-in. tubing is small, all 8-in. size might be more practical for Main *A*.

Example 13.3 From Fig. 13.8 determine the clay tile size for line *A*1. The soil is clay loam and all surface water from the 8-acre drainage area is to be removed through the surface inlet. Field crops are to be grown.

Solution. From Table 13.2 select $\frac{3}{8}$- and $\frac{3}{4}$-in. D.C. for normal conditions and surface inlets, respectively. The maximum drainage area for *A*1 above station 5 + 62 is the area drained by laterals *A*1.1, *A*1.2, and *A*1.3, which is $(3 \times 60 \times 726)/43\,560$ or 3.0 acres. From Fig. 13.6*b* read a minimum size of 4 in. The 8-acre drainage area to the surface inlet with a D.C. of $\frac{3}{4}$ in. is equivalent to 16 acres at $\frac{3}{8}$ in. The total drainage area for line *A*1 must include a 60-ft strip below the surface inlet, which is $(562 \times 60)/43\,560$ or 0.8 acre. From Fig. 13.6*b* read a tile size of 8-in. for 16.8 acres, slope of 0.13 percent, and D.C. of $\frac{3}{8}$ in. The total tile for *A*1 is 562 ft of 8-in. and 164 ft of 4-in.

Note that the area drained by the three laterals in the depression is not considered in determining the drainage area for line *A*1 below the surface inlet. To do so would result in using the same area twice. The assumption is made that the pipe size will be adequate if the design is based on the D.C. for surface inlets.

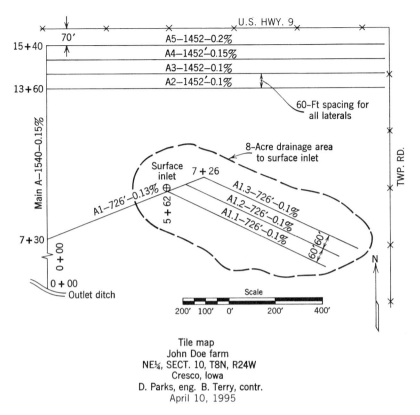

FIGURE 13.8 A map for a subsurface drainage system.

13.13 Envelope Filters

For drainage of irrigated land in the West, gravel envelopes are placed completely around the drain. These envelopes are installed (1) to prevent the inflow of soil into the drain and (2) to increase the effective pipe diameter, which increases the inflow rate. Because of practical considerations and cost, the envelope is usually limited to one gradation of material, which is selected from local naturally occurring deposits. The minimum thickness of the filter is 3 in. In humid areas this practice is normally not necessary. However, initial backfilling (blinding) with 6 to 12 in. of topsoil directly over the pipe is desirable. Such a practice will prevent misalignment and increase inflow. So-called "blinding" can be done with corncobs, cinders, or other porous material.

For keeping out fine, uniform sands, thin filter materials (called *geotextiles*) have been commercially developed. These materials are normally placed around the tubing at the extruding plant. These textiles may be made from nylon, polypropylene, and similar products in a variety of thicknesses and opening sizes. One type called a bean sock stretches giving variable sizes of openings. The opening size in the filter material should allow some of the fine soil particles to pass through so as not to plug the filter but retain the larger particles that tend to deposit in the drain. The maximum opening size is related to the soil particle size distribution, but the relationship has not yet been fully established for all soils.

13.14   Accessories

As shown in Fig. 12.6, a surface inlet is an intake structure for the removal of surface water from potholes, road, ditches, farmsteads, and the like. Surface drains (Chapter 12) are preferred, where possible, because of their greater capacity to remove flood water.

To allow surface water to reach the drain more quickly, short lengths of the trench may be filled with various gradations of sand and gravel. The coarsest material is placed immediately over the pipe, and the size is gradually reduced toward the surface. Short lengths backfilled in this manner are known as blind inlets.

Relief pipes and breathers are small vertical risers extending from the drain line to above the surface. The riser should be made of steel pipe or mortared bell and spigot tile and should be located at fence lines where they are not likely to be damaged. Relief pipes serve to relieve the excess water pressure in the pipe during periods of high outflow, thus preventing blowouts. A relief pipe should be installed where a steep section of a main changes to a flat section, unless the capacity of the flat section exceeds the capacity of the steep section by 25 percent.

13.15   Selection of Drain Pipe

Drain pipes are made from clay or concrete in short lengths and smooth plastic, steel, or corrugated plastic perforated tubing. Good quality concrete tiles are very resistant to freezing and thawing but may be subject to deterioration in acid and alkaline soils. In these soils, concrete tiles should be used only if they are approved by local recommendations. Clay tiles are not affected by acid or alkaline soils. When subjected to frequent alternate freezing and thawing conditions, concrete tile is safer to use, although most clay tile are resistant to frost damage.

Good clay or concrete drain tile should have the following characteristics: (1) resistance to weathering and deterioration in the soil, (2) sufficient strength to support static and traffic loads under conditions for which they are designed, (3) low water absorption, that is, a high density, (4) resistance to alternate freezing and thawing, (5) relative freedom from defects, such as cracks and ragged ends, and (6) uniformity in wall thickness and shape. Drain tiles that meet current specifications of the American Society for Testing Materials (ASTM) have the essential qualities listed previously. Specifications have been prescribed for three classes of drain tile, namely, standard, extra-quality, and special-quality (concrete only) or heavy-duty (clay only). Standard-quality tile is satisfactory for drains of moderate size and depth found in most farm drainage work.

Corrugated plastic tubing, first manufactured in the United States about

1967, increased in usage to 95 percent of all agricultural drains by 1983. It is lightweight and unaffected by all soil chemicals. The weight of 4-in. diameter plastic tubing is only about $\frac{1}{25}$ of that for clay or concrete tile. Tubing is made in sizes varying from 1 to 36 in. in diameter as well as in black, white, yellow, and other colors. Small size tubing is coiled in lengths up to several thousand feet. The most common resins are high density polyethylene (HDPE) and polyvinyl chloride (PVC). In the United States and Canada, HDPE has the greater share of the market, whereas in Europe, PVC is more common. Tubing should meet ASTM Standards F405 or F667 for deflection and elongation. These standards specify either standard or heavy-duty quality.

13.16   Allowable Drain Pipe Depths

Frequently, deep trenches are necessary for mains, and the soil load on the pipe may cause failure. Allowable depths for standard and extra-quality clay or concrete tile are given in Table 13.3.

The width of the trench at the top of the pipe governs the load, regardless of the trench width above the top of the pipe. Thus, a trench 2 ft wide at the top of the pipe and 5 ft wide at the surface would result in a load on the pipe equal to that from a trench 2 ft wide all the way to the surface. Under

TABLE 13.3 Allowable Clay or Concrete Tile Depths to the Bottom of the Trench in Feet

Tile Size (in.)	ASTM Class	Crushing Strengths (lbs/lin ft)	Width of Trench at Top of Tile (in.)						
			15	18	21	24	27	30	36
4 or 5	Standard	1200	a	8.0	7.3	7.3	7.3	7.3	7.3
	Extra-quality	1650	a	a	10.5	9.8	9.8	9.8	9.8
6	Standard	1200	a	8.9	6.3	6.3	6.3	6.3	6.3
	Extra-quality	1650	a	a	10.6	8.3	8.3	8.3	8.3
8	Standard	1200	a	9.2	6.4	5.4	5.4	5.4	5.4
	Extra-quality	1650	a	a	10.8	7.8	7.1	7.1	7.1
10	Standard	1200	a	9.5	6.6	5.4	4.8	4.8	4.8
	Extra-quality	1650	a	a	11.0	8.0	6.5	6.2	6.2
12	Standard	1200	a	9.7	6.8	5.6	5.0	4.7	4.7
	Extra-quality	1650	a	a	11.2	8.2	6.7	5.8	5.8
15	Extra-quality	1650	—	—	11.5	8.5	7.0	6.0	5.2
18	Extra-quality	1800	—	—	14.8	9.8	8.0	7.0	5.6
21	Extra-quality	2100	—	—	—	13.8	9.7	8.5	6.7

[a]Any depth is permissible at this width or less.

Assumptions: (1) Crushing strengths given are averages in pounds per linear foot based on sandbearing method. Specifications for Drain Tile ASTM C4 and C412. (2) These values allow a safety factor of 1.5. (3) Loadings were computed for wet clay soil at 120 lb per cu ft. Somewhat greater depths are permissible for lighter soils. (4) Ordinary laying whereby the underside of the tile is well bedded on soil for 60°–90° of the circumference.

conditions to the left and below the heavy line in Table 13.3, tile may be installed at any depth without the danger of breakage. To reduce the possibility of failure, tile should be selected that is uniform in quality. Where the depth of cut is greater than allowed, the trench width may be reduced, or higher strength pipe, corrugated metal pipe, and other more rigid materials may be installed.

Allowable depths for corrugated plastic tubing are given in Table 13.4. These depths are based on an allowable deflection of 20 percent of the nominal tubing diameter, although tubing may withstand up to 30 percent deflection without collapsing. Good side support is an important factor in preventing deflection. For tubing installed with less than 2 ft of cover, gravel backfill should be placed around the tubing and to a depth equal to the pipe diameter above the tubing. For further information see Schwab et al. (1993).

13.17   Layout and Design Procedure

The first step in design is to make a preliminary survey to evaluate the feasibility of drainage. The kind of information needed is indicated in Fig. 13.9. Drain layout can be made directly in the field or on a topographic map. If the field is flat, several elevations should be taken with the level to locate the outlet and plan the type of system best suited to the area. As shown in Fig. 13.7, the total

TABLE 13.4   Allowable Corrugated Plastic Tubing Depths to the Bottom of the Trench in Feet

Nominal Tubing Diameter (in.)	Tubing Quality (ASTM)	Trench Width at Top of Tubing (in.)			
		12	16	24	32 or more
4	[a]Standard	12.7	7.0	5.5	5.2
	[b]Heavy-Duty	c	9.8	7.0	6.3
6	[a]Standard	10.1	6.9	5.4	5.2
	[b]Heavy-Duty	c	9.6	6.7	6.1
8	[a] Standard	10.3	7.1	5.6	5.3
	[b]Heavy-Duty	c	9.8	6.8	6.2
10	b	—	9.2	6.6	6.2
12	b	—	8.8	6.6	6.2
15	b	—	—	6.8	6.3

Source: Fenemor et al. (1979).

[a]Pipe stiffness 13 psi at 20 percent deflection.

[b]Pipe stiffness 18 psi at 20 percent deflection.

[c]Any depth is permissible at this width or less and for 8-inch trench width for all sizes.

Assumptions: (1) Tubing buried in loose, fine-textured soil with density of 109 lb/cu ft. (2) Modulus of soil reaction 50 psi. (3) Deflection lag factor 3.4. (4) Bedding angle 90°. (5) Vertical deflection 110 percent of horizontal deflection.

Note: Differences in commercial tubing from several manufacturers, including corrugation design and pipe stiffness, and differences in soil conditions may change the assumptions; therefore, the maximum depths may be more or less than stated in the table. These depths are based on limited research and should be used with CAUTION.

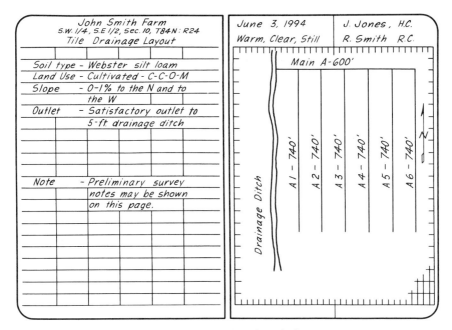

FIGURE 13.9   Field notes for a preliminary subsurface drainage system survey.

length of pipe required and the amount of large pipe can be reduced by using a layout with long laterals and a short main. This reduction is due to a smaller area that is drained both by the main and by the laterals. This double-drained area for the field in Fig. 13.7 is reduced from 1.5 acres with short laterals to 0.5 acre for the long-lateral layout (Fig. 13.7a). By joining the laterals to the main at an angle of 90°, the double-drained area is kept at a minimum. Such a layout also facilitates installation. Smaller angles are permissible, provided the flow from the lateral is directed downstream.

After the final plan has been selected, the drain lines are then staked out by placing hub stakes and guard stakes (station markers) at each 50- or 100-ft station. The stakes must be offset about 5 ft from the center line of the trench so that they will not be removed by the excavation. The contractor can advise as to the offset distance and to which side of the trench to set the stakes. Laser grade control reduces staking and computations.

After the drain line has been laid out, the elevations of the hub stakes are determined by profile leveling, and the notes are recorded as shown in Fig. 13.10. This form of field notes is convenient for simple drainage systems, as the cuts can be computed in the field and given to the contractor. On the left-hand side of Fig. 13.10, a blank line is provided between 100-ft station numbers. This method of spacing provides the proper scale for plotting the profile shown on the right-hand side of the page. The blank line also permits the recording of intermediate stations without confusing the data, for example, see station 2 + 50. The hub elevations in column five on the left-hand side of the page are then plotted on the right-hand side to determine the ground profile. The elevation of the outlet

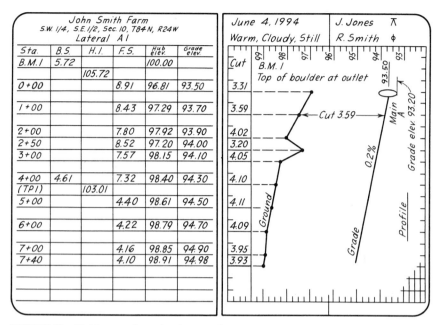

Sta.	B.S.	H.I.	F.S.	Hub elev.	Grade elev.
B.M.1	5.72			100.00	
		105.72			
0+00			8.91	96.81	93.50
1+00			8.43	97.29	93.70
2+00			7.80	97.92	93.90
2+50			8.52	97.20	94.00
3+00			7.57	98.15	94.10
4+00	4.61		7.32	98.40	94.30
(TP1)		103.01			
5+00			4.40	98.61	94.50
6+00			4.22	98.79	94.70
7+00			4.16	98.85	94.90
7+40			4.10	98.91	94.98

John Smith Farm — S.W. 1/4, S.E. 1/2, Sec. 10, T84N, R24W — Lateral A1

June 4, 1994 — J. Jones — Warm, Cloudy, Still — R. Smith

FIGURE 13.10 Field notes for selecting grades and calculating cuts for a subsurface drainage lateral.

drain line or ditch is also plotted on the profile, and the grade line for the pipe is selected so that cuts at each station will not be too deep. The field notes for each lateral or main should be recorded on a separate page of the field book.

The grade elevation at $0+00$ on the lateral is then determined from the grade elevation previously computed for the main at the junction. When the main and lateral are the same diameter, the grade elevation of the lateral should be 0.1 ft above the grade elevation of the main at the junction. In Fig. 13.10, a rise of 0.3 ft (93.50–93.20) was allowed for convenience. However, the pipes are connected in the normal manner and the extra fall is adjusted in the first few feet of the lateral. This additional fall at the junction causes the velocity to increase and reduces sedimentation.

Where the main is larger than the lateral, the grade line of the lateral must be higher so that the center line of the main and the center line of the lateral intersect. This difference in elevation can be computed by subtracting half the outside diameter of the lateral from half the outside diameter of the main.

As illustrated in Fig. 13.10, cuts are computed for each station by the subtraction of the grade elevation from the hub elevation. Grade elevations for each station above $0+00$ are computed by adding the rise (distance times percent slope) to the grade elevation of the previous station.

Where subsurface drains are to be installed with trenching machines or with drain plows equipped with laser grade control systems, the procedure of staking and determining cuts at 100-ft intervals is not required. Instead, it is necessary to locate only the drain line and the points where changes in grade are required. The rotating laser beam from the command post can be set level or on a sloping plane for reference. A detector on the machine picks up the beam and automatically keeps the machine on the same grade as the plane produced by the laser beam. When a change in grade is desired, some machines are equipped with a "grade breaker," which will allow the machine to be controlled at a grade different from that of the laser beam. Sufficient survey points must be taken along the drain line to determine the desired grade and points of change to keep the cuts within limits of the equipment.

Example 13.4   Determine the minimum grade elevation and cut at the outlet of a 4-in. diameter lateral to be connected to a 10-in. clay main if the hub stake elevation at the junction $(0 + 00)$ is 96.81 as shown in Fig. 13.10.

Solution.   The outside diameter (o.d.) of the 10-in. main is about 1.0 ft (o.d. in tenths of feet is numerically about equal to the inside tile diameter in inches for sizes up to 12 in.) and 0.4 ft for the 4-in. lateral. For the two lines to connect center to center

$$\text{rise in elevation} = \frac{1.0}{2} - \frac{0.4}{2} = 0.3 \text{ ft}$$

Minimum grade elevation of the 4-in. lateral is $93.20 + 0.3 = 93.50$ ft, and the cut on the lateral is $96.81 - 93.50 = 3.31$ ft, as shown in Fig. 13.10.

13.18   Installation and Maintenance

Wheel and endless-chain trenchers and drain plows (trenchless) are the most common installation machines. Backhoes may require hand shaping of the trench bottom because accurate grade control devices have not been developed. By setting targets at each station or by using the laser plane of light at the desired grade (Fig. 13.11), a line parallel to the grade line is established from which the desired depth of cut and grade are maintained. The vertical distance

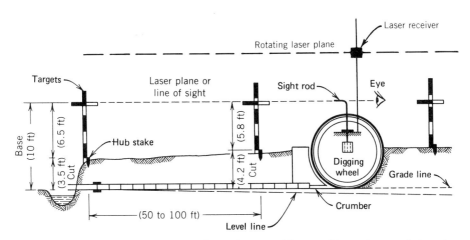

FIGURE 13.11   Establishing a grade line for a pipe trenching machine using grade targets or a sloping laser beam. (The laser beam is a plane of light and is parallel to the line of sight.) (Source: Schwab, G. O. et al., 1993.) (Reprinted and revised with permission of John Wiley & Sons, Inc.)

(base height) from the sight bar or laser detector on the machine to the bottom of the digging mechanism is fixed. In Fig. 13.11 this base height is 10 ft. Subtracting the cut at each station from the base will give the height of the target above the hub stake. For example, at the first station in Fig. 13.11, the target height is 6.5 ft $(10 - 3.5)$. The allowable variation from the true grade should not be more than 0.01 ft per in. of pipe diameter in 100 ft.

The crack spacing between individual tile (1-ft lengths) should not be more than $\frac{1}{8}$ in. for sandy soils and $\frac{1}{4}$ in. for clay. In unstable soils, tiles should be laid as close together as possible. For equivalent drainage, perforated tubing should have about 28 and 42 perforations per foot ($\frac{1}{4}$ in. diameter) to compare to $\frac{1}{8}$- and $\frac{1}{4}$-in. cracks, respectively. Perforating ordinary tile will give some increase in flow, but increasing the crack width would give the same effect.

Poor outlets and failure to make timely repairs are major maintenance problems. Drains that are designed and installed properly require little maintenance. Roots of brush and trees, particularly willow, elm, cottonwood, soft maple, and eucalyptus, will grow into the line and obstruct the flow, especially if the drain is fed by springs supplying water far into the dry season. Under such conditions, this vegetation should be removed if it is within 100 ft of a line. If this is not convenient, a sealed pipe may be installed. In a few areas, minerals, such as iron and manganese, may be deposited in the openings, thus sealing the drains. Iron ochre is common in some areas, especially in peat and muck soils. It is a gelatinous material, caused by chemical oxidation or bacteria. Its removal or control is a difficult problem, for which there is not a satisfactory solution.

13.19   Mapping the Drainage System

After the drainage system has been installed, a map should be made and filed with the deed to the property. This map may be made with suitable survey equipment, or, if the system is simple (such as the one shown in Fig. 13.8), it may be made by measuring a few distances and angles with a steel tape. The drainage system can best be shown on an aerial photo or a topographical map. The owner of the farm should retain a copy of the map, as it is useful in locating the lines at a later date. This map is likewise evidence that the drains have been installed and is especially valuable to future owners.

13.20   Estimating Cost of Drainage Systems

The principal items of cost for a subsurface drainage system are installation, engineering, pipe, and accessories, such as outlet tubes and surface inlets. Pipe prices vary from one region to another and with the distance from the plant. Most ditching-machine contractors charge a fixed price per unit length. Additional charges may be made for cuts below a specified depth, for large pipe, and for backfilling. Engineering and supervision costs may vary from 5 to 10 percent of the total. In the Midwest, the cost for materials is about the same as the trenching cost; but in California, where depths are 6 ft or more, installation is about twice the cost of the pipe.

REFERENCES

American Society of Agricultural Engineers (ASAE) (1993) *Design and Construction of Subsurface Drains in Humid Areas.* EP260.4. ASAE, St. Joseph, Michigan.

Drablos, C. J. W., and R. C. Moe (1984) "Illinois Drainage Guide," *University Illinois Ext. Cir. 1226.*

Fenemor, A. D., B. R. Bevier, and G. O. Schwab (1979) "Prediction of Deflection for Corrugated Plastic Tubing," *Trans. ASAE,* 22(6):1338–1342.

Manson, P. W., and C. O. Rost (1951) "Farm Drainage—An Important Conservation Practice," *Agr. Engin.,* 32:325–327.

Ritzema, H. P., and H. M. H. Braun (1994) "Environmental Aspects of Drainage," Chapter 25. In *Drainage Principles and Applications,* International Institute for Land Reclamation (ILRI) Publ. 16, 2nd ed. Wageningen, The Netherlands.

Schwab, G. O., D. D. Fangmeier, W. J. Elliot, and R. K. Frevert (1993) *Soil and Water Conservation Engineering.* 4th ed. Wiley, New York.

Skaggs, R. W. (1980) *A Water Management Model for Artificially Drained Soils,* Tech. Bull. 267. University North Carolina Water Resources Res. Inst.

Skaggs, R. W., M. A. Breve, and J. W. Gilliam (1994) "*Hydrologic and Water Quality Impacts of Agricultural Drainage,*" *Critical Reviews in Environmental Science and Tech.,* 24(1):1–32. CRC Press, Inc., Boca Raton, Florida.

Smedema, L. K., and D. W. Rycroft (1983) *Land Drainage.* Cornell University Press, Ithaca, New York.

U.S. Bureau of Reclamation (USBR) (1993) *Drainage Manual.* Government Printing Office, Washington, D.C.

U.S. Soil Conservation Service (SCS) (1973) *Drainage of Agricultural Land.* Water Information Center, Inc., Port Washington, D.C.

U.S. Soil Conservation Service (SCS) (1979) *Engineering Field Manual,* (Lithographed). SCS, Washington, D.C.

PROBLEMS

13.1 A pipe drainage system draining 12 acres flows at design capacity for 2 days following a storm. If the system is designed using a D.C. of $\frac{1}{2}$ in., how many cubic feet of water will be removed during this period?

13.2 What size pipe (plastic tubing) is required to remove the surface water through a surface inlet if the runoff accumulates from 20 acres of mineral soil and the slope in the pipe is 0.3 percent? Design for field crops in your area.

13.3 Determine the total drainage area in acres for the pipe system in Fig. 13.9, assuming that Main *A* drains only a 50-ft strip on one side of the main and the drain spacing is 100 ft.

13.4 Determine the plastic tubing size at the outlet of a 10-acre drainage system (a) if the D.C. is $\frac{3}{8}$ in. and the grade is 0.3 percent, (b) if the grade is reduced to 0.1 percent.

13.5 Determine the design drain flow from 10 000 ft of pipe at a spacing of 200 ft, following local recommendations for irrigated conditions. What size clay tile is required for the main at the outlet if the slope is 0.25 percent?

13.6 For the pipe system shown in Fig. 13.8, determine the size of clay tile required for Main *A* and lateral *A1* if the D.C. is $\frac{3}{8}$ in. for normal conditions and $\frac{3}{4}$ in. for surface inlets. Assume that the location is in a humid area and that all surface water from the 8 acres must be removed through the surface inlet.

13.7 For the pipe system shown in Fig. 13.8, determine the length and size of all plastic tubing required. Assume that the drainage coefficients are to be at the lowest recommended rate for truck crops. The field is in a humid area, and (mineral) clay loam soil predominates. All surface water from the 8-acre depression is to be removed through the surface inlet.

13.8 Solve Problem 13.6, assuming that the surface inlet is not to be installed. Surface water is to be removed with a surface drain.

13.9 What is the maximum depth that a 10-in. standard-quality clay tile will withstand in a trench 24 in. wide? Extra-quality?

13.10 Six-inch concrete tile is to be installed in a trench 21 in. wide and 9.5 ft deep. Will standard-quality tile support the soil load? If not, what could be done to prevent breakage?

13.11 Using a topographic map supplied by your instructor, select a depth, a spacing, a drainage coefficient for the crop, and the layout for the subsurface drainage system. By estimating the grade from the elevations on the map, determine the size and length of plastic tubing required, including junctions and other accessories.

13.12 The elevation of a hub stake on a 10-in. tile main is 86.40. If the cut on the main at this station is 3.40 ft, what is the lowest grade elevation at which a 4-in. lateral should be connected? Assume that a 10×4-in. manufactured T-junction is to be installed and that the wall thickness of both sizes is 1 in.

13.13 If the sight bar on a trenching machine is mounted 9 ft above the bottom of the digging wheel, how high should the sighting target be set above the hub stake if the cut is 3.56 ft?

13.14 Using Fig. 13.10, compute the cuts for lateral *A1* if the slope in the drain had been 0.3 percent up to stations 3 + 00 and 0.16 percent above this point.

13.15 Estimate the cost of the drainage system in Fig. 13.7*a*. Assume the average cut for the main is to be 4.5 ft and for the laterals 4.0 ft or less. *Installation cost:* 4.0 ft or less, $30 per 100 ft; overcut, $1.00 per 0.1 ft per 100 ft. *Materials:* Pipe prices to be supplied by your instructor. Add 5 percent of length for breakage for clay or concrete, but no extra cost for plastic; pipe junctions at five times the cost per foot of pipe of the largest size. *Engineering:* 5 percent of total material and installation cost. Show bill of materials and compute the cost per acre.

13.16 What is the maximum depth at which 8-in. standard-quality corrugated plastic tubing can be installed in a 24-in.-wide trench to prevent a deflection not greater than 20 percent of the tubing diameter? What is the maximum depth at which heavy-duty tubing can be installed?

13.17 Four-inch-diameter corrugated plastic tubing is to be installed in a trench 24 in. wide and 6.0 ft deep. Will standard quality tubing provide sufficient support to prevent a deflection of less than 20 percent? If not, what could be done to reduce the deflection?

13.18 The grade elevation at the outlet of a 600-ft lateral is 83.50 ft. The hub elevations at each 100-ft station, starting at the outlet, are 86.81, 87.29, 87.92, 88.15, 88.40, 88.61, and 88.79 ft. Select a uniform slope to provide an average cut of about 4 ft. Tabulate all grade elevations and compute the cuts for each station.

CHAPTER 14 _____
SOIL WATER RELATIONS AND EVAPOTRANSPIRATION

The soil is a reservoir for water and chemicals, including plant nutrients, and provides a medium to support the plants. Water is removed from the soil reservoir by evapotranspiration (evaporation plus transpiration). The rate at which water is removed from the soil by plants and the amount of water stored in the soil by rainfall or irrigation determine the type of plants to be grown, plant spacing, yield, and general management criteria. Knowledge of soil water relations and evapotranspiration is essential for managing soil water systems, particularly for irrigation management. For irrigation, the water-holding capacity (or the quantity of water in the soil water reservoir) and the rate of water removal by plants must be considered. Although this chapter will emphasize irrigation, the principles apply to all crop production systems.

14.1 The Soil Water Reservoir

In planning and managing irrigation, the soil's capacity to store available water can be thought of as the soil water reservoir, which must be filled periodically by irrigation or rainfall. It is slowly depleted by evapotranspiration. Water application in excess of the reservoir capacity is wasted unless it is required for leaching (Section 15.5) or to meet a specific management need. Irrigation must be scheduled to prevent the soil water reservoir from becoming so low as to inhibit plant growth.

Representative means and ranges of soil physical properties including water holding abilities are listed by soil texture in Table 14.1. Field capacity (FC) is the water content after a soil is wetted and allowed to drain 1 to 2 days. It represents the upper limit of water available to plants, usually defined as 0.1 to 0.3 bars tension. Permanent wilting point (PWP) represents the lower limit of water available to plants, usually defined as 15 bars tension. Neither FC nor PWP can be precisely defined, since both vary with soils and plants. The difference between FC and PWP is the available water (AW) and can be estimated from

$$AW = (FC_v - PWP_v)\, D_r/100 \tag{14.1}$$

TABLE 14.1 Representative Physical Properties of Soils

Soil Texture	Saturated Hydraulic Conductivity,[a] k_3 (mm/h)	(in./h)	Total Pore Space (% by vol)	Apparent Specific Gravity (A_s)	Field Capacity fc (% by vol)	Permanent Wilting pwp (% by vol)	Total Available Water		
							Percent by Volume (v)	(mm/cm)	(in./ft)
Sandy	50 (25–250)	2 (1–10)	38 (32–42)	1.65 (1.55–1.80)	15 (10–20)	7 (3–10)	8 (6–10)	0.8 (0.7–1.0)	1.0 (0.8–1.2)
Sandy Loam	25 (12–75)	1 (0.5–3)	43 (40–47)	1.50 (1.40–1.60)	21 (15–27)	9 (6–12)	12 (9–15)	1.2 (0.9–1.5)	1.4 (1.1–1.8)
Loam	12 (8–20)	0.5 (0.3–0.8)	47 (43–49)	1.40 (1.35–1.50)	31 (25–36)	14 (11–17)	17 (14–20)	1.7 (1.4–1.9)	2.0 (1.7–2.3)
Clay Loam	8 (3–5)	0.3 (0.1–0.6)	49 (47–51)	1.35 (1.30–1.40)	36 (31–42)	18 (15–20)	18 (16–22)	1.9 (1.7–2.2)	2.3 (2.0–2.6)
Silty Clay	3 (0.25–5)	0.1 (0.01–0.2)	51 (49–53)	1.30 (1.25–1.35)	40 (35–46)	20 (17–22)	20 (18–23)	2.1 (1.8–2.3)	2.5 (2.2–2.8)
Clay	5 (1–10)	0.2 (0.05–0.4)	53 (51–55)	1.25 (1.20–1.30)	44 (39–49)	21 (19–24)	23 (20–25)	2.3 (2.0–2.5)	2.7 (2.4–3.0)

Source: Hansen, V. E. et al. (1980). Reprinted with permission of John Wiley & Sons, Inc.

Note: Normal ranges are shown in parentheses.

[a]Saturated hydraulic conductivities vary greatly with soil structure and structural stability, even beyond the normal ranges shown.

where FC_v and PWP_v = the volumetric field capacity and permanent wilting point percentages, respectively

D_r = depth of the root zone or depth of a layer of soil within the root zone *(L)*

AW = depth of water available to plants *(L)*

Since dry weight water contents are easier to obtain, it is useful to relate volumetric water content to dry weight content by

$$P_v = (P_d)(A_s) \tag{14.2}$$

where P_v = volumetric water content

P_d = dry weight water content

A_s = the soil apparent specific gravity

Plants can remove only a portion of the available water before growth and yield are affected. This portion is readily available water (RAW) and for most crops ranges between 40 and 65 percent of AW in the root zone. Readily available water can be estimated from

$$RAW = (MAD)(AW) \tag{14.3}$$

in which MAD is the management allowed deficiency or the portion (decimal) of the available water that management determines can be removed from the root zone without adversely affecting yield and/or economic return. Table 14.2 provides suggested values for MAD and typical root zone depths when soil conditions do not limit root penetration.

Example 14.1.   Determine the readily available water for alfalfa grown on a sandy loam soil.

Solution.   From Table 14.1, the available water in a sandy loam soil is 1.4 in./ft. From Table 14.2, a rooting depth of 5 feet is selected. The available water is 1.4 in./ft $\times$ 5 ft or 7.0 in. The readily available water is obtained from Eq. 14.3 with MAD = 0.55 from Table 14.2.

$$RAW = 0.55 \times 7.0 \text{ in.} = 3.9 \text{ in. (99 mm)}$$

14.2   Soil Water Measurement

Measurement of the soil water content in the crop root zone is useful for determining when to irrigate and/or how much water to apply. A soil sample is needed for some methods and can be obtained with a shovel. Special augers or probes can obtain samples from any depth in the root zone.

The feel and appearance of the soil is the simplest indicator of soil water depletion. A small sample of soil is squeezed to make a ball or pressed between the thumb and forefinger. Table 14.3 gives the criteria for estimating the soil

TABLE 14.2 Typical Management Allowed Deficits (MAD) and Rooting Depths of Various Crops

Crop	MAD	Rooting Depth	
		ft	*m*
Alfalfa	0.55	3–10	1.0–3.0
Banana	0.35	2–3	0.5–0.9
Barley	0.55	3–5	1.0–1.5
Beans	0.45	1–2	0.5–0.7
Beets	0.5	2–3	0.6–1.0
Cabbage	0.45	1–2	0.4–0.5
Carrots	0.35	1–3	0.5–1.0
Celery	0.2	1–2	0.3–0.5
Citrus	0.5	4–5	1.2–1.5
Clover	0.35	2–3	0.6–0.9
Corn (maize)	0.6	3–5	1.0–1.7
Cotton	0.6	3–5	1.0–1.7
Cucumber	0.5	2–4	0.7–1.2
Dates	0.5	5–8	1.5–2.5
Flax	0.5	3–5	1.0–1.5
Grains, small	0.55	3–5	0.9–1.5
Grapes	0.35	3–6	1.0–2.0
Grass	0.5	2–5	0.5–1.5
Lettuce	0.3	1–2	0.3–0.5
Melons	0.35	3–5	1.0–1.5
Olives	0.65	4–5	1.2–1.7
Onions	0.25	1–2	0.3–0.5
Peas	0.35	2–3	0.6–1.0
Peppers	0.25	2–3	0.5–1.0
Pineapple	0.5	1–2	0.3–0.6
Potatoes	0.25	1–2	0.4–0.6
Safflower	0.6	3–6	1.0–2.0
Sorghum	0.55	3–6	1.0–2.0
Soybeans	0.5	2–4	0.6–1.3
Spinach	0.2	1–2	0.3–0.5
Strawberries	0.15	0.5–1	0.2–0.3
Sugarbeets	0.5	2–4	0.7–1.2
Sugarcane	0.65	4–6	1.2–2.0
Sunflower	0.45	3–5	0.8–1.5
Sweet potatoes	0.65	3–5	1.0–1.5
Tomatoes	0.4	2–5	0.7–1.5
Vegetables	0.2	1–2	0.3–0.6
Wheat	0.55	3–5	1.0–1.5

Source: Doorenbos, J. and W. O. Pruitt (1977). Reprinted with permission from Food and Agriculture Organization of the United Nations.

TABLE 14.3 Guide for Estimating Soil Water Deficiency

Soil Water Deficiency	Feel or Appearance of Soil and Water Deficiency in Inches of Water Per Foot of Soil			
	Coarse Texture	Moderately Coarse Texture	Medium Texture	Fine and Very Fine Texture
0% (Field capacity)	Upon squeezing, no free water apears on soil but wet outline of ball is left on hand 0.0	Upon squeezing, no free water appears on soil but wet outline of ball is left on hand 0.0	Upon squeezing, no free water appears on soil but wet outline of ball is left on hand 0.0	Upon squeezing, no free water appears on soil but wet outline of ball is left on hand 0.0
0–25%	Tends to stick together slightly, sometimes forms a very weak ball under pressure 0.0 to 0.2	Forms weak ball, breaks easily, will not slick 0.0 to 0.4	Forms a ball, is very pliable, slicks readily if relatively high in clay. 0.0 to 0.5	Easily ribbons out between fingers, has slick feeling 0.0 to 0.6
25–50%	Appears to be dry, will not form a ball with pressure 0.2 to 0.5	Tends to ball under pressure but seldom holds together 0.4 to 0.8	Forms a ball somewhat plastic, will sometimes slick slightly with pressure. 0.5 to 1.0	Forms a ball, ribbons out between thumb and forefinger 0.6 to 1.2
50–75%	Appears to be dry, will not form a ball with pressure[a] 0.5 to 0.8	Appears to be dry, will not form a ball[a] 0.8 to 1.2	Somewhat crumbly but holds together from pressure. 1.0 to 1.5	Somewhat pliable, will ball under pressure[a] 1.2 to 1.9
75–100% (100% is permanent wilting)	Dry, loose, single-grained, flows through fingers 0.8 to 1.0	Dry, loose, flows through fingers 1.2 to 1.5	Powdery, dry, sometimes slightly crusted but easily broken down into powder 1.5 to 2.0	Hard, baked, cracked, sometimes has loose crumbs on surface 1.9 to 2.5

Source: Hansen, V. E. et al. (1980). Reprinted with permission of John Wiley & Sons, Inc.

[a]Ball is formed by squeezing a handful of soil very firmly.

water deficiency in inches per foot. This method requires judgment and experience for best results.

The gravimetric sampling method requires that soil samples be collected in the field and placed in a tight container. The container is weighed, placed in an oven to dry, and then weighed again. This method is labor intensive, but can be used to calibrate other methods.

Soil tensiometers consist of a tube with a ceramic tip at one end and a suction gage at the other. The ceramic tip is carefully placed in the soil, and the soil water tension is detected by the gage. These devices operate up to tensions of 0.7 bars, which is in the wet range. Calibration curves are used to convert tension to water content; however, with experience, the tension reading can be used to indicate the need for irrigation.

Electrical resistance measurement of gypsum blocks is an indirect method for estimating soil water tension, which can be converted to soil water content through soil water release curves or by calibration. Water permeates the gypsum at about the same tension as the surrounding soil. The resistance between two electrodes in the gypsum is measured with an ohmmeter or with a meter supplied by the manufacturer of the blocks. Salt content of the soil water can affect the readings. The resistance is often nonlinear with water content, and blocks should be calibrated for the specific soil.

Heat dissipation in a soil has been found to be related to water content. Thus, porous blocks have been developed with a small heat source and temperature sensor. The temperature rise is related to the water content of the block, which is proportional to the water in the surrounding soil. These devices are more accurate than gypsum blocks, but require more expensive meters to provide power to the heat source and measure the temperature.

Thermocouple psychrometers determine the total water potential of the liquid phase of the soil water by measurement of the vapor phase. These devices have been influenced by temperature fluctuations and gradients, vapor pressure gradients, and the design of the sensor. Recent improvements have overcome some of these problems.

Neutron probes can accurately measure soil water content. The hydrogen nuclei of water molecules in the soil slows and scatters the neutrons from a fast neutron source. A sensor in the neutron probe detects and counts the thermalized neutrons. Manufacturers provide a standard calibration curve; however, for more accurate measurements the probe should be calibrated for each soil. Since a radiation source is used, licensing is required, film badges are essential for operators, and maintenance should be done by trained operators.

Time domain reflectrometry (TDR) measures the propagation of an electrical pulse that is dependent on the soil dielectric constant. Soil water content and electrical conductivity directly affect the soil dielectric constant. A voltage pulse is applied to a pair of rods placed in the soil. The propagation velocity and the amplitude of the signal generated by the pulse are used in the estimation of the water content. With proper analysis, the signals also can be related to soil bulk density and salinity.

14.3   Crop Water Requirements

The water requirements and time of maximum demand vary with different crops. Although growing crops are continuously using water, the rate of water use depends on (1) the kind of crop, (2) the degree of maturity, and (3) atmospheric conditions, such as radiation, temperature, wind, and humidity. The rate of growth at different soil water contents varies with different soils and crops. Some crops are able to withstand drought or high water content better than others. During the early stages of growth the water needs are generally low, but they increase rapidly during the maximum growing period to the fruiting stage. During the later stages of maturity, water use decreases as the crops ripen.

To make maximum use of available water supplies, managers must have a knowledge of the total seasonal water requirements of crops and how water use varies during the growing season. The seasonal requirement is necessary to select crops and areas that match the available water supply. Knowledge of the variation during the season aids in scheduling irrigations. Table 14.4 illustrates seasonal evapotranspiration and water requirements for crops grown near Deming, New Mexico. Expected effective rainfall is considered in determining the field irrigation requirement. The duration and length of periods of inadequate precipitation during the growing season in humid and subhumid regions largely determine the economic feasibility of irrigation. In the Northern Hemisphere, water deficiency during the months of June, July, and August is more serious than in earlier or later months. Extensive data on crop-water requirements are sometimes available locally. Examples for daily, semimonthly, and seasonal water use for alfalfa and wheat are shown in Fig. 14.1 for central Arizona.

TABLE 14.4   Seasonal Evapotranspiration and Irrigation Requirements for Crops Near Deming, New Mexico[a]

Crop	Length of Growing Season (Days)	Evapo-transpiration Depth (in.)	Effective Rainfall Depth (in.)	ET Less Rainfall (in.)	Water Application Efficiency (%)	Irrigation Requirement Depth (in.)
Alfalfa	197	36.0	6.0	30.0	70	42.9
Beans (dry)	92	13.2	4.0	9.2	65	14.1
Corn	137	23.1	5.3	17.8	65	27.4
Cotton	197	26.3	6.0	20.3	65	31.3
Grain (spring)	112	15.6	1.3	14.3	65	22.0
Sorghum	137	21.6	5.3	16.3	65	25.1

Source: Jensen, M. E. (ed.) (1973). Reprinted with permission of American Society of Civil Engineers.

[a]Average frost-free period is April 15 to October 29. Irrigation prior to the frost-free period may be necessary for some crops.

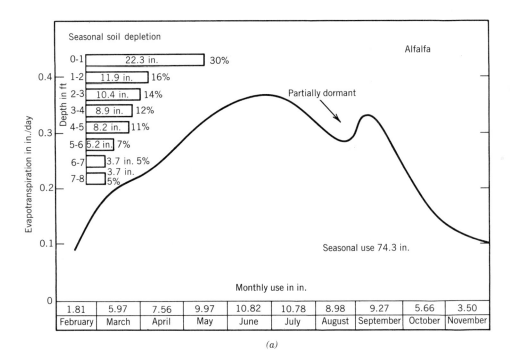

(a)

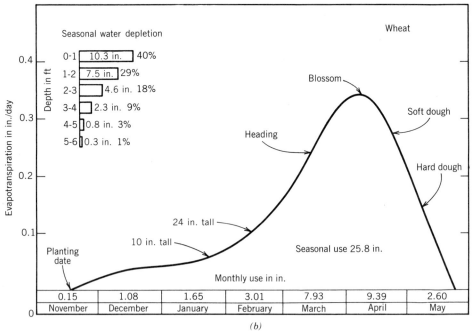

(b)

FIGURE 14.1. Average evaportranspiration and seasonal soil water depletion with depth for (a) alfalfa and (b) wheat at Mesa and Tempe, Arizona. (Redrawn and adapted from Erie et al., 1982.)

EVAPOTRANSPIRATION

For convenience, evaporation and transpiration are combined into evapotranspiration *(ET)* to define the water use by crops. The various methods for determining evapotranspiration include (1) tank and lysimeter experiments; (2) field experimental plots in which the quantity of water applied is kept small to avoid deep percolation losses and surface runoff is avoided or measured; (3) soil water studies based on a large number of samples taken at various depths in the root zone; (4) analysis of climatological data; (5) integration methods in which the water used by plants and evaporation from the water and soil surfaces are combined for the entire area involved; and (6) inflow-outflow method for large areas in which yearly inflow into the area, annual precipitation, yearly outflow from the area, and the change in ground-water level are evaluated.

Many practical applications can be made of evapotranspiration estimates, but the principal use is to predict soil water deficits for irrigation. Analyzing weather records and estimating evapotranspiration rates, drought frequencies, and excess water periods can show potential needs for irrigation and drainage.

14.4 Blaney-Criddle Method

The Blaney-Criddle equation is one of the simplest for estimating evapotranspiration (consumptive use). The equation is

$$u = \frac{ktp}{100} = kf \qquad (14.4)$$

where u = monthly evapotranspiration in inches

k = monthly evapotranspiration (consumptive use) coefficient (determined for each crop and location from experimental data)

t = mean monthly temperature in degrees Fahrenheit

p = monthly percent of annual daytime hours (Table 14.5)

$f = (tp)/100$ = monthly evapotranspiration (consumptive use) factor

For the entire growing season the following equation is more convenient

$$U = KF = K\Sigma f = \Sigma kf \qquad (14.5)$$

where U is the seasonal water use in inches, K is the seasonal evapotranspiration coefficient, and F = sum of the monthly evapotranspiration (consumptive use) factors f for the period. Mean monthly temperatures can be determined from local weather data. Evapotranspiration coefficients are given in Table 14.6 or can be obtained from local sources (i.e. Erie et al., 1982).

Example 14.2. Determine the evapotranspiration and irrigation requirements for wheat (small grain) near Plainview, Texas (34 degrees north latitude), if the water application efficiency is 65 percent.

Solution.

Month	Temperature[a] °F	Daytime Hours[b] %	f[c]	Effective Rainfall[a] (in.)
April	59.2	8.8	5.21	1.92
May	67.5	9.72	6.56	2.58
June	75.6	9.70	7.33	3.04
		Total 19.10		7.54

[a]Mean monthly temperature and rainfall can be obtained from local weather data.
[b]Monthly percent of daytime hours of the year from Table 14.5.
[c]Monthly evapotranspiration factor (tp/100).

From Eq. 14.5 and Table 14.6, $K = 0.80, U = 0.80 \times 19.1 = 15.28$ in.

irrigation water required $= (15.28 - 7.54)/0.65 = 11.9$ in. (300 mm)

The original Blaney-Criddle formula was developed primarily to obtain monthly or seasonal estimates of crop water use. The SCS (1970) modified the

TABLE 14.5 Average Monthly Percentage of Annual Daylight Hours for Various Latitudes

Latitude 0° North	Jan.	Feb.	Mar.	Apr.	May	June	July	Aug.	Sept.	Oct.	Nov.	Dec.
0	8.50	7.66	8.49	8.21	8.50	8.22	8.50	8.49	8.21	8.50	8.22	8.50
5	8.32	7.57	8.47	8.29	8.65	8.41	8.67	8.60	8.23	8.42	8.07	8.30
10	8.13	7.47	8.45	8.37	8.81	8.60	8.86	8.71	8.25	8.34	7.91	8.10
15	7.94	7.36	8.43	8.44	8.98	8.80	9.05	8.83	8.28	8.26	7.75	7.88
20	7.74	7.25	8.41	8.52	9.15	9.00	9.25	8.96	8.30	8.18	7.58	7.66
25	7.53	7.14	8.39	8.61	9.33	9.23	9.45	9.09	8.32	8.09	7.40	7.42
30	7.30	7.03	8.38	8.72	9.53	9.49	9.67	9.22	8.33	7.99	7.19	7.15
32	7.20	6.97	8.37	8.76	9.62	9.59	9.77	9.27	8.34	7.95	7.11	7.05
34	7.10	6.91	8.36	8.80	9.72	9.70	9.88	9.33	8.36	7.90	7.02	6.92
36	6.99	6.85	8.35	8.85	9.82	9.82	9.99	9.40	8.37	7.85	6.92	6.79
38	6.87	6.79	8.34	8.90	9.92	9.95	10.10	9.47	8.38	7.80	6.82	6.66
40	6.76	6.72	8.33	8.95	10.02	10.08	10.22	9.54	8.39	7.75	6.72	6.52
42	6.63	6.65	8.31	9.00	10.14	10.22	10.35	9.62	8.40	7.69	6.62	6.37
44	6.49	6.58	8.30	9.06	10.26	10.38	10.49	9.70	8.41	7.63	6.49	6.21
46	6.34	6.50	8.29	9.12	10.39	10.54	10.64	9.79	8.42	7.57	6.36	6.04
48	6.17	6.41	8.27	9.18	10.53	10.71	10.80	9.89	8.44	7.51	6.23	5.86
50	5.98	6.30	8.24	9.24	10.68	10.91	10.99	10.11	8.46	7.45	6.10	5.65
52	5.77	6.19	8.21	9.29	10.85	11.13	11.20	10.12	8.49	7.39	5.93	5.43
54	5.55	6.08	8.18	9.36	11.03	11.38	11.43	10.26	8.51	7.30	5.74	5.18
56	5.30	5.95	8.15	9.45	11.22	11.67	11.69	10.40	8.53	7.21	5.54	4.89
58	5.01	5.81	8.12	9.55	11.46	12.00	11.98	10.55	8.55	7.10	5.31	4.56
60	4.67	5.65	8.08	9.65	11.74	12.39	12.31	10.70	8.57	6.98	5.04	4.22

Source: SCS (1970).

TABLE 14.6 Seasonal Evapotranspiration Crop Coefficients K for the
Blaney-Criddle Formula

Crop	Length of Normal Growing Season or Period[a]	Evapotranspiration Coefficient K[b]
Alfalfa	Between frosts	0.80 to 0.90
Bananas	Full year	0.80 to 1.00
Beans	3 months	0.60 to 0.70
Cocoa	Full year	0.70 to 0.80
Coffee	Full year	0.70 to 0.80
Corn (maize)	4 months	0.75 to 0.85
Cotton	7 months	0.60 to 0.70
Dates	Full year	0.65 to 0.80
Flax	7 to 8 months	0.70 to 0.80
Grains, small	3 months	0.75 to 0.85
Grain, sorghums	4 to 5 months	0.70 to 0.80
Oilseeds	3 to 5 months	0.65 to 0.75
Orchard crops		
Avocado	Full year	0.50 to 0.55
Grapefruit	Full year	0.55 to 0.65
Orange and lemon	Full year	0.45 to 0.55
Walnuts	Between frosts	0.60 to 0.70
Deciduous	Between frosts	0.60 to 0.70
Pasture crops		
Grass	Between frosts	0.75 to 0.85
Ladino white clover	Between frosts	0.80 to 0.85
Potatoes	3 to 5 months	0.65 to 0.75
Rice	3 to 5 months	1.00 to 1.10
Soybeans	140 days	0.65 to 0.70
Sugar beet	6 months	0.65 to 0.75
Sugarcane	Full year	0.80 to 0.90
Tobacco	4 months	0.70 to 0.80
Tomatoes	4 months	0.65 to 0.70
Truck crops, small	2 to 4 months	0.60 to 0.70
Vineyard	5 to 7 months	0.50 to 0.60

Source: U.S. Soil Conservation Service (1970).

[a]Length of season depends largely on the variety and the time of year when the crop is grown. Annual crops grown during the winter period may take much longer than if they are grown in the summertime.

[b]The lower values of K for use in the Blaney-Criddle formula, $U = KF$, are for the more humid areas, and the higher values are for the more arid climates.

method by including both climatic and crop coefficients in k. Based on climatic data from 13 sites around the world, Doorenbos and Pruitt (1977) revised the Blaney-Criddle method (FAO Blaney-Criddle) to determine grass reference crop ET_r so that the method was more compatible with other methods for estimating ET.

14.5 Penman Method

With the development of reliable equipment for measuring climatic data, it is now feasible to use the Penman equation for estimating ET. Penman (1948, 1956) combined energy balance and mass transfer into an equation for evapotranspiration that uses meteorological data. The equation for well-watered grass or reference ET_O (Penman, 1963), given in SI units (Jensen et al., 1990), is

$$\lambda ET_o = \frac{\Delta}{\Delta + \gamma} (R_n - G) + \frac{\gamma}{\Delta + \gamma} 6.43(1.0 + 0.53 v_2)(e_s - e_d) \qquad (14.6)$$

where λET_o = reference ET for a well-watered grass expressed as latent heat flux density, MJ m^{-2} d^{-1}

λ = latent heat of vaporization of water in MJ/kg

Δ = slope of the saturation vapor pressure curve in kPa/°C

γ = psychrometric constant in kPa/°C

R = net radiation in MJ m^{-2} d^{-1}

G = heat flux density to the soil in MJ m^{-2} d^{-1}

v_2 = average wind speed at a height of 2 m in m/s

e_s = saturated vapor pressure at mean air temperature in kPa

e_d = saturated vapor pressure at mean dew point temperature in kPa (also e_s × relative humidity)

The following equations and constants were summarized from Jensen et al. (1990). Values for Δ can be obtained from

$$\Delta = 0.20(0.00738T + 0.8072)^7 - 0.000116 \qquad (14.7)$$

where T is the mean air temperature in °C.
The psychrometric constant in kPa/°C is

$$\gamma = 0.00163 \, P/\lambda \qquad (14.8)$$

$$P = 101.3 - 0.01055EL \qquad (14.9)$$

$$\lambda = 2.501 - 2.361 \times 10^{-3}T \qquad (14.10)$$

where P = estimated atmospheric pressure in kPa

EL = elevation in m

T = mean air temperature in °C

The net radiation can be calculated from

$$R_n = (1 - \alpha)R_s - \sigma(Ta)^4[0.34 - 0.139(e_d)^{0.5}](0.1 + 0.9 \, n/N) \quad (14.11)$$

where R_s = the solar radiation received at the earth's surface in MJ m^{-2} d^{-1}

α = the radiation reflection coefficient or albedo with values near 0.25 for green crops

σ = Stefan-Boltzman constant (4.903 × 10^{-9} MJ m^{-2} d^{-1} K^{-4})

Ta = air temperature in °K (°C + 273)

n/N = ratio of actual to possible hours of sunshine

If solar radiation is not measured, it can be obtained from

$$R_s = (0.35 + 0.61 \ n/N)R_{so} \qquad (14.12)$$

where R_{so} = the mean solar radiation for cloudless skies in MJ m^{-2} d^{-1} from Table 14.7

The soil heat flux G is small and often assumed as zero.
The saturation vapor pressure is calculated from

$$e_s = \exp\left(\frac{16.78T - 116.9}{T + 237.3}\right) \qquad (14.13)$$

where T is the mean air temperature in °C. Equation 14.13 also can be used to determine e_d by substituting the dew-point temperature for T.

Since Eq. 14.6 estimates reference ET as latent heat flux density for a well-watered grass, ET_o in mm/day is obtained from

$$ET_o = \lambda ET_o/\lambda \qquad (14.14)$$

where ET_o = the estimated grass reference ET.

The actual ET for various crops is estimated with crop coefficients from

$$ET_c = (K_c)ET_o \qquad (14.15)$$

TABLE 14.7   Mean Solar Radiation for Cloudless Skies

Month	Degrees North Latitude					
	0	10	20	30	40	50
	Mean Solar Radiation MJ m^{-2}d^{-1}					
Jan.	28.18	25.25	21.65	17.46	12.27	6.70
Feb.	29.18	26.63	25.00	21.65	17.04	11.43
Mar.	30.02	29.43	28.18	25.96	22.90	18.55
Apr.	28.47	29.60	31.14	29.85	28.34	25.83
May	26.92	29.60	31.40	32.11	32.11	30.98
June	26.25	29.31	31.82	33.20	33.49	33.08
July	26.27	29.46	31.53	32.66	32.66	31.53
Aug.	27.76	28.76	31.14	30.44	29.18	26.67
Sept.	29.60	29.60	28.47	26.67	23.73	20.10
Oct.	29.60	28.05	25.83	22.48	18.42	13.52
Nov.	28.47	25.83	22.48	18.30	13.52	8.08
Dec.	26.80	24.41	20.50	16.04	10.76	5.44

Source: Jensen et al. (1990).

TABLE 14.8   Approximate Crop Coefficients for a Grass Reference Crop, ET_o

Date		Corn (Grain)	Cotton	Potatoes	Soybeans
			Location		
		Midwest	Southwest	Northwest	Midwest
Apr	1–15				
	16–30		0.1		
May	1–15	0.2	0.2	0.1	0.1
	16–31	0.4	0.3	0.4	0.3
June	1–15	0.7	0.6	0.7	0.6
	16–30	0.9	0.9	0.9	0.8
July	1–15	1.0	1.2	1.1	1.0
	16–31	1.0	1.2	1.1	1.0
Aug.	1–15	1.0	1.2	1.1	1.0
	16–31	0.9	1.2	1.0	1.0
Sept.	1–15	0.6	1.1	0.8	0.8
	16–30		0.8		0.6
Oct.	1–15		0.5		
	16–31				

Source: Jensen, M. E. et al. (eds.) (1990). Adapted from, and with permission of, American Society of Civil Engineers.

where ET_c = the estimated ET for a crop, and K_c = crop coefficient for the specific crop and location. Approximate values of crop coefficients for selected crops are given in Table 14.8. Specific values for these and other crops can be obtained from Doorenbos and Pruitt (1977), Jensen et al. (1990), Hoffman et al. (1990), Cuenca (1989), or local data. Values of K_c vary with local conditions and the equation used to determine reference ET.

Example 14.3.   Compute average daily ET_o for June 16 to 30 near Bakersfield, California, 35° north latitude, using the Penman equation. Assume the following average data for the period: mean maximum temperature is 36°C, mean minimum temperature is 22°C, mean dew-point temperature is 9°C, mean wind speed at 2 m is 1.5 m/s, mean percent of possible sunshine is 94 percent, elevation is 50 m, and assume $G = 0$.

Solution.

1. Calculate the mean temperature $T = (36 + 22)/2 = 29$, then calculate Δ from Eq. 14.7:

$$\Delta = 0.2[0.00738(29) + 0.8072]^7 - 0.000116 = 0.232 \text{ kPa/°C}$$

2. Calculate P from Eq. 14.9:

$$P = 101.3 - 0.01055(50) = 100.8 \text{ kPa}$$

3. Calculate the latent heat of vaporization, Eq. 14.10:

$$\lambda = 2.501 - 0.002361(29) = 2.43 \text{ MJ/kg}$$

4. Calculate the psychrometric constant, Eq. 14.8:

$$\lambda = 0.00163(100.8)/2.43 = 0.0676 \text{ kPa/°C}$$

5. Calculate the saturation vapor pressure at mean air temperature and at dew-point temperature from Eq. 14.13:

$$e_s = \exp\left[(16.78(29) - 116.9)/(29 + 237.3)\right] = 4.008 \text{ kPa}$$

$$e_d = \exp\left[(16.78(9) - 116.9)/(9 + 237.3)\right] = 1.149 \text{ kPa}$$

6. Calculate the climatic constants:

$$\frac{\Delta}{\Delta + \gamma} = \frac{0.232}{0.232 + 0.0675} = 0.774$$

$$\frac{\gamma}{\Delta + \gamma} = 1.0 - \frac{\Delta}{\Delta + \gamma} = 1.0 - 0.774 = 0.226$$

7. From Table 14.7, determine the mean cloudless solar radiation $= (33.20 + 33.49)/2 = 33.34 \text{ MJ m}^{-2} \text{ d}^{-1}$, then find R_s with Eq. 14.12:

$$R_s = [0.35 + 0.61(94/100)]\,33.34 = 30.79 \text{ MJ m}^{-2}\text{d}^{-1}$$

8. Calculate net radiation, Eq. 14.11:

$$R_n = (1.0 - 0.25)30.79 - 4.903 \times 10^{-9}(29 + 273)^4[0.34 - 0.139(1.112)^{0.5}]$$
$$\times (0.1 + 0.9(94/100)) = 15.63 \text{ MJ m}^{-2}\text{d}^{-1}$$

9. Calculate λET_o from Eq. 14.6:

$$\lambda ET_o = 0.774(15.63) + 0.226(6.43)[1.0 + 0.53(1.5)]\,(4.008 - 1.149)$$
$$= 19.5 \text{ MJ m}^{-2} \text{ d}^{-1}$$

10. Calculate ET_o from Eq. 14.14 (Note: 1 kg of water $= 1 \text{ mm} \times 1 \text{ m}^2$ and 1 in. $= 25.4$ mm)

$$ET_o = 19.5/2.43 = 8.0 \text{ mm/d or } 0.32 \text{ in./day}$$

Example 14.4. Compute the estimated ET for cotton for June 16 to 30 near Bakersfield, California, 35° north latitude, using the Penman equation.

Solution. From Example 14.3 the reference $ET = 0.32$ in./day and from Table 14.8 the crop coefficient $= 0.9$, then calculate the ET for cotton from Eq. 14.15:

$$ET_c = 0.9 \times 0.32 = 0.29 \text{ in./day}$$

For 15 days the estimated ET for cotton is 4.35 in. (110 mm).

14.6. Jensen-Haise Method

Jensen and Haise (1963) and Jensen (1966) presented an energy-balance (solar radiation) approach to estimate evapotranspiration that is simpler in application than Penman's equation. Based on extensive field data, they developed a method for well-watered alfalfa reference ET_r

$$\lambda ET_r = C_t (T - T_x) R_s \qquad (14.16)$$

where λET_r = alfalfa-based reference ET expressed as latent heat flux density in MJ m^{-2} d^{-1}

R_s = solar radiation as measured or defined by Eq. 14.12 in MJ m^{-2} d^{-1}

T = mean air temperature for the period of calculation in °C

$$C_t = 1.0/(C_1 + 7.3 C_H) \qquad (14.17)$$

$$C_1 = 38 - (2EL/305), EL \text{ is elevation in m} \qquad (14.18)$$

$$C_H = 5.0/(e_2 - e_1) \qquad (14.19)$$

$$T_x = -2.5 - 1.4(e_2 - e_1) - EL/550 \qquad (14.20)$$

where e_2 and e_1 are the saturation vapor pressures in kPa for the mean maximum and mean minimum temperatures, respectively, for the warmest month of the year in the area and are calculated with Eq. 14.13.

Crop ET is obtained from Eq. 14.15 using a crop coefficient K_c for alfalfa-based reference ET. Note that grass and alfalfa-based crop coefficients are not identical. The coefficients vary with the stage of growth of the plant as given in Table 14.8. Jensen and Haise (1963), Wright (1982), and others have developed alfalfa-based coefficients for many crops and locations.

Example 14.5. Compute the alfalfa-based reference evapotranspiration for south central Idaho for July 10 to 15 using the solar radiation method. The location is 43° north latitude, elevation is 700 m, average temperature for July 10 to 15 is 22°C, mean maximum and minimum temperatures for the warmest month are 30°C and 17°C, respectively, and 90 percent of the possible sunshine is received.

Solution.

1. Calculate C_1 from Eq. 14.18:

$$C_1 = 38 - 2(700)/305 = 33.41$$

2. Calculate e_2 and e_1 from Eq. 14.13:

$$e_2 = \exp\left[(16.78(30) - 116.9)/(30 + 237.3)\right] = 4.24 \text{ kPa}$$

$$e_1 = \exp\left[(16.78(17) - 116.9)/(17 + 237.3)\right] = 1.94 \text{ kPa}$$

3. Calculate C_H from Eq. 14.19:

$$C_H = 5.0/(4.24 - 1.94) = 2.17$$

4. Calculate T_x from Eq. 14.20:

$$T_x = -2.5 - 1.4(4.24 - 1.94) - (700/550) = -6.99$$

5. Read cloudless radiation $R_{so} = 32.32$ from Table 14.6, then determine R_s from Eq. 14.12:

$$R_s = [0.35 + 0.61(80/100)] \, 32.66 = 29.06 \text{ MJ m}^{-2}\text{d}^{-1}$$

6. Calculate C_t from Eq. 14.17:

$$C_t = 1.0/[33.41 + 7.3(2.17)] = 0.0203$$

7. Substitute values into Eq. 14.16 to obtain ET_r:

$$\lambda ET_r = 0.0203(22 + 6.99) \, 29.06 = 17.1 \text{ MJ m}^{-2}\text{ d}^{-1}$$

8. Calculate the latent heat of vaporization from Eq. 14.10:

$$\lambda = 2.501 - 0.002361(22) = 2.44 \text{ MJ/kg}$$

9. Calculate ET_r from $ET_r = \lambda ET_r / \lambda$:

$$ET_r = 17.1/2.44 = 0.28 \text{ in./day } (7.0 \text{ mm/day})$$

Example 14.6. Compute the estimated ET for potatoes in south central Idaho for July 10 to 15 using the solar radiation method. Planting is May 1 and full cover is July 10.

Solution. From Table 14.9 estimate the average crop coefficient for July 10 to 15 to be 0.75 and $ET_r = 0.28$ from Example 14.5, calculate ET_c for the period from $ET_c = ET_r K_c$ (alfalfa-based)

$$ET_c = (0.28 \text{ in./day} \times 0.75) \, 5 \text{ days} = 1.0 \text{ in. } (26 \text{ mm}).$$

Because these equations for ET require extensive climatic data, several states have developed weather station networks, operated by trained professionals, to obtain the climatic data. In addition, these networks calculate the reference ET for various locations and make it available on a computer network or the local media such as newspapers, radio, or television. This system is one of many tools that water managers can use in their conservation practices.

14.7 Pan Evaporation

Pan evaporation data can be used to estimate reference ET using a pan coefficient K_p. Pans can be placed in or near the fields and measurements taken at regular intervals of the evaporation. Extreme care must be taken in interpreting pan data to obtain ET_o. The evaporation from a pan is influenced by the environment, type of pan, and climate. For irrigation purposes, it is better to place

TABLE 14.9 Approximate Crop Efficients K_c for Use with Alfalfa ET_r for Irrigated Crops Grown in an Arid Region with a Temperate Intermountain Climate

	Percent Time From Planting to Effective Cover										
Crop	0	10	20	30	40	50	60	70	80	90	100
Spring grain[a]	0.15	0.15	0.15	0.20	0.25	0.40	0.50	0.65	0.80	0.95	1.00
Peas	0.15	0.15	0.15	0.20	0.20	0.30	0.40	0.45	0.65	0.80	0.90
Sugar beets	0.15	0.15	0.15	0.15	0.15	0.15	0.20	0.30	0.40	0.70	1.00
Potatoes	0.15	0.15	0.15	0.15	0.15	0.20	0.30	0.45	0.60	0.70	0.75
Corn	0.15	0.15	0.15	0.15	0.15	0.20	0.25	0.40	0.55	0.75	0.95
Beans	0.15	0.15	0.15	0.20	0.20	0.35	0.45	0.60	0.75	0.90	0.90
Winter wheat	0.15	0.15	0.15	0.30	0.55	0.80	0.95	1.00	1.00	1.00	1.00

	Days After Effective Cover										
Crop	0	10	20	30	40	50	60	70	80	90	100
Spring grain[a]	1.00	1.00	1.00	1.00	0.90	0.40	0.15	0.10	0.05	—	—
Peas	0.90	0.90	0.70	0.50	0.30	0.15	0.05	0.05	—	—	—
Sugar beets	1.00	1.00	1.00	1.00	1.00	0.90	0.85	0.80	0.75	0.70	0.65
Potatoes	0.75	0.75	0.75	0.70	0.65	0.65	0.60	0.50	0.20	0.10	0.10
Field corn	0.95	0.95	0.95	0.90	0.85	0.85	0.75	0.70	0.30	0.20	0.15
Sweet corn	0.95	0.90	0.90	0.90	0.80	0.70	0.50	0.25	0.15	—	—
Beans	0.90	0.90	0.85	0.65	0.30	0.10	0.05	—	—	—	—
Winter wheat	1.00	1.00	1.00	1.00	0.95	0.50	0.20	0.10	0.05	—	—

	Percent Time from New Growth or Harvest to Harvest										
Crop	0	10	20	30	40	50	60	70	80	90	100
Alfalfa (first)[b]	0.40	0.50	0.60	0.80	0.90	0.95	1.00	1.00	1.00	0.95	0.95
(Intermediate)	0.25	0.30	0.40	0.70	0.90	0.95	1.00	1.00	1.00	0.95	0.95
(Last)	0.25	0.30	0.40	0.50	0.55	0.50	0.40	0.35	0.30	0.25	0.25

Source: Wright, J. L. (1982) and Jensen, M. E. et al. (eds.) (1990). Adapted from, and with permission of, American Society of Civil Engineers.

[a]Spring grain includes wheat and barley.

[b]First denotes first harvest, intermediate harvest may be one or more depending on season length.

The last harvest is when crop becomes dormant in cool weather. Cultivar used was Ranger.

the pan within the irrigated area, because the evaporation is greatly affected by upwind conditions. For details on this method see Jensen et al. (1990), Hoffman et al. (1990), or Doorenbos and Pruitt (1977).

REFERENCES

Burman, R. D., P. R. Nixon, J. L. Wright, and W. O. Pruitt (1983) "Water Requirements." In Jensen, M. E. (ed.) *Design and Operation of Farm Irrigation Systems.* American Society of Agricultural Engineers, St. Joseph, Michigan.

Cuenca, R. H. (1989) *Irrigation System Design: An Engineering Approach.* Prentice–Hall, Englewood Cliffs, New Jersey.

Doorenbos, J., and W. O. Pruitt (1977) *Guidelines for Predicting Crop Water Requirements,* FAO Irrig. and Drainage Paper No. 24, 2nd ed. FAO, Rome, Italy.

Erie, L. J., O. F. French, D. A. Bucks, and K. Harris (1982) "Consumptive Use of Water by Major Crops in the Southwestern United States," USDA-ARS Cons. Res. Rpt. No. 29. USDA, Washington, D.C.

Hansen, V. E., O. W. Israelsen, and G. E. Stringham (1980) *Irrigation Principles and Practices,* 4th ed. Wiley, New York.

Hoffman G. J., T. A. Howell, and K. H. Solomon (eds.) (1990) *Management of Farm Irrigation Systems,* Monograph. American Society of Agricultural Engineers, St. Joseph, Michigan.

Jensen, M. E. (1966) "Empirical Methods of Estimating or Predicting Evapotranspiration Using Radiation." In ASAE Proceedings, *Evapotranspiration and Its Role in Water Resources Management,* Dec., pp. 49–53, 64, St. Joseph, Michigan.

Jensen, M. E. (1973) *Consumptive Use of Water and Irrigation Water Requirements.* American Society of Civil Engineers, New York.

Jensen, M. E., R. D. Burman, and R. G. Allen (eds.) (1990) *Evapotranspiration and Irrigation Water Requirements.* American Society of Civil Engineers, New York.

Jensen, M. E., and H. R. Haise (1963) "Estimating Evapotranspiration from Solar Radiation," *Proc. ASCE J. Irrigation and Drainage Div.,* 89(IR4):15–41.

Penman, H. L. (1948) "Natural Evapotranspiration from Open Water, Bare Soil, and Grass," *Proc. R. Soc. London,* 193:120–145.

Penman, H. L. (1956) "Estimating Evapotranspiration," *Trans. Am. Geophys. Union,* 37:43–46.

Penman, H. L. (1963) *Vegetation and Hydrology,* Tech. Communication No. 53. Commonwealth Bureau of Soils, Harpenden, England.

U.S. Soil Conservation Service (SCS) (1970) *Irrigation Water Requirements,* Tech. Release No. 21. Washington, D.C.

Wright, J. L. (1982) "New Evapotranspiration Crop Coefficients," *ASCE J. of Irrigation and Drainage Div.,* 108:57–74.

PROBLEMS

14.1 Determine the available water in loam soil for a crop with a 3-ft root zone. What is the readily available water if wheat is being grown?

14.2 Estimate how much water has been depleted from 3 ft of a loam soil if the dry-weight water content is 15 percent?

14.3 Cotton is being grown in July on a field where the top 2 ft of soil is a clay loam and the next 2 ft is a sandy loam. Determine the available water and readily available water. If cotton uses 0.3 in./day, how often should irrigation occur if no rainfall occurs?

14.4 Corn is being grown on a clay loam soil in the Midwest. Estimate the readily available water, assuming the plants are starting to tassel.

14.5 Determine the seasonal water requirement for a crop in your area using the Blaney-Criddle method.

14.6 Determine ET_o for your area for July 1.

14.7 What is the estimated water use for soybeans in eastern Nebraska for June 1, assuming ET_o is 0.28 in./day?

14.8 Using data for a location given by the instructor, determine the alfalfa-based reference ET_r.

14.9 Determine the alfalfa-based crop coefficient for field corn on June 30 and August 10 if planting is May 1, and effective cover occurs July 10.

CHAPTER 15

IRRIGATION PRINCIPLES

Human dependence on irrigation can be traced to the earliest biblical references. Irrigation in very early times was practiced by the Egyptians, the Asians, and the Indians of North and South America. For the most part, water supplies were available to these people only during the periods of heavy runoff. Sediment in the runoff carried beneficial soil and nutrients for some fields, but the sediment filled canals and limited irrigation, which reduced food supplies and led to the decline of civilizations (Lowdermilk, 1953).

Current concepts of irrigation were made possible by the application of efficient power sources to deep well pumps and by the storage of large quantities of water in reservoirs. Thus, by using either underground or surface reservoirs, it is possible to bridge over the months and years and provide consistent water supplies.

Benefits from proper irrigation include (1) increased yields, (2) improved crop quality, particularly from vegetable and fruit crops, (3) controlled time of planting and harvesting so as to obtain a more favorable market price, (4) reduced damage from freezing temperatures, (5) reduced damage by control of high air temperatures, (6) increased efficiency of fertilizers and reduced cost of application, and (7) a stabilized farm income.

Disadvantages of irrigation include (1) increased costs of operation, not only for the irrigation system, but increased needs for fertilizer, pesticides, seeds, and more field operations, (2) an increase in weed populations, (3) potential increase in waterborne diseases to animals and humans, (4) concentration of contaminants in the root zone, (5) need for artificial subsurface drainage for water table management and leaching requirements, (6) detrimental effects on ground-water and downstream water quality, and (7) increased and conflicting demands on limited water resources.

In semihumid and humid areas where irrigation primarily supplements natural precipitation, the prospective irrigator should answer positively the following questions before considering irrigation: (1) Is the water supply adequate and of good quality? (2) Is sufficient labor available to operate and manage the irrigation system? (3) Is capital available to purchase the necessary equipment? (4) Will irrigation sufficiently increase crop yields over a period of years to justify the cost? Investment in alternative means of increasing production or conservation should also be considered if these practices might be more advantageous than irrigation. Not only is capital required for the irrigation system, but capital for additional fertilizer and seed is also needed. Higher fertility levels and higher planting rates are usually necessary to obtain maximum returns from the water applied.

15.1   Irrigation Effects on the Environment

Although irrigation has greatly increased the availability of food supplies and reduced their costs, it has a significant effect on the environment. Some early civilizations flourished because of irrigation, but also perished because they did not or could not control the effects of salinity. Salinity problems are typically associated with high water tables. The combination of salinity and high water tables is prevalent in many irrigated regions around the world. It is unclear whether today's technology is adequate to overcome water table and salinity problems. Assuming it is, there is no assurance that economic and political commitments are adequate to provide implementation.

Crop production by irrigation produces other effects on the environment because of the chemicals applied. Surface runoff from irrigated fields may contain fertilizers or pesticides. In addition, some chemicals are injected into the irrigation water for application to the field and carried in the runoff. Most irrigation waters contain dissolved salts, which remain in the soil after water use by evapotranspiration. These salts must be removed to maintain crop production, and natural or artificial drainage is needed to remove excess water and associated salts from the plant root zone. This water is obviously of lower quality than the applied water and is frequently returned to a stream or underground water supply, degrading the quality. Dissolved nitrogen from fertilizer applications is another source of contamination. The nitrogen may be in surface runoff and combined with salts in the drainage water.

The adverse effects of salinity and nitrogen have long been known. Only recently, however, has the potential impact of trace elements been recognized. These trace elements, including arsenic, boron, cadmium, chromium, lead, molybdenum, and selenium, originate mainly in the geologic materials on site instead of the irrigation water. This impact adds a new dimension to irrigation and drainage management, since both salinity and toxic trace elements must be considered when planning the disposal system. Selenium is a prime example.

The contamination of the Kesterson National Wildlife Refuge in the San Joaquin Valley of California by selenium has created awareness of the trace element problem. The environmental issues and impacts are documented by the National Research Council (1989). The wildlife ponds were built in 1971 and received fresh water until 1982. By 1981, the supply was entirely drainage water. Less than 2 years later, in 1982, reproductive failures and deaths of some aquatic organisms and waterfowl were reported. The contamination was caused by increased irrigation development, subsequent installation of subsurface drains, and failure to install an adequate disposal system. It is important to recognize that this event is not isolated and future planning must include measures to alleviate or prevent additional occurrences. As a result of public concern, efforts were initiated to eliminate drainage flows into Kesterson.

15.2 Irrigation Efficiencies

Various definitions of irrigation efficiencies have been formulated to aid in defining the performance of various systems and their operation. Efficiency is an output divided by an input and is usually expressed as a percentage. An efficiency figure is only meaningful when the output and input are clearly defined. Three basic irrigation efficiency concepts follow.

1. Water-conveyance efficiency (E_c),

$$E_c = 100 W_d / W_i \tag{15.1}$$

where W_d = water delivered by a distribution system
W_i = water introduced into the distribution system

The water-conveyance efficiency definition can be applied along any reach of a distribution system. For example, a water-conveyance efficiency could be calculated from a pump discharge to a given field or from a major diversion work to a farm turnout.

2. Water-application efficiency (E_a),

$$E_a = 100 W_s / W_d \tag{15.2}$$

where W_s = water stored in the soil root zone by irrigation
W_d = water delivered to the area being irrigated

This efficiency may be calculated for an individual furrow or border, for an entire field, or for an entire farm or project. When it is applied to areas larger than a field, it overlaps the definition of conveyance efficiency.

3. Water-use efficiency (E_u),

$$E_u = 100 W_u / W_d \tag{15.3}$$

where W_u = water beneficially used
W_d = water delivered to the area being irrigated

The concept of beneficial use differs from that of water stored in the root zone in that leaching water would be considered beneficially used, even though it moved through the soil water reservoir. Sometimes water-use efficiency is based on plant yield produced by a unit volume of water.

Another useful measurement of the effectiveness of irrigation is the uniformity coefficient *(UC)*,

$$UC = 1 - \frac{y}{d} \tag{15.4}$$

where y = average of the absolute values of the deviations in depth of water stored

d = average depth of water stored

This coefficient indicates how uniformly the infiltrated water has been distributed throughout the field. Each value of infiltrated water should represent an equal area; and when the deviation from the average depth is 0, the uniformity coefficient is 1.0. *UC* values above 0.8 are acceptable.

A second measure of effectiveness is the distribution uniformity *(DU)*,

$$DU = \frac{\text{average low-quarter depth of water infiltrated or caught}}{\text{average depth of water infiltrated or caught}} \quad (15.5)$$

The average low-quarter depth is the average of the lowest one fourth of all values where each value represents an equal area. *DU* values above 0.8 are considered acceptable.

Example 15.1. If 150 cfs are delivered to a distribution canal 4 miles long, and 140 cfs are delivered at the end of the canal, what is the conveyance efficiency of this system?

Solution. Substituting into Eq. 15.1:

$$E_c = 100 W_d / W_i - (100)140/150 = 93 \text{ percent}$$

Example 15.2. Delivery of 6 cfs to an 80-acre field is continued for 3 days. Soil water measurements indicate 4.1 inches of water was stored in the root zone. Compute the application efficiency.

Solution. Apply Eq. 15.2:

$$W_d = (6 \text{ cfs})(3600 \text{ sec/h})(24\text{h/day})(3 \text{ days})/43560 \text{ ft}^2/\text{ac} = 35.7 \text{ ac-ft}$$

$$W_i = (80 \text{ ac})(4.1 \text{ in.})/12 \text{ in./ft} = 27.3 \text{ ac-ft}$$

$$E_a = (100)27.3/35.7 = 76 \text{ percent}$$

Example 15.3. The depths of water infiltrated along the length of a border strip taken at 100-ft stations were found to be 4.5, 4.7, 4.4, 4.2, 3.8, 3.5, 4.3 in. Compute the uniformity coefficient.

Solution. Apply Eq. 15.4:

$$d = (4.5 + 4.7 + 4.4 + 4.2 + 3.8 + 3.5 + 4.3)/7 = 29.4/7 = 4.2 \text{ in.}$$

$$y = (0.3 + 0.5 + 0.2 + 0 + 0.4 + 0.7 + 0.1)/7 = 2.2/7 = 0.31$$

$$UC = 1 - (0.31 / 4.2) = 0.93$$

Example 15.4. A solid-set sprinkler system with 40 by 40-ft head spacing was operated for 4 h. Sixteen catch cans were placed under the system on a 10 by 10-ft spacing and the following depths of water were measured in the cans immediately after irrigation stopped. Assuming evaporation was equal from all cans, determine the distribution uniformity.

		depths (in.)	
2.5	2.3	2.2	2.3
2.2	2.1	1.9	2.2
2.3	2.1	1.8	2.0
2.3	2.2	2.0	2.2

Solution.   Apply Eq. 15.5:

Average low-quarter depth $= (1.8 + 1.9 + 2.0 + 2.0)/4 = 7.7/4 = 1.92$ in.

Average depth $= (2.5 + 2.3 + \cdots + 2.0 + 2.2)/16 = 34.6/16 = 2.16$ in.

$$DU = 1.92/2.16 = 0.89$$

SALINITY

15.3 Soil Salinity

The presence of soluble salts in the root zone can be a serious problem in arid regions. In subhumid regions, where irrigation is provided on a supplemental basis, salinity is usually of little concern because rainfall is sufficient to leach any accumulated salts; however, all water from surface and underground sources contains dissolved salts. The salt applied to the soil with irrigation water remains in the soil unless it is flushed out in drainage water or is removed in the harvested crop. Usually the quantity of salt removed by crops is so small that it will not make a significant contribution to salt removal or enter into determination of leaching requirements.

Salt-affected soils may be classified as saline, sodic, or saline-sodic soils. The classification depends on the electrical conductivity, the exchangeable sodium percentage, and the pH. The electrical conductivity EC is given in deciSiemens/m (dS/m) or millimhos/cm (mmho/cm) (1 dS/m $= 1$ mmho/cm) and is obtained from a saturation extract of the soil. The degree of saturation of the soil-exchange complex with sodium is the exchangeable sodium percentage ESP and defined

$$ESP = \frac{\text{Exchangeable sodium (milliequivalents per 100-g soil)}}{\text{Cation-exchange capacity (milliequivalents per 100-g soil)}} \times 100 \quad (15.6)$$

Saline soils have an $EC > 4$ dS/m, $ESP < 15$ percent, and pH < 8.5. These soils contain sufficient soluble salts to interfere with the growth of most plants. Sodium salt concentrations are relatively low in comparison with calcium and magnesium salts. Saline soils often are recognized by the presence of white crusts on the soil, spotty stands, and by stunted and irregular plant growth. Saline soils generally are flocculated, and the permeability is comparable to that of similar nonsaline soils.

The principal effect of salinity is to reduce the availability of water to the plant. In cases of extremely high salinity, there may be curling and yellowing of the leaves, firing of the margins of the leaves, or actual death of the plant. Long before such effects are observed, the general nutrition and growth physiology will have been altered and the yield potential reduced.

Sodic soils have an $EC < 4$ dS/m, an $ESP > 15$ percent, and a pH between 8.5 and 10. These soils are relatively low in soluble salts, but contain sufficient exchangeable (adsorbed) sodium to interfere with the growth of most plants. Exchangeable sodium is adsorbed on the surfaces of the fine soil particles. It is not leached readily until displaced by other cations, such as calcium and magnesium.

As the proportion of exchangeable sodium increases, soils tend to become dispersed, less permeable to water, and of poorer tilth. High-sodium soils usually are plastic and sticky when wet and are prone to form clods and crusts on drying. These conditions result in reduced plant growth, poor germination, and, because of inadequate water penetration, poor root aeration and soil crusting.

Saline-sodic soils have an $EC > 4$ dS/m, $ESP > 15$ percent and pH > 8.5. These soils contain sufficient quantities of both total soluble salt and adsorbed sodium to reduce the yields of plants. As long as excess soluble salts are present, the physical properties of these soils are similar to those of saline soils. If excess soluble salts are removed, these soils may assume the properties of sodic soils.

Both sodic and saline-sodic soils may be improved by the replacement of the excess adsorbed sodium by calcium or magnesium. This usually is done by applying soluble amendments that supply these cations. Acid-forming amendments, such as sulfur or sulfuric acid, may be used on calcareous soils since they react with limestone (calcium carbonate) to form gypsum, a more soluble calcium salt.

With the necessity of using additional water beyond the needs of the plant to provide sufficient leaching, it is imperative to have adequate drainage of the root zone. Natural drainage through the underlying soil may be adequate. In cases where subsurface drainage is inadequate, open or pipe drains must be provided.

Water will rise 2 to 5 ft or more in the soil above the water table by capillarity. The height to which water will rise above a free-water surface depends on soil texture, structure, and other factors. Water reaching the surface evaporates, leaving a salt deposit typical of saline soils.

To prevent the water table from rising too near the soil surface, a drainage system must be installed. In addition to lowering the water table, the drainage system carries excess salts away. The drainage water is carried to an outlet for disposal. Many times this outlet is another water supply and the salinity level of that water is then increased for downstream users.

Some plants can tolerate large amounts of salt. Others are more easily injured. The relative salt tolerance of a number of crops is shown in Table 15.1. The tolerance of crops listed may vary somewhat, depending on the variety, the cultural practices, and climatic factors (see Ayers and Westcot, 1985 for more

TABLE 15.1 Soil (EC_e) and Irrigation Water (EC_w) Salinity Tolerance Levels for Crops[a]

| | Yield Potential | | | | |
| | 100% | | 90% | | 0% |
Crop	EC_e	EC_w	EC_e	EC_w	Max EC_e
Field crops					
Barley[b]	8.0	5.3	10.0	6.7	28
Corn	1.7	1.1	2.5	1.7	10
Cotton	7.7	5.1	9.6	6.4	27
Sorghum	4.0	2.7	5.1	3.4	18
Soybeans	5.0	3.3	5.5	3.7	10
Wheat[b]	6.0	4.0	7.4	4.9	20
Vegetable crops					
Beans	1.0	0.7	1.5	1.0	7
Lettuce	1.3	0.9	2.1	1.4	9
Potato, sweet potato	1.6	1.1	2.5	1.7	10
Tomato	2.5	1.7	3.5	2.3	13
Forage crops					
Alfalfa	2.0	1.3	3.4	2.2	16
Bermuda grass	6.9	4.6	8.5	5.7	23
Sudan grass	2.8	1.9	5.1	3.4	26
Fruit crops					
Date palm	4.0	2.7	6.8	4.5	32
Grape	1.5	1.0	2.5	1.7	12
Orange, grapefruit, lemon	1.7	1.1	2.3	1.6	8

Source: Ayers, R. S. and D. W. Westcot (1985). Reprinted with permission from Food and Agriculture Organization of the United Nations.

[a]All values are in dS/m at 25°C.

[b]During germination and seedling stage, EC_e should not exceed 4 or 5 dS/m. Data may not apply to new semi-dwarf varieties of wheat.

details). The term EC_e in the table denotes the electrical conductivity of the saturated extract of the soil in deciSiemens per meter (dS/m) at 25°C. EC_w is the electrical conductivity of the irrigation water in dS/m at 25°C. One dS/m represents about 670 ppm of dissolved salts. In general, a yield reduction of 10 percent (as a result of salinity) is considered acceptable since there are many other possible factors that might limit the maximum yield of a given crop. Similar data for additional crops are available from Ayers and Westcot (1985) and Tanji (1990).

15.4 Irrigation Water Quality Criteria

The chemical quality of water largely determines its suitability for irrigation (Tanji, 1990; and Ayers and Westcot, 1985). The most important characteristics of irrigation water are (1) total concentration of soluble salts, (2) proportion of sodium to other cations, (3) concentration of potentially toxic elements, and

(4) bicarbonate concentration as related to the concentration of calcium plus magnesium.

Total soluble salts are commonly indexed by the electrical conductivity of the water expressed in dS/m. The major ions causing salinity of water are the cations Na, Ca, and Mg and the anions Cl, SO_4, and HCO_3. Other ions are usually minor components of salinity.

Sodium is a significant hazard in irrigation water; however, it does not impair water uptake by plants but reduces the infiltration of water into the soil. The reduction in infiltration is generally attributed to dispersion and movement of clays into the soil pores, surface crusting, and swelling of expandable clays. The effect of sodium is determined by the other cations in the water and soil. The proportion of sodium to other cations or the sodium hazard of the water is indicated by the sodium-adsorption ratio, or SAR, calculated from

$$SAR = \frac{Na^+}{\sqrt{(Ca^{2+} + Mg^{2+}) \, / \, 2}} \tag{15.7}$$

where Na^+, Ca^{2+}, and Mg^{2+} represent the concentration in milliequivalents per liter of the respective ions. Organic matter, pH, and minerals in the soil also affect the influence of sodium. Recent recommendations are to evaluate the potential sodium hazard for each specific soil and water.

Boron, though essential to normal growth of plants, is toxic under some conditions in concentrations as low as 0.33 parts per million. High concentrations of bicarbonate ions may result in precipitation of calcium and magnesium bicarbonates from the soil solution, increasing the relative proportions of sodium and thus the sodium hazard. USDA Handbook No. 60 (USDA, 1954), Tanji (1990), or Ayres and Westcot (1985) should be consulted if there are potential problems with toxic elements or high bicarbonate ion concentration.

15.5 Leaching

Leaching is the only way the salts added to the soil by the irrigation water can be removed satisfactorily. Sufficient water must be applied to dissolve the excess salts and carry them away by subsurface drainage. The traditional concept of leaching involves the application of water to remove salts from the entire root zone; however, Ayers and Westcot (1985) have shown that salt accumulation can take place for short periods in the lower root zone without adverse effects. As can be noted in Fig. 14.1, most of the water transpired by the plant is taken from the upper portion of the root zone. Salts in the upper root zone will be leached to a considerable degree by normal applications of irrigation water and by rainfall. The same amount of water, when applied with more frequent irrigations, is more effective in removing salts from this critical upper portion of the root zone than from the lower root zone. Thus, high-frequency sprinkler irrigation, microirrigation, or surface irrigation should be effective for salinity control.

Other concepts that may be helpful in controlling salinity are the use of soil or water amendments, deep tillage, and irrigation before planting. Some cultural practices such as alternate furrow irrigation, planting configuration, etc. can aid in reducing the effects of salinity.

Ayers and Westcot (1985) summarized several methods for estimating the leaching requirement for specific crops. One method is

$$LR = EC_w / [5(EC_e) - EC_w]$$ (15.8)

where LR = leaching requirement expressed as a portion of the infiltrated water

EC_w = salinity of the irrigation water in dS/m

EC_e = average soil salinity tolerated by the specific crop from Table 15.1 in dS/m

LR represents that portion of the infiltrated water that must pass through the soil and percolate below the root zone. Thus, the depth of water infiltrated must equal $(1 + LR)$ times the soil water deficit to maintain a salt balance in the soil appropriate to the crop and the accepted potential yield reduction. With some irrigation applications, sufficient water moves through the root zone to leach the salts, and additional water is not needed. It is good practice to monitor the actual salt content of the soil through chemical analyses of soil samples to ensure that neither inadequate nor excessive quantities of water are being applied.

IRRIGATION MANAGEMENT

Irrigation management requires determining when to irrigate and how much water to apply at each irrigation. The objective is to manage the production system for profit and optimal crop production per unit of water without compromising the environment. In addition to knowing soil water relations discussed in Chapter 14, the irrigation manager needs to consider soil salinity, irrigation water quality, and the effects of rainfall. The irrigation manager also must know the capabilities and limitations of the system and how it should be operated and maintained for optimum efficiency. The characteristics of an irrigation system can be defined by its efficiency, however, poor management can greatly reduce the efficiency of a well-designed system. A comprehensive review of irrigation management is given by Hoffman et al. (1990).

15.6 Effective Rainfall

Rainfall adds water to the soil and must be considered when determining irrigation water requirements and when scheduling irrigations. Only a portion of the measured precipitation is effective in meeting the plant water requirement. The effectiveness of rainfall depends on the following: (1) evaporative demand, (2) soil water status, (3) crop characteristics, (4) rainfall amount and intensity, and (5) cropping practices. Effective rainfall P_e can be estimated using procedures developed by the SCS (1970) or measured under local conditions. In general, small depths of rainfall are not effective and only a portion of larger depths of rainfall are effective.

15.7 Irrigation Water Requirements

The total seasonal irrigation requirement, the flow rate required to meet the peak water demand, and the flow rate and time to apply a specific quantity of water are important factors in irrigation system management and design. The total seasonal irrigation requirement *(IR)* is the total amount of water that must be supplied over a growing season to a crop that is not limited by water, fertilizer, salinity, or diseases. In some instances, deficit irrigation is practiced where less water than needed to meet the full seasonal requirement is applied to optimize the use of limited irrigation water supplies and rainfall. The irrigation requirement is obtained from

$$IR = [(ET - P_e)(1 + LR)]/E_a \tag{15.9}$$

where IR = seasonal irrigation requirement, in.

ET = seasonal evapotranspiration, in.

P_e = effective rainfall, in.

LR = leaching requirement as defined by Eq. 15.8

E_a = application efficiency (decimal)

Additional water may be needed to meet the needs for cooling or frost protection and added to *IR*. Equation 15.9 assumes that the soil water contents at the beginning and end of the season are similar. In regions that receive off-season precipitation, it is often desirable to allow soil water to decline to low but safe levels at the end of the irrigation season. The soil water deficit may be refilled during the off-season by precipitation and the irrigation requirement reduced. Precipitation during the off-season can aid leaching, which reduces the water required.

Example 15.5. Corn is grown and sprinkler irrigated with water having an electrical conductivity of 1.2 dS/m; the average depth of irrigation is 2 in., and the application efficiency is 70 percent. The monthly *ET* and effective rainfall are as follows:

Month	ET (in.)	Effective Rainfall (in.)
May	4	2
June	6	4
July	8	3
August	8	2
September	3	1
Total	29	12

Compute the seasonal irrigation requirement.

Solution. Assume no yield reduction is preferred and apply Eq. 15.8. Read $EC_e = 1.7$ from Table 15.1, then:

$$LR = 1.2/[5(1.7) - 1.2] = 0.16$$

Now solve Eq. 15.9:

$$IR = [(29 - 12.0)(1 + 0.16)]/0.7 = 28.2 \text{ in.}$$

15.8 Irrigation Flow Management Equation

Irrigators must be able to utilize their water supply effectively over the field area to be irrigated. The following equation, which gives the relationship among flow rate, time, area, and depth of water to apply, is a valuable tool for irrigation management.

$$qt = cad \tag{15.10}$$

where q = the flow rate available for irrigation, cfs or L/s

t = the time that the flow is delivered to the area being irrigated, hours

a = the area irrigated, acres or ha

d = the total depth to be applied, inches or cm (typically the total depth of water applied is the soil water deficit divided by the application efficiency)

c = 1.0 for q in cfs, a in acres, and d in inches (valid because 1 cfs for 1 h equals 1 acre-in. with less than 1 percent error)

c = 28 for q in L/s, a in ha, and d in cm

This equation is very useful as an aid to irrigators since it can be used to determine (1) how long to irrigate a field, (2) the flow required, (3) the area that can be irrigated, or (4) the depth that can be applied.

Example 15.6. A flow of 2 cfs is available to irrigate a 30-acre field. The soil water deficit is 3 in. and the irrigation efficiency is 70 percent. Determine the time required for irrigation.

Solution. Apply Eq. 15.10:

$$\text{total depth } d = 3/0.70 = 4.3 \text{ inches}$$
$$\text{time } t = 30(4.3)/2 = 64.5 \text{ hours.}$$

15.9 Irrigation Scheduling

Irrigations must be scheduled according to water availability and crop need. If adequate water supplies are available, irrigations are usually provided to obtain optimum or maximum yield. Overirrigation should be avoided, however, as it can decrease yields by reducing soil aeration and leaching fertilizers while increasing water and energy costs. In addition, overirrigation can contribute to high water tables and water pollution. If water supplies are limited and/or expensive, the irrigation scheduling strategy becomes one of maximizing economic return. In practice, much irrigation water is applied on a routine schedule based on the experience of the farm manager.

Irrigation scheduling requires knowing when to irrigate and how much water to apply. When to irrigate can be determined on the basis of plant or soil indicators or water balance techniques. How much water to apply can be based on soil water measurements or water balance techniques.

Soil indicators are based on soil water measurement techniques as discussed in Chapter 14. Irrigation is scheduled when the readily available water has been depleted. With adequate water supplies, the soil water deficit is refilled.

Since the objective of irrigation is to provide water for plant growth, plant indicators can directly show the need for water. Growth and appearance are visual indicators of water status. Slow growth of leaves and stems may indicate water stress. Irregular growth patterns may suggest excessive wet and dry cycles. Plant wilting and dark color are also indicators of water stress. When visual indications of plant stress occur, the yield potential has usually been reduced. Earlier indications of plant stress are leaf and canopy temperatures, which can be easily measured with infrared thermometers. Increases in leaf temperatures relative to air temperature indicate transpiration is slowing, hence leaves are hotter than those of well-watered plants. Canopy temperatures can be obtained with thermal scanners from aircraft and analyzed with computers to show temperature variations over a field. Leaf water potentials become lower (more negative) when plants are water stressed; however, these measurements are difficult. Care must be taken when interpreting these signs, since diseases or nutrient deficiencies may cause similar reactions.

Water balance techniques can provide a record of the estimated water in the root zone. When the plants have removed the Readily Available Water (RAW), irrigation is scheduled. Calculation of RAW includes the irrigation manager's decision on Management Allowed Depletion (MAD) based on suggested values in Table 14.2. Crop and soil characteristics, water availability and costs, irrigation system capabilities, and other factors influence the values selected for MAD and RAW. The soil water balance may be determined from soil water measurements. If estimates of ET are available, the soil water balance can be calculated from

$$D_i = D_{i-1} + (ET - P_e)_i \qquad (15.11)$$

where D_i and D_{i-1} = the total depth of water deficit in the soil root zone at the end of days i and $i-1$, respectively, in.

ET = the evapotranspiration calculated from climatic data for day i, in.

P_e = the effective rainfall for day i, in.

Irrigation is normally scheduled when D_i equals RAW. Historic data can be used to estimate ET and P_e and a projected irrigation schedule developed either for the next irrigation or for the season. The water balance can be updated from current climatic data also.

Example 15.7. Determine the date and amount of water to apply to a field having a soil water deficit of 1.25 in. at the end of the day on June 14. The soil is a loam, the effective root zone depth is 3 ft, and MAD is 0.5.

Solution. From Table 14.1 read $AW = 2$ in./ft, then determine RAW from Eqs. 14.1 and 14.3. $AW = 2$ in./ft $\times$ 3 ft root zone = 6 in.

$$RAW = AW \times MAD = 6 \text{ in.} \times 0.5 = 3 \text{ in.}$$

For the following ET and P_e values, apply Eq. 15.11 to calculate D_i, beginning with $i = 15$ for June 15. For example, $D_{15} = 1.25 + 0.3 - 0 = 1.55$, $D_{16} = 1.55 + 0.27 - 0.2 = 1.62$, etc.

Date (June)	ET (in.)	P_e (in.)	D_i (in.)
15	0.3	0	1.55
16	0.27	0.2	1.62
17	0.33	0	1.95
18	0.3	0	2.25
19	0.32	0.4	2.17
20	0.35	0	2.52
21	0.35	0	2.87
22	0.3	0	3.17

On June 22 the calculated deficit exceeds *RAW*, therefore, irrigation should begin early June 22 with a net application of 2.9 in. If this were a large field requiring several days to irrigate, irrigation should begin earlier so the last set is irrigated before the soil water deficit in that set exceeds *RAW*.

In some cases, irrigation districts, government agencies, or irrigation consultants provide irrigation water management services to farm operators. These services can provide the irrigation manager with recommendations for effective management of irrigation activities. Evapotranspiration rates also may be available from government agencies. Computer programs may be used to facilitate data acquisition and handling. Programs are available that collect current climatic data or use historic data to calculate evapotranspiration, compute the soil water balance, and project when to irrigate.

IRRIGATION METHODS

Many factors should be considered when determining if an irrigation system should be installed and which system is best suited. A comparison of the different types of irrigation systems in relation to various site and situation factors is summarized in Table 15.2. The methods of applying water may be classified as subirrigation, surface irrigation, sprinkler irrigation, and microirrigation.

15.10   Subirrigation

In special situations, water may be applied below the soil surface by developing or maintaining a water table that allows water to move up through the root zone by capillary action. This practice is essentially the same as the controlled drainage discussed in Chapter 13. Controlled drainage becomes subirrigation if water must be supplied to maintain the desired water level. Water may be introduced into the soil profile through open ditches, mole drains, or pipe drains. The open ditch method is most widely used. In some river valleys and near lakes, subirrigation is a natural process. Water table maintenance is suitable where the soil in the plant root zone is relatively permeable and there is either a continuous impermeable layer or a natural water table below the root zone. Since subirrigation allows no opportunity for leaching and establishes an upward movement of water, salt accumulation is a hazard. The salt content of the water should be low or must be closely monitored, and a period of drainage included in the annual operation.

15.11   Surface Irrigation

By far the most common method of applying irrigation water, especially in arid regions, is by flooding the surface. Surface methods include wild flooding, where the flow of water is essentially uncontrolled, and surface applications, where the flow is controlled by furrows, corrugations, border dikes, contour dikes, or basins. Except in the case of wild flooding, the land should be carefully prepared before irrigation water is applied. To conserve water, the rate of water application should be measured and controlled, and the land properly graded (Chapter 16).

Efficient surface irrigation requires grading of the land surface to control the flow of water. The grading required depends on the topography. In some soil and topographical situations, the presence of unproductive subsoils may make grading for surface irrigation unfeasible. The utilization of level basins where large streams of water are available generally provides high irrigation efficiencies.

15.12   Sprinkler Irrigation

Lightweight portable pipes with slip joint connections were once common for water distribution; however, in view of the high labor cost in moving these systems, such applications are becoming limited to high-value crops. Mechanical-move systems are now widely accepted. These may be either intermittent or continuous. Solid-set and permanent systems are suitable for intensively cultivated areas growing a high-income crop, such as flowers, fruits, or vegetables.

Sprinkler irrigation systems provide reasonably uniform application of water. On coarse-textured soils, water application efficiency may be twice as high as with surface irrigation.

Sprinkler irrigation can be used for temperature moderation. Irrigation water, especially if supplied from wells, is often considerably warmer than the soil and air near the surface under frost conditions. The heat of fusion released by water freezing on plant parts keeps the temperature from falling below freezing. Conversely, irrigation may be used for cooling, particularly when germination occurs under high temperatures, or to delay premature blossoming of fruit trees when warm weather occurs before the frost danger has passed. The cooling effect of evaporation lowers the temperature of plant parts (Chapter 17).

Sprinkler irrigation is particularly adaptable to hilly land where grading for surface irrigation is not feasible. It is appropriate for most circumstances where the soil infiltration rate exceeds the rate of water application. With sprinkler irrigation, the rate of water application can be easily controlled. Sprinkler irrigation systems usually have a moderate initial cost. With mechanical-move systems, labor can be substantially reduced. In some cases, disease problems have resulted from moistened foliage. Evaporation losses with sprinkler irrigation are not excessively high, even in arid regions. A well-designed sprinkler irrigation system can provide a high efficiency of water application.

15.13   Microirrigation

Increasing use is being made of microirrigation (trickle or drip) systems that apply water at very low rates, often to individual plants. Such rates are achieved through the use of specially designed emitters or perforated tubes. A typical emitter might apply water at from 0.5 to 4 gal/h, and it is usually installed on or just below the soil surface. Perforated or porous tubes apply 0.2 to 0.6 gal/min per 100 ft of tubing and are often installed 2 to 12 in. below the soil surface. These systems provide an opportunity for efficient use of water because of minimum evaporation losses and because irrigation is limited to the root zone. Microirrigation is generally limited to high-value crops, because of its high costs. Plugging is a potential problem and requires careful management of physical and chemical constituents in the water. Since the distribution pipes are usually at or near the surface, operation of field equipment is difficult. Both sprinkler and microirrigation systems are well adapted to application of agricultural chemicals, such as fertilizers and pesticides, with the irrigation water (Chapter 18).

A well-designed microirrigation system can provide a high efficiency of water application. It is especially well-suited to tree fruit and high-value crops. Water must be clean and uncontaminated, usually achieved by a filtration system. Microirrigation lends itself well to automation and has a low labor requirement. Since microirrigation systems usually operate at low pressure, energy requirements are generally lower than with sprinkler systems; however, some low-pressure sprinklers operate at pressures comparable to those of microirrigation systems.

TABLE 15.2 Comparison of Irrigation Systems in Relation to Site and Situation Factors

Site and Situation Factors	Improved Surface Systems		Sprinkler Systems			Microirrigation Systems
	Redesigned Surface Systems	Level Basins	Intermittent Mechanical-Move	Continuous Mechanical-Move	Solid-Set and Permanent	Emitters and Porous Tubes
Infiltration rate	Moderate-to-low	Moderate	All	Medium-to-high	All	All
Topography	Moderate slopes	Small slopes	Level to rolling	Level to rolling	Level to rolling	All
Crops	All	All	Generally shorter crops	All but trees and vineyards	All	High value required
Water supply	Large streams	Very large streams	Small streams nearly continuous	Small streams nearly continuous	Small streams	Small streams, continuous and clean
Water quality	All but very high salts	All	Salty water may harm plants	Salty water may harm plants	Salty water may harm plants	All—can potentially use high salt waters
Efficiency	Average 60–70%	Average 80%	Average 70–80%	Average 80%	Average 70–80%	Average 80–90%
Labor requirement	High, training required	Low, some training	Moderate, some training	Low, some training	Low-to-seasonal high, little training	Low to high, some training

Capital requirement	Low-to-moderate	Moderate	Moderate	Moderate	High	High
Energy requirement	Low	Low	Moderate to high	Moderate-to-high	Moderate	Low-to-moderate
Management skill	Moderate	Moderate	Moderate	Moderate-to-high	Moderate	High
Machinery operations	Medium-to-long fields	Short fields	Medium field length, small interference	Some interference circular fields	Some interference	May have considerable interference
Duration of use	Short-to-long	Long	Short-to-medium	Short-to-medium	Long term	Long term, but durability unknown
Weather	All	All	Poor in windy conditions	Better in windy conditions than other sprinklers	Windy conditions reduce performance, good for cooling	All
Chemical application	Fair	Good	Good	Good	Good	Very good

Source: Fangmeier and Biggs (1986).

15.14   Automation

With high labor costs, much attention is given to minimizing labor requirements through automation. Mechanical-move and solid-set sprinkler systems
lend themselves well to automatic controls. The application of automatic controls to surface irrigation is also feasible in some cases. Electrically triggered,
mechanically operated control gates and valves can provide a considerable reduction in labor for most irrigation systems. Automation is also advantageous if
irrigation efficiency can be improved and water is saved.

REFERENCES

Ayers, R. S., and D. W. Westcot (1985) *Water Quality for Agriculture,* FAO Irrigation and
     Drainage Paper No. 29, Rev. 1. Food and Agricultural Organization of the United
     Nations, Rome.

Fangmeier, D. D., and E. N. Biggs (1986) *Alternative Irrigation Systems,* Rpt. 8555. Coop.
     Ext. Service, University of Arizona, Tucson, Arizona.

Hansen, V. E., O. W. Israelsen, and G. E. Stringham (1980) *Irrigation Principles and Prac-
     tices,* 4th ed. Wiley, New York.

Hoffman G. J., T. A. Howell, and K. H. Solomon (eds.) (1990) *Management of Farm Irriga-
     tion Systems,* Monograph. ASAE, St. Joseph, Michigan.

Jensen, M. E. (ed.) (1983) *Design and Operation of Farm Irrigation Systems,* Monograph No.
     3. ASAE, St. Joseph, Michigan.

Lowdermilk, W. C. (1953) *Conquest of the Land Through 7,000 Years,* Agr. Infor. Bul. No.
     99. U.S. Soil Conservation Service, U.S. Government Printing Office, Washing-
     ton, D.C.

National Research Council (1989) *Irrigation-Induced Water Quality Problems.* National Aca-
     demic Press, Washington D.C.

Stegman, E. C. et al. (1977) *Crop Curves for Water Balance Irrigation Scheduling in S.E. North
     Dakota.* North Dakota Agr. Expt. Sta. Res. Rep. 66, Fargo, North Dakota.

Tanji, K. K. (ed.) (1990) *Agricultural Salinity Assessment and Management.* American Soc. of
     Civil Engineers, New York.

U.S. Soil Conservation Service (SCS) (1970) *Irrigation Water Requirements,* Tech. Release
     No. 21. USDA, Washington, D.C.

U.S. Department of Agriculture (USDA) (1954) *Diagnosis and Improvement of Saline and
     Alkali Soils,* Handbook No. 60. U.S. Government Printing Office, Washington, D.C.

PROBLEMS

15.1 How long will it take to irrigate 50 acres with a flow of 4 cfs if 3.5 in. of water is
to be applied?

15.2 An irrigator has a stream of 750 gpm for 2 days to irrigate 25 acres. Determine
the average depth of water delivered.

15.3 What area can be irrigated with a flow of 900 gpm in 30 h if 3 in. is to be applied?

15.4 Determine the leaching requirement and depth of application to a soybean field if no yield reduction is acceptable, the salinity of the irrigation water is 1.5 dS/m, 15 days have elapsed since the last irrigation, and the average ET rate is 0.3 in./day. Assume surface irrigation and no precipitation.

15.5 Determine the water application efficiency, uniformity coefficient, and distribution uniformity if a stream of 3 cfs was delivered to a field for 2 h, runoff averaged 1.4 cfs for 1 h, and depth of water penetration varied linearly from 5 ft at the upper end of the field to 3 ft at the lower end of the field. The root zone depth is 5 ft.

15.6 Determine the irrigation requirement of wheat using the evapotranspiration data from Fig. 14.1*b*. Assume the only effective rainfall is 2 in. in March. The irrigation water has an $EC_w = 1.3$ dS/m.

15.7 Determine the date and amount of the next irrigation for cotton if the last irrigation was July 31, evapotranspiration is 0.3 in./day, effective rainfall is 0.5 in. on August 6, soil is a clay loam 4 ft deep, and MAD is 0.55 percent of the available water.

CHAPTER 16
SURFACE IRRIGATION

Surface irrigation is the application of water to the soil by allowing the water to flow over the soil surface. Levees, dikes, or furrows are constructed to control the flow and aid in distributing the water over a field. It is generally believed that surface irrigation is the least efficient method, however, recent innovations in water control have greatly improved potential efficiencies.

Surface irrigation is the predominant method of irrigation in the United States and in most other countries with large irrigation areas. A 1993 U.S. irrigation survey (1994 *Irrigation Jour.*) reported that 55 percent of the irrigation was accomplished with surface methods.

Surface irrigation is better suited to certain conditions, which are:

1. Level topography—When the land is flat, leveling costs are low and surface irrigation efficiencies are often higher. (See Chapter 12 for land-leveling procedures.)

2. Uniform deep soils of medium texture—These soils have high water-holding capacities, consequently larger depths of water can be applied at each irrigation. Surface irrigation systems are least efficient when the depth of application is small. Also, medium-textured soils have moderate infiltration rates in which surface irrigation application efficiencies are higher than for soils with very high or very low infiltration rates. Since land leveling may remove some of the topsoil, the impact on crop production is reduced with deep soils.

3. Large streams—Since the water is allowed to flow over the soil surface, a large stream is needed to cover the area quickly. However, if not managed carefully, it is easy to apply excess water.

4. Experienced labor—Irrigators must balance the soil and flow characteristics to provide optimum efficiency. The knowledge to manage the system effectively is improved by experience.

In the western states, where surface irrigation predominates, the major water supply is surface runoff, usually stored in reservoirs. Since this water must be conveyed for considerable distances over rough terrain, conveyance canals and control structures are key parts of most irrigation systems in arid regions. These systems must be capable of carrying hundreds or even thousands of cubic feet per second (cfs). Ground water also provides an important source of water for surface irrigation. Frequently, wells for ground-water pumping are located on the farm and conveyance systems are small.

DISTRIBUTION OF WATER ON THE FARM

The farm water supply is normally delivered either by conveyance ditches from surface storage or from irrigation wells. Sometimes surface storage and underground water supplies are combined to provide an adequate water supply to the farm. Mixing waters of different qualities may permit the utilization of poor quality waters.

16.1   Surface Ditches

A system of open ditches often distributes water from the source on the farm to the field as shown in Fig. 16.1. These ditch systems should be carefully designed to provide adequate head (elevation) and capacity to supply water to all areas to be irrigated. The amount of land that can be irrigated is often limited by the quantity of water available and by the design and location of the ditch system. Where irrigation is used as an occasional supplement to rainfall, these ditches may be temporary. To minimize water losses, there is an increasing tendency to line ditches with impermeable materials. This practice is particularly applicable where soils are sandy and in more arid regions where irrigation water supplies are limited and crop needs are dependent largely on irrigation water.

16.2   Devices to Control Water Flow

Control structures, such as illustrated in Fig. 16.2, are essential in open ditch systems to (1) divide the flow into two or more ditches, (2) lower the water elevation without erosion, (3) raise the water level in the ditch so that it will have adequate head for removal, and (4) measure the flow. Various devices are used to divert water from the irrigation ditch and to control its flow to the appropriate basin, furrow, or border. Valves or gates may be installed in the side or bottom of the ditch during construction. Other devices shown in Fig. 16.3 are spiles, gate turnouts, and siphon tubes. Siphons, usually plastic or aluminum, carry the water from the ditch to the surface of the field and have the advantage of metering the quantity of water applied. Figure 16.4 gives the rate of flow that can be expected from siphons of various diameters and heads (water surface elevation differences between the inflow and outflow ends of the tubes).

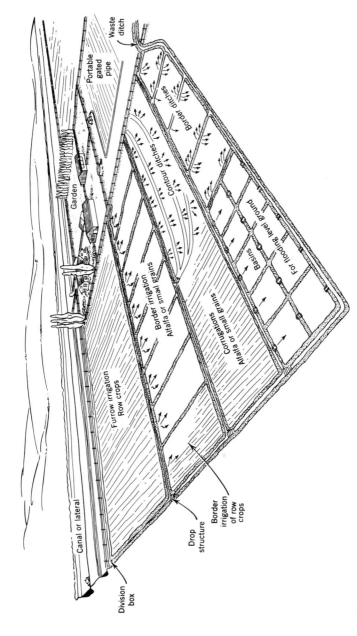

FIGURE 16.1 Surface methods of applying irrigation water to field crops. (Redrawn from U.S. Soil Conservation Service, 1947.)

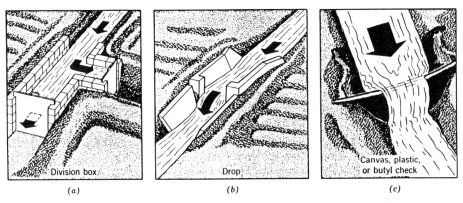

FIGURE 16.2 Devices to control water flow in irrigation ditches: *(a)* division box, *(b)* drop, *(c)* canvas, plastic, or butyl check dam. (Adapted from U.S. Soil Conservation Service and U.S. Bureau of Reclamation, 1959, and U.S. Soil Conservation Service, 1967.)

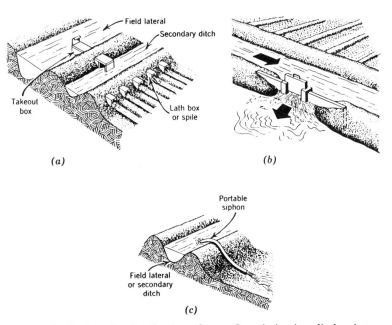

FIGURE 16.3 Devices for distribution of water from irrigation ditches into fields: *(a)* spile or lathe box, *(b)* border takeout, *(c)* siphon. (Adapted from U.S. Soil Conservation Service and U.S. Bureau of Reclamation, 1959.)

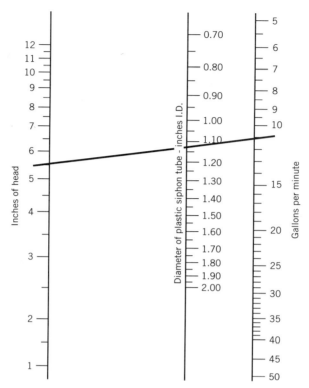

FIGURE 16.4 Discharge versus head for plastic siphon tubes. (Source: Hansen et al., 1980.) Reprinted with permission of John Wiley & Sons, Inc.

16.3   Underground Pipe

Although open-channel distribution systems provide high capacities at lower costs than pipe systems, they have continuing maintenance problems, constitute an obstruction to farming operations, may pose a safety hazard, and are subject to surface evaporation losses. Since underground pipe distribution systems (Fig. 16.5a) are easier to manage, they are becoming increasingly popular. In these systems, water flows from the distribution pipe upward through riser pipes and irrigation valves (often called alfalfa valves, Fig. 16.5a) to the desired basins, borders, or furrows. In Fig. 16.5b, a multiple-outlet riser controls the distribution of water to several furrows.

16.4   Portable Pipe

Surface irrigation with portable aluminum pipe or large-diameter plastic tubing may be advantageous, particularly where water is applied infrequently. Lightweight gated pipe (Fig. 16.5c) provides a convenient and portable method of applying water to furrows.

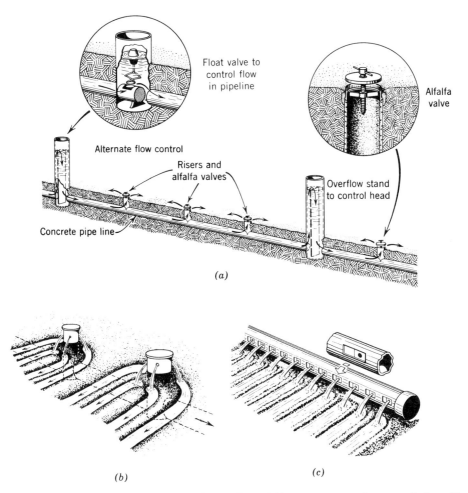

Float valve to
control flow
in pipeline

Alfalfa
valve

Alternate flow control

Risers and
alfalfa valves

Overflow stand
to control head

Concrete pipe line

(a)

(b)

(c)

FIGURE 16.5 Methods of distribution of water from (a) low-pressure underground pipe, (b) multiple-outlet risers, and (c) portable gated pipe. (Adapted from U.S. Soil Conservation Service and U.S. Bureau of Reclamation, 1959, and U.S. Soil Conservation Service, 1967.)

SURFACE IRRIGATION METHODS

The various surface irrigation methods for applying water to field crops shown in Fig. 16.1 include flooding, borders, furrows, and basins.

16.5 Flooding

Flooding is the application of irrigation from field ditches that may be nearly on the contour or up and down the slope. After the water leaves the ditches, no attempt is made to control the flow by means of levees or other methods of restricting water movement. For this reason, flooding is frequently referred to as "wild flooding." Although the initial cost for land preparation is low, labor requirements are usually high and the efficiency of water application is low. Flooding is most suitable for close-growing crops, particularly where slopes are steep, such as mountain meadows. Contour ditches are spaced 50 to 150 ft apart, depending on the slope, texture, and depth of the soil, stream size, and crop grown. This method may be selected on rolling land where borders, basins, and furrow are not feasible and adequate water supplies are available.

16.6 Graded Borders

Graded borders are typically used for close-growing crops such as grains, alfalfa, and grasses. The graded border method of irrigation consists of dividing the field into a series of strips separated by low ridges or levees. Normally, the direction of the strip is in the direction of greatest slope, but in some cases the borders are placed nearly on the contour. Slopes less than 0.5 percent are most suitable for border irrigation; however, slopes up to 4 percent may be used if erosion can be controlled. The strips usually vary from 30 to 60 ft in width and are 300 to 1300 ft in length.

Water is delivered to the upper end of the border strip and floods the surface. Inflow is usually cut off before the stream advances to the lower end of the field. After cutoff, the stream continues to advance down the field. Thus, the cutoff time should be selected to allow adequate irrigation of the lower end of the field with minimun runoff. Typical cutoff times are 2 to 4 h.

Ridges between borders should be sufficiently high to prevent overtopping during irrigation. To prevent water from concentrating on either side of the border, the land should be level perpendicular to the flow. Where row crops are grown in the border strip, furrows confine the flow and eliminate this difficulty. The efficiencies of well-designed, well-managed border irrigation systems are typically 60 to 70 percent.

16.7 Level Borders and Basins

The layout of level borders is similar to that of graded borders, except that the surface is leveled within the area to be irrigated. If the areas are nearly square, they are called basins and may range in size from 1 tree to 10 or more acres. Modern laser leveling techniques have the precision to form the smooth level surface required for this method. With relatively large flow rates, the field can be quickly covered, and application efficiencies of 80 to 90 percent are typical. Because of the high flow rates, special flow control and energy dissipation devices are needed at the field inlets. These systems are more expensive than sloping fields because leveling costs are higher, however, they can be automated because water is applied on a volume basis. Level fields are adaptable to most crops, except those susceptible to diseases when flooded. Also, if excess water is applied, prolonged ponding can increase the potential for crop damage.

16.8 Furrows

Although in the border and basin methods water covers the entire soil surface, irrigation by furrows submerges only $\frac{1}{5}$ to $\frac{1}{2}$ the surface, resulting in less puddling of the soil, less evaporation, and allows cultivation sooner after irrigation. Furrows vary in size and may be up and down the slope or on the contour. Small, shallow furrows, called corrugations, are particularly suitable for relatively irregular topography and close-growing crops, such as meadows and small grains. Furrows 3 to 8 in. deep are especially suited to row crops, since the furrow can be constructed with normal tillage. Slopes less than 2 percent are best suited for this method; however, contour furrow irrigation may be practiced on slopes up to 12 percent, depending on the crop, the erodibility of the soil, and the size of the irrigation stream. To minimize erosion in furrows, the irrigation stream size should not be larger than the value calculated from the following equation.

$$Q = 10/S \qquad (16.1)$$

where Q is the maximum flow rate per furrow in gpm and S is the furrow slope in percent.

Zero furrow slopes are allowable and desirable in arid and semiarid areas if the crops permit impoundment of water around the root system. In humid and subhumid areas, a minimum slope of 0.05 percent should be provided to ensure surface drainage. The slope perpendicular to the direction of the furrow should not result in either irrigation water or storm runoff overtopping the ridges. The maximum recommended furrow slope for erodible soils is given in Table 16.1. This is based on the 30-min rainfall with a 2-year frequency (Fig. 16.6).

The length of the furrows depends on the infiltration capacity, the slope, and the size of the stream. Excessively long furrows result in deep percolation

TABLE 16.1 Maximum Recommended Furrow Slopes for Erodible Soils[a]

30-min Rainfall (in.)	Maximum Furrow Slope (percent)
0.4	3.3
.6	1.9
.8	1.3
1.0	1.0
1.2	.8
1.4	.65
1.6	.55
1.8	.50

Source: SCS (1984).

[a]Less erodible soils may exceed these limits by approximately 25 percent.

FIGURE 16.6 Contours of 2-year, 30-min rainfall in inches for the United States. (From U.S. Soil Conservation Service, 1984.)

losses and erosion at the upper ends of the furrows. Suggested lengths of furrows for various slopes and soil textures are given in Table 16.2.

Furrow irrigation efficiencies typically range from 40 to 60 percent. Because only a portion of the soil surface is wetted, water must be applied for long periods of time, which causes deep percolation or runoff. To reduce the runoff, the inflow stream to the furrow can be cut back (reduced) after water reaches the end of the field and the application efficiency greatly improves. This procedure increases the labor requirement because cutback must be made during the irrigation. Unless the supply flow can be decreased, additional labor is re-

TABLE 16.2 Lengths of Run for Furrows and Corrugations[a]

	Lengths of Furrows or Corrugations (ft)			
Slope (percent)	Loamy Sand and Coarse Sandy Loams	Sandy Loams	Silt Loams	Clay Loams
0–2	250–400	300–660	660–1320	880–1320
2–5	200–300	200–300	300–660	400–880
5–8	150–200	150–250	200–300	250–400
8–15	100–150	100–200	100–200	200–300

[a]From USBR, 1951.

quired to use the flow remaining from the cutback stream. An alternative to cutback is installation of a runoff reuse system.

In cases where water advance is too slow, for example when infiltration rates are high, surge irrigation may be advantageous. Here intermittent, rather than continuous, streams are delivered to furrows. Between surges all the water infiltrates. The next surge advances faster across the previously wetted portion because the infiltration rate is lower and roughness may be reduced. Fast advance improves uniformity, and, with proper design and management, runoff and deep percolation can be reduced, increasing application efficiencies to near 80 percent. Special surge valves are available to facilitate surge irrigation. These valves alternate the flow between two sets of furrows to form the surges. The water is delivered to the furrows through gated pipe connected to the surge valves.

16.9   Runoff Reuse Systems

Runoff reuse or tailwater systems collect the runoff from graded border or furrow systems. These systems may simply deliver the water to another field or collect and store the water in a sump or reservoir. A pump installed in the sump and a pipeline allow delivery of the runoff to the same or another field. If a small sump is installed, pumps must cycle frequently, but not more than 15 cycles per hour to avoid damaging the motor. With a large sump, sufficient water may be stored to irrigate a regular set. Also, the sump and pump can supply water to the normal stream until water reaches the end of the furrow, then the pump is stopped to give a cutback stream. Runoff reuse systems can increase farm application efficiencies to 80 percent.

MANAGEMENT AND EVALUATION

A recognition and understanding of the variables involved in the hydraulics of surface irrigation are essential to effective management and evaluation. Hansen et al. (1980) listed the pertinent variables as (1) size of stream, (2) rate of advance, (3) length of run and time involved, (4) depth of flow, (5) infiltration rate, (6) slope of land surface, (7) surface roughness, (8) erosion hazard, (9) shape of flow channel, (10) depth of water to be applied, and (11) fluid characteristics. These variables are illustrated in the schematic view of the flow in Fig. 16.7.

Irrigators should understand the interrelationships among these variables to efficiently apply water. Time of application is the easiest variable to control since it can be changed during the irrigation. The stream size is next easiest to adjust. Even if a constant stream is available, the effective stream size may be adjusted by changing the number of furrows or borders irrigated in a set. To determine if adjustments are necessary or beneficial, evaluation of the current irrigation practices is very helpful.

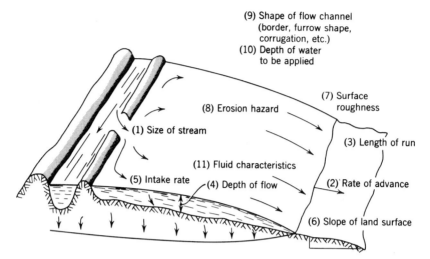

FIGURE 16.7 Schematic view of flow in surface irrigation indicating the variables involved. (Source: Hansen et al., 1980.) Reprinted with permission of John Wiley & Sons, Inc.

16.10 Evaluation of Existing Systems

Irrigation managers must continually assure that irrigation systems are operated efficiently. This is achieved by regularly evaluating the performance of a water application. Many older systems were designed before the interaction of the variables affecting surface irrigation was understood, and when water supplies were not as limited. Fields or portions of fields that do not receive sufficient water have limited production potential. Excessive irrigation not only wastes water, but leaches water-soluble nutrients and may cause drainage problems. In addition, runoff may flood adjacent fields and roads as well as carry pesticides and weed seeds to these fields or to downstream water supplies.

The evaluation of existing systems can be approached in a number of ways. They may be very simple or require elaborate measurements of a complete irrigation. The following describes the various procedures and measurements that can be taken. The simplest evaluation can be obtained from the procedure given under Soil Water Penetration. Additional measurements can be added in the order presented, except soil water content measurements can replace soil water penetration data.

SOIL WATER PENETRATION The simplest measurement, but very valuable, is to probe the soil the day after an irrigation. The purpose is to identify the depth of the wetting front and indicate if the field has been adequately irrigated. A simple probe can be constructed with a $\frac{1}{4}$- or $\frac{3}{8}$-in. rod. One end of the rod should be expanded so that when the rod is pushed into the soil it will form a hole slightly larger than the rod. A handle is attached to the other end of the rod. The rod will penetrate easily to the depth that the soil is well wetted, but will be stopped by dry soil. Thus, it will easily show areas that were underirrigated. The irrigation uniformity can be estimated from the depths of penetration. If the rod penetrates its full length everywhere, it probably indicates overirrigation. This method cannot detect if the soil was wet from a previous irrigation. It does not consider differences in soil type, and the probe will also be stopped by rocks. While the irrigator is walking the field probing the soil, observations can be made of plant growth, diseases, and insects. This additional knowledge can be beneficial for overall crop management.

WATER MEASUREMENT Flow measurement devices should be installed permanently to measure the inflow for every irrigation so the irrigator can determine the average depth applied. If such a device has not been installed, the appropriate device should be installed for the evaluation of the system. An additional measuring device should be installed to measure the runoff. The difference between inflow and runoff is the water infiltrated, which should be compared to the soil water deficit prior to irrigation. From these measurements, the irrigator can make an estimate of application efficiency.

SOIL WATER CONTENT MEASUREMENT Since the objective of an irrigation is to replace the soil water deficit, the deficit should be known before irrigation. The soil water deficit can be estimated from *ET* calculations or more accurately obtained by soil measurements (Chapter 14). These measurements will include the effects of previous irrigations and differences in soil type. Soil water measurements should be taken before and 1 to 2 days after an irrigation. The difference between the after and before irrigation water contents will be the depths of water stored in the soil by irrigation; these depths will be used to estimate uniformity. Furthermore, these depths combined with the flow measurements described previously will permit calculation of the application efficiency.

ADVANCE AND RECESSION MEASUREMENT When water is introduced into a field, a front of water advances across the field. To obtain advance data, stakes are usually placed at 100-ft intervals down the field, and the time that the average waterfront reaches that distance recorded. After the inflow is cut off, the water infiltrates and/or runs off the field. When water is no longer visible on about $\frac{3}{4}$ of the soil surface adjacent to each stake, this *recession* time is recorded, indicating water has receded from the field surface. The difference between the advance and recession times is the time water was on the soil surface for infiltration, hence this time is known as the infiltration opportunity time. When the advance and recession curves are plotted, the irrigator has a graphic representation of the uniformity of the irrigation (Fig. 16.8). Parallel curves occur for a uniform irrigation, assuming uniform soils and infiltration rates. With experience, the advance and recession curves reveal much about the performance of an irrigation. Soil surface elevations are frequently obtained near the stakes, which may aid in interpreting the results.

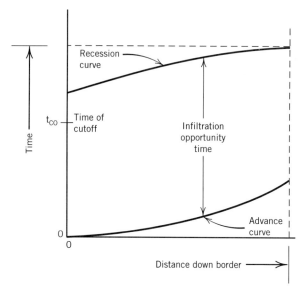

FIGURE 16.8   Advance and recession curves for a border irrigation.

INFILTRATION MEASUREMENT   Infiltration is usually obtained from a steel cylinder driven into the soil in borders and basins. Water is carefully added to the cylinder, and the change in water depth recorded with time. Cylinders may be used in furrows or a short section blocked with plates and water ponded between the plates. It is important to keep the depth of water constant between the plates and at about the same depth as the irrigation flow so that the wetted areas are similar. Since the water is ponded, these measurements of infiltration are frequently lower than those obtained with flowing water. The accumulated depth of infiltration should be plotted versus time. Depths of infiltration for various locations in the field can be obtained from the infiltration opportunity times.

Based on the preceding measurements, analysis of the evaluation should determine first if the optimum application time, and second if the optimum stream size were achieved since these variables are most easily changed by the irrigator. If slopes and lengths are not appropriate, the system may need to be redesigned. For more details on evaluating farm irrigation systems, see Merriam and Keller (1978).

REFERENCES

Hansen, V. E., O. W. Israelsen, and G. E. Stringham (1980) *Irrigation Principles and Practices,* 4th ed. Wiley, New York.

"1993 Irrigation Survey" (1994) *Irrigation Jour.,* 44(1):41.

Merriam, J. L., and J. Keller (1978) *Farm Irrigation System Evaluation: A Guide for Management.* Utah State Univ., Logan, Utah.

U.S. Bureau of Reclamation (1951) *Irrigation Advisor's Guide.* U.S. Dept. of Interior, Washington, D.C.

U.S. Soil Conservation Service (SCS) (1947) *First Aid for the Irrigator,* USDA Misc. Pub. 624.

U.S. Soil Conservation Service (SCS) (1967) "Planning Farm Irrigation Systems." In *National Engineering Handbook,* Sect. 15, Chapter 3, Washington, D.C.

U.S. Soil Conservation Service (SCS) (1974) "Border Irrigation." In *National Engineering Handbook,* Sect. 15, Chapter 4, Washington, D.C.

U.S. Soil Conservation Service (SCS) (1984) "Furrow Irrigation," In *National Engineering Handbook,* Sect. 15, Chapter 5, Washington, D.C.

U.S. Soil Conservation Service (SCS) and U.S. Bureau of Reclamation (USBR) (1959) *Irrigation on Western Farms,* Agr. Inf. Bull. 199. Washington, D.C.

PROBLEMS

16.1 How many 2-in. diameter siphon tubes are required to supply 1 cfs to a border strip if the head on the tube is 12 in. (1 cu. ft = 7.5 gal)?

16.2 If a flow of 18 gpm is to be supplied to a furrow by two 1-in. diameter siphon tubes, how far below the water level in the ditch should the tube outlets be placed?

16.3 What is the maximum flow rate in a furrow on a 0.5 percent slope that will cause minimum erosion?

16.4 Determine the maximum recommended furrow slope for eastern Nebraska.

16.5 A 10-acre field is irrigated with a 2 cfs stream for 24 h. Runoff was measured and averaged 0.5 cfs for 6 h. (a) Based on only these two measurements, what is the maximum possible application efficiency? (b) If soil water content measurements before and after the irrigation show 3.5 in. of water was stored in the root zone, what is the application efficiency?

16.6 The distribution of infiltrated depth after irrigation of a border varies linearly from 8 in. at the upper end of the field to 4 in. at the lower end of the field. Determine the distribution uniformity based on the low quarter.

16.7 Assume the border in Problem 16.6 is 800 ft long and 50 ft wide, and the soil water deficit before irrigation was 4 in. A flow of 2 cfs was delivered for 3 h. What is the application efficiency?

CHAPTER 17 _____
SPRINKLER IRRIGATION

Sprinkler irrigation is a versatile means of applying water to any crop, soil, and topographical condition. With this method, water is sprayed into the air and falls (much like rainfall) on the crop and soil. It is popular because surface ditches are not necessary, prior land preparation is minimal, and pipes are easily transported and provide no obstruction to farm operations when irrigation is not needed. Sprinkler irrigation is well suited for sandy soils or any other soil and topographical condition in which surface irrigation may be inefficient or expensive, or where erosion may be hazardous. Low amounts and rates of water may be applied, such as are required for seed germination, frost protection, delay of fruit budding, land application of wastewater, and cooling of crops in hot weather. Fertilizers and soil amendments may be dissolved in the water and applied through the irrigation system. Sprinklers are not suited to windy conditions that distort the sprinkler patterns and reduce efficiency and uniformity. Also, sprinkling with salty water may result in reduced yields because salts remain on the leaves as the water evaporates and sodium and chloride ions may be absorbed through the plant leaves. Other major concerns of sprinkler systems are energy and investment costs, and labor requirements.

According to the *Irrigation Journal* (1994) survey, over 42 percent of the irrigated land in the United States is irrigated by sprinkler systems. About 50 percent of this land is irrigated by center-pivot systems shown in Fig. 17.1. The leading states with sprinkler irrigation (in decreasing order) are Nebraska, California, Idaho, and Texas. Sprinkler irrigation is increasing and is expected to continue increasing.

17.1 Sprinkler Systems

In general, sprinkler systems are described according to the method of moving the lateral lines on which the sprinklers are attached. These systems are identified and compared in Fig. 17.2 and Table 17.1. During normal operation, laterals may be hand-moved or mechanically moved. The sprinkler system may apply water to only a small part of a field at one time or be a solid-set system in which sprinklers are placed over the entire field. With the solid-set system, part or all of the sprinklers may be operated at the same time. Many sprinklers rotate and wet a full circle. Some may be set to operate for any portion of a circle (part-cir-

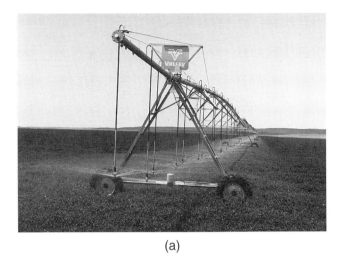

(a)

(b)

FIGURE 17.1 *(a)* Center-pivot sprinkler system. *(b)* Crop pattern of several adjacent systems. (Courtesy Valmont Irrigation Industries, Inc., Valley, Nebraska.)

cle) and are often placed near a field boundary and adjusted to prevent wetting adjacent fields. Nonrotating heads are also available and may be either full- or part-circle. Perforated pipes are suitable for distributing water to small acreages on high-value crops or plants, such as a nursery, where a rectangular pattern is desired.

The development of aluminum lateral pipe and rotating sprinklers led to the early use of sprinkler irrigation. These hand-move systems are suitable for low-growing crops, rather than tall crops, because of the adverse conditions for

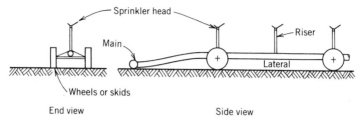

(a) End-pull lateral

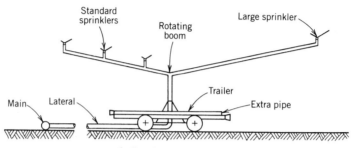

(b) Rotating boom-type unit

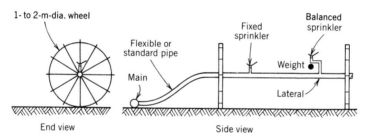

(c) Side-roll lateral (hand or mechanical move)

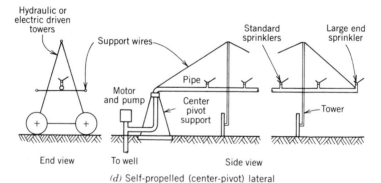

(d) Self-propelled (center-pivot) lateral

FIGURE 17.2   Mechanical-move sprinkler systems.

TABLE 17.1   Comparison of Sprinkler Irrigation Equipment

Type of System	Relative Investment Cost per Acre[a]	Relative Labor Cost	Practical Hours of Operation per Day
Hand-move laterals (standard sprinklers)	0.4	5.0	16
Hand-move laterals (giant sprinklers)	0.5	4.0	12–16
End-pull laterals (tractor tow)	0.5	1.4	16
Boom-type sprinklers (trailer-mounted)	0.6	3.7	12–16
Side-roll laterals (powered-wheel move)	0.7	1.7	18–20
Self-propelled (center pivot)	1.0	1.0	24
Solid set	3.0–5.0	1.0	24

[a]Based on a 160-acre field, 1000 gpm from pump, and 80% application efficiency. (Based on data from Berge and Groskopp, 1964.)

moving the pipe. As shown in Table 17.1, hand-move laterals have the lowest investment cost but the highest labor requirement. The labor requirement can be reduced with giant sprinklers, but application rates are higher, droplet size is larger, and higher pressures are required, which increase the pumping cost. These sprinklers are pulled or transported from one location to another or moved continuously.

With the end-pull systems shown in Fig. 17.2a, the lateral is moved to the next position by pulling one end with a tractor to the opposite side of the main. The lateral is pulled at a diagonal to move the entire length down the field one half the lateral spacing with each direction pulled. This system is suitable for low-growing crops, tree crops, and where adequate moving space is available to prevent damage to the crop. Labor is greatly reduced (Table 17.1) compared to hand-move systems.

The rotating boom-type system (Fig. 17.2b) operates with one trailer unit per lateral at spacings up to 350 ft. The trailer and boom unit is moved to the next position along the lateral with a tractor or winch. The lateral line is added or picked up as the trailer is moved progressively through the field, away from or toward the main line. The boom-type unit will cover about the same area as a giant sprinkler, but the pressure required is not as high.

The side-roll lateral system (Fig. 17.2c) uses the irrigation pipe as the axle for large diameter wheels that are spaced 30 to 40 ft apart. The lateral is moved to the next position by a small gasoline engine mounted at the midpoint of the line. The side-roll lateral is limited to crops that will not interfere with the movement of the pipe. Laterals are frequently connected to the main with a flexible hose, which must be disconnected for each move. Labor requirements are about the same as for the end-pull system, but much less than for hand-move systems (Table 17.1). Sprinklers that stay in a vertical position, regardless of the position of the wheels (Fig. 17.2c), eliminate the necessity of having the sprinkler on top, as required with fixed-position sprinklers.

The self-propelled center-pivot system (Figs. 17.1 and 17.2d) consists of a radial pipeline supported at a height of 10 to 12 ft by towers at intervals ranging from 140 to 170 ft. The radial line rotates slowly around a central pivot by either electric motors, water pressure, or oil hydraulic motors. The towers are supported by wheels and are kept in alignment by switches or wires. The nozzles may increase in size and increase or decrease in spacing from the pivot to the end of the line to account for the increased area irrigated with widening diameter. In addition, a large sprinkler may be installed at the end of the line to obtain the maximum diameter of coverage. The nozzles are selected to provide a uniform depth of application, varying from $\frac{1}{2}$ to 4 in. per revolution. The depth of application is determined by the speed of rotation. The system is best suited to sandy soils, but it will operate in heavier soils if the depth of application is greatly reduced. A common system designed for a quarter section (160 acres) is about 1300 ft in length. Such a system will irrigate about 135 acres of the 160 acres. The remaining 25 acres in the corners are not covered and may serve as homestead sites, woodlands, and wildlife habitats. Normally, center-pivot systems remain on a single field. Some can be towed, however, from the tower end to another field by rotating the wheels parallel to the pipe. The major advantages of the center-pivot are the savings in labor, and ease of adjusting the application depth. The major disadvantage is the investment cost.

Linear-move laterals use hardware similar to that of center-pivot systems, but move in a straight line across large rectangular fields. Water is supplied along the entire length of a field by a ditch or pipeline. If a ditch is used, an engine with pump and generator is mounted on a carriage unit beside the ditch. This unit pumps water from the ditch, generates electrical energy for the drive motors, and guides the system across the field. These units are typically 1300 or 2600 ft long and irrigate 160 or 320 acres.

Solid-set systems may be operated by changing the flow from one lateral to the next or by sequencing sprinklers (one operating on each lateral) along the lateral lines. For frost protection and cooling, all sprinklers are operated at the same time. Operation can be automated with timers and solenoid valves. As with self-propelled systems, the only labor required is for setup and maintenance. Because of high investments costs (Table 17.1), solid-set systems are used primarily for high-value crops and for turf.

17.2 Components of Sprinkler Systems

The major components of a hand-move portable sprinkler system are shown in Fig. 17.3. Basically the same components are required for the end-pull, rotating boom, and side-roll systems shown in Fig 17.2. With the rotating-boom or giant sprinkler systems, the lateral spacing and the distance between sprinklers along the lateral are much greater than for standard sprinklers. Latches, gaskets, and

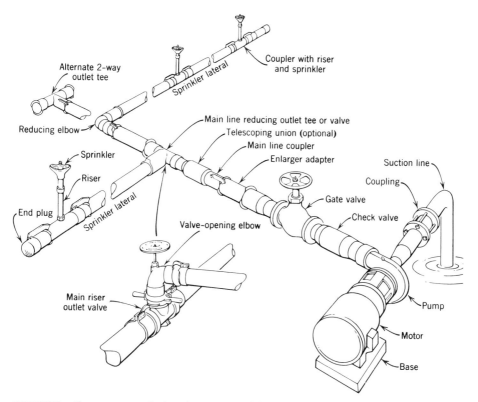

FIGURE 17.3   Components of a hand-move portable sprinkler system.

other details of these components vary considerably, depending on the manufacturer.

PUMPS   The first component, a pump, is required to lift water from the source, push it through the distribution system, and spray it over the area. Types of pumps used for irrigation vary with the rate of flow, the discharge pressure, and the vertical distance to the source of water. The pump should have adequate capacity for present and future needs. Centrifugal and vertical turbine pumps are suitable for sprinkler systems (see Chapter 10). If pumping is from a stream, pond, pit, or lake, an intake screen should be provided to avoid clogging and excessive wear of the sprinkler nozzles.

MAIN LINES AND LATERALS   The second component of the system, the main line, may be either moveable or permanent, and either aluminum or plastic; however, aluminum lines should not be buried unless coated. Moveable mains generally have a lower first cost and can be more easily adapted to a variety of conditions, whereas permanent mains offer a savings in labor and reduced obstruction to field operations. Water is taken from the main, either through valves placed at each junction with a lateral or, in some cases, through either an ell or a tee section that has been substituted for one of the standard couplings in the main (Fig. 17.3).

Moveable laterals usually consist of 20- or 30-ft lengths of 3- or 4-in.-diameter aluminum tubing with attached couplers. Two examples of quick-connecting couplers are shown in Fig. 17.4a. The riser and sprinkler head may be placed on the coupler or near the center of the pipe, to facilitate movement of the pipe.

SPRINKLER HEADS AND NOZZLES The operating pressure for most rotating sprinklers ranges from 30 to 100 lb per sq in. and most have two nozzles (Fig. 17.4b). One nozzle applies water at a considerable distance from the sprinkler (range nozzle) and the other covers the area near the sprinkler (spreader nozzle). Many sprinklers are rotated by the taps of a small hammer that is activated by the force of the water striking against a small vane connected to the hammer. Other sprinklers, particularly for turf, are gear-driven and powered by a small water turbine.

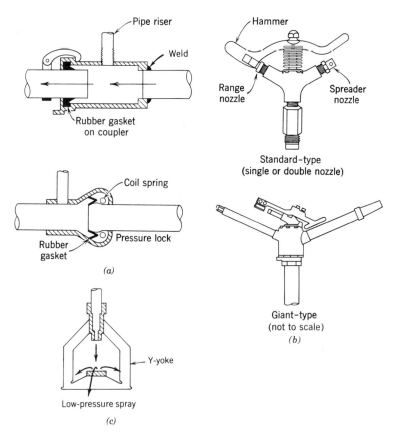

FIGURE 17.4 Examples of (a) quick-connecting couplers, (b) rotating sprinkler heads, and (c) low-pressure spray head.

A number of sprinkler heads are available for special purposes. Some provide low-angle jets for orchards. Some operate at pressures as low as 5 lb per sq in., and some operate only in a part-circle. Giant sprinklers may discharge 300 to 500 gpm at pressures over 100 lb per sq in. and spray the water up to 450 ft. These high pressures result in high pumping costs; however, because of the large area covered, a smaller number of moves is required.

Low-pressure sprays (Fig. 17.4c), which operate at 10 to 30 lb per sq in., were developed primarily for use on center-pivot and linear-move systems to reduce the pressure requirement. Droplet sizes are small, but application rates tend to be high unless small trusses or booms are attached nearly perpendicular to the lateral. Spray heads are mounted on the booms, both ahead and behind the lateral, to increase the wetted area and decrease the application rate. Pressure regulators may be installed with each head to assure uniform pressure, which increases the uniformity of water application.

17.3   Evaporation Losses

In sprinkler application of irrigation water, evaporation occurs from the spray and from the wet foliage surfaces. Frost (1963) developed relationships between spray losses and atmospheric parameters. He compared the evapotranspiration rates of various crops under nonsprinkling conditions to the evapotranspiration rates of the same foliage as it was being wetted by sprinkler water. It was found that the evaporation from wet foliage was essentially the same as for dry foliage. In those situations where wet-leaf loss was less than dry-leaf loss, energy available for evaporation of water was not reaching the wet leaves at the rate it reached dry leaves. The energy difference was absorbed by the spray. Thus, over a vegetated surface, the net loss of water is negligible in humid regions and 3 to 10 percent in arid regions. Losses may be higher with bare soils and windy conditions. Since winds carry fine droplets, spray heads, which have more fine droplets, have higher evaporation losses than other heads.

17.4   Distribution Pattern of Sprinklers

Most sprinklers have a circular wetting pattern; however, some sprinklers are designed for other patterns. The diameter of the pattern is affected by the nozzle diameter, pressure, and sprinkler design. The shape of the pattern also is affected by wind and somewhat by sprinkler wear. A typical distribution pattern showing the effect of wind for a single sprinkler is illustrated in Fig. 17.5. Since one sprinkler does not apply water uniformly over the area, sprinkler patterns are overlapped to provide more uniform coverage. The distribution pattern in Fig. 17.6 illustrates how the overlapping patterns combine to give a relatively uniform distribution between sprinklers. Although Fig. 17.6 shows a relatively uniform distribution over the area, wind will skew the pattern and give a less uniform distribution. A schematic of the overlapping patterns and the relative depth of water applied at various points between four adjacent sprinklers is given in Fig. 17.7.

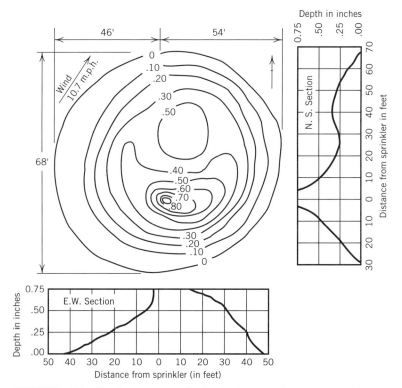

FIGURE 17.5 Distribution pattern of a single, intermediate-pressure, double-nozzle sprinkler showing the effect of wind. (Source: SCS, 1983.)

The factors that influence the distribution pattern of sprinklers are nozzle pressure, wind velocity, and speed of rotation. Too low a pressure will result in a "ring-shaped" distribution and a reduction in the area covered; high pressures produce smaller drops with high application rates near the sprinkler. Wind will cause a variable diameter of coverage and also somewhat higher rates near the sprinkler. A high speed of rotation of the sprinkler reduces the area covered and causes excessive wear of the sprinkler. The variation in speed of rotation is not due to the wind but to changes in frictional resistance attributed to lack of

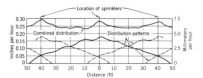

FIGURE 17.6 Sprinkler distribution pattern overlapping to give relatively uniform combined distribution.

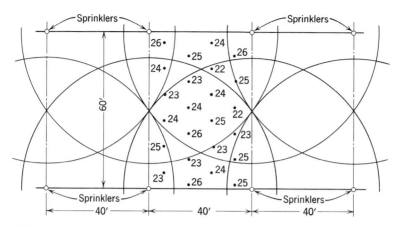

FIGURE 17.7 Relative depth distribution of water between sprinklers with overlapping patterns.

precision in manufacture and to wear. Uniformity of application of a sprinkler system can be expressed by either of the uniformity coefficients given in Chapter 15. The uniformity may be improved in windy areas by reducing the sprinkler spacing, by irrigating at night when the wind is low, or by using only the range nozzle on two-nozzle sprinklers. Another option with some systems is to offset the laterals midway between the positions used during the previous irrigation.

17.5   Size of Laterals and Mains

As water is forced through the irrigation system, head or pressure is lost because of friction of the water against the sides of the conduit. The amount of head lost depends largely on the rate of flow, the inside diameter of the pipe, the roughness of the inner surface of the pipe, and the number and abruptness of the turns.

Irrigation pipelines are sized to carry the required flow at minimum cost. Small pipes are less expensive, but have more energy loss from friction than large pipes with the same flow. Thus, pipe sizing is based on the minimum cost of the pipe and energy costs. The friction loss in pipes can be determined from the Hazen-Williams equation:

$$H_f = 10.46 \left(\frac{Q}{C}\right)^{1.85} \frac{L}{D^{4.87}} \tag{17.1}$$

where H_f = friction loss in feet
  Q = flow rate in the pipe in gpm
  L = pipe length in feet
  D = actual inside pipe diameter (not nominal diameter) in inches
  C = Hazen-Williams coefficient from Table 17.2.

TABLE 17.2   Hazen-Williams Coefficients for Equation 17.1.

Pipe Material	Values of C
Aluminum with couplers	120
Aluminum without couplers	140
Polyethylene (PE) and polyvinyl chloride (PVC)	140
Cast iron	100
Copper	130
Steel	100

The inside diameter of pipes can be obtained from tables in handbooks or from many pipe dealers. Likewise, tables are available that give the friction losses in pipes for various pipe materials, sizes, and flow rates.

Example 17.1.   Calculate the head loss in 300 ft of 2 in. schedule 40 PVC pipe (inside diameter = 2.067 in.) for a flow of 40 gpm.

Solution.   From Table 17.2 read $C = 140$, then use Eq. 17.1 to calculate the head loss.

$$H_f = 10.46 \times \left(\frac{40}{140}\right)^{1.85} \times \frac{300}{2.067^{4.87}}$$

$$H_f = 10.46 \times 0.098 \times 8.74 = 9.0 \text{ ft}$$

The pipe diameter should provide the required rate of flow with a reasonable head loss. In the case of a lateral, the sections at the distant end of the line will have less water to carry and may therefore be smaller. A disadvantage of tapering pipe diameters in laterals is that the various pipe sizes must be kept in the same relative positions. The system also may be less adaptable to other fields and situations. For permanent systems, pipe tapering is acceptable and reduces the initial cost.

The total pressure variation in laterals for intermittent-move systems should not be more than 20 percent of the average pressure in the lateral. If the lateral runs uphill or downhill, allowance for this difference in elevation should be made in determining the variation in head. If the water runs uphill, a greater head will be required, whereas if it runs downhill, there will be a tendency to balance the loss of head caused by friction.

The diameter of the main should provide the required pressure for the laterals in each of their positions. The rate of flow for each lateral may be determined by multiplying the number of sprinklers by the flow. The position of the laterals that gives the highest friction loss should be used for design purposes. Allowable friction loss in the main will vary with the cost of power and price differential between different diameters of pipe and fittings. The most economical size can be determined by balancing the increase in pumping cost against the amortized cost difference in the pipe sizes. Pumping against friction presents a continuing cost for as long as the system is operated.

The friction loss in laterals is more difficult to determine because the flow rate decreases at each sprinkler. Special tables have been developed (Pair et al., 1983) to adjust for the changing flow and are based on the number of sprinklers on the lateral.

INTERMITTENT-MOVE SPRINKLER SYSTEMS

Well-designed intermittent-move systems have application efficiencies of 70 to 80 percent and uniformities of 80 percent or higher. Winds can reduce these values significantly.

17.6   System Layouts of Mains and Laterals

The number of possible arrangements for the mains, laterals, and sprinklers is practically unlimited. The arrangement selected should allow a minimal investment in irrigation pipe, have a low labor requirement, and provide uniform application of water over the total area in the required time period. The most suitable layout can be determined only after a careful study of the conditions to be encountered. The choice will depend on the location of the water supply, crop, and the types and capacities of the sprinklers and their operating pressures. For medium-pressure systems, laterals are moved 40 to 60 ft between sets and the sprinklers are spaced 30 to 40 ft along each lateral.

Typical layouts for sprinkler systems are shown in Fig. 17.8. The layout in Fig. 17.8a is suitable where the water supply can be obtained from a stream or canal alongside the field to be irrigated. This arrangement either eliminates the main line or requires a relatively short main, depending on the number of moves of the pump. The layouts illustrated in Figs. 17.8b and 17.8c are suitable where the water supply is from a well or pond. In Fig. 17.8b, the two laterals are started at opposite ends of the field and moved in opposite directions. Since the farther half of the main supplies a maximum of one lateral at a time, the diameter of this section can be reduced. This arrangement is well suited to day and night operation when the required amount of water can be applied in 6 to 10 h. The system in Fig. 17.8c can be designed for continuous operation. While line A is in operation, the operator moves line B. When the required amount of water has been applied, line B is turned on and then line A is moved. With this procedure the capacity of the pumps needs to be adequate to supply only one lateral.

Some general rules that may be helpful for the layout of systems are as follows:

1. Sprinkler spacings both along the lateral and between laterals should be as wide as possible to reduce moving costs. Wider spacings require sprinklers with higher pressures, which produce higher application rates.
2. Mains should be laid uphill and downhill.
3. Laterals should be laid across the slope or nearly on the contour.

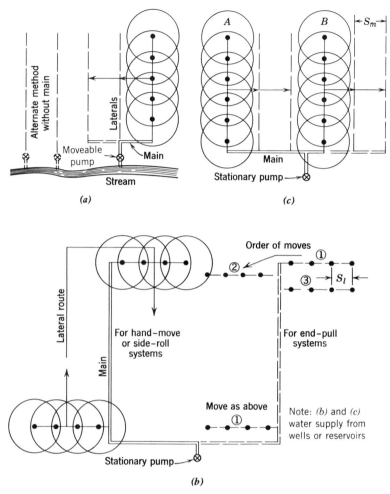

FIGURE 17.8 Field layouts of main and laterals for hand-move, side-roll, and end-pull sprinkler systems: *(a)* fully portable, *(b)* portable or permanent (buried) main and portable laterals, *(c)* portable main and laterals.

4. Whenever possible, laterals should be perpendicular to the prevailing wind direction.
5. Lateral pipe sizes should be limited to not more that two different diameters.
6. For a more balanced layout, lateral operation should conform to that shown in Fig. 17.8*b*.
7. Layout should facilitate and minimize lateral movement during the season.
8. Whenever possible, the water supply should be located near the center of the irrigated area.
9. The number of sprinklers operating for the various setups should be nearly the same.

17.7   Application Rate

The application rate of the sprinkler system should not exceed the infiltration rate of the soil. The application rate for the layout shown in Fig. 17.7 can be calculated from the following equation:

$$r = \frac{q \times 96.3}{S_l \times S_m} \tag{17.2}$$

where r = the application rate, in./h
q = the discharge rate of the sprinklers, gpm
S_l = the sprinkler spacing along the lateral, ft
S_m = the lateral spacing along the main, ft

To prevent runoff on steep slopes, such as highway embankments, the application rate should be lower than with the same soil on a level field.

17.8   Sprinkler Spacing

The spacing of sprinklers along the lateral S_l and along the main S_m is a function of the sprinkler discharge rate, the wetted diameter, wind speed, and the infiltration rate of the soil. Wetted diameters and discharge rates are provided by sprinkler manufacturers for each type, size, and operating pressure. An example of the characteristics of one sprinkler with five nozzle combinations is given in Table 17.3.

The uniformity of sprinkler application is determined by the amount of

TABLE 17.3   Manufacturer's Sprinkler Characteristics

	*Nozzle Diameters in Inches*								
Nozzle Pressure (psi)	$\frac{5}{32} \times \frac{1}{8}$		$\frac{5}{32} \times \frac{5}{32}$		$\frac{3}{16} \times \frac{5}{32}$		$\frac{7}{32} \times \frac{5}{32}$		$\frac{1}{4} \times \frac{5}{32}$	
	Dia.[a]	GPM[b]	Dia.[a]	GPM[b]	Dia.[a]	GPM[b]	Dia.[a]	GPM[b]	Dia.[a]	GPM[b]
20	77	4.6	77	5.6	78	6.6	81	8.2	82	9.7
25	80	5.3	80	6.2	81	7.6	85	9.2	86	11.0
30	83	5.8	83	6.8	86	8.3	88	10.2	91	12.1
35	85	6.3	85	7.4	89	9.1	92	11.1	96	13.2
40	87	6.8	87	8.0	93	9.7	96	11.9	101	14.2
45	89	7.2	89	8.5	96	10.3	101	12.7	105	15.0
50	92	7.5	92	9.0	99	10.8	105	13.4	111	15.8
55	94	7.9	94	9.4	101	11.4	100	13.9	115	16.6
60	97	8.3	97	9.8	103	11.8	112	14.5	119	17.4

Pressures to left and below dashed line recommended for best breakup of stream.

[a]Diameter of coverage in feet.

[b]The discharge for other nozzle sizes not shown may be computed from the formula, $q = 29.85\, C\, d^2\, P^{1/2}$, where q is the discharge in gallons per minute, C is the discharge coefficient, d is the nozzle diameter in inches, and P is the pressure in pounds per square inch (psi) using $C = 0.87$.

TABLE 17.4   Maximum Spacings for Low- or Medium-Pressure Sprinklers[a]

Wind Velocity (mph)	Lateral Spacings in Percent of the Diameter of Coverage	
	S_1, Along the Lateral	S_m, Along the Main
0	50	65
≤5	45	60
6–10	40	50
>10	30	40

[a]For laterals normal to the wind direction only.

overlap of the wetting patterns. Approximate maximum sprinkler spacings for satisfactory uniformity of application are given in Table 17.4 for a distribution pattern from a single sprinkler that is triangular in cross section or for one similar to that shown in Fig. 17.6. The spacings are functions of the wetted diameter and the wind speed. A better method to determine the overlap is to measure the actual wetting pattern of the sprinkler selected and evaluate uniformity of various sprinkler spacings. The spacing with the highest uniformity is selected or another sprinkler may need to be tested.

CONTINUOUS-MOVE SYSTEMS

These systems are typically more efficient and have higher uniformities than intermittent-move systems. Since these systems are continuously moving, there is no overlapping of patterns in the direction of movement. With center-pivot and linear-move systems, there is only one lateral and the spacing between sprinklers can be designed for close spacings for high uniformities.

17.9   Center-Pivot and Linear-Move Systems

These systems are usually designed by the manufacturer and sold as a complete unit. Center-pivot systems may have constant sprinkler spacings along the lateral with variable discharge or with a combination of spacings and discharges. Linear-move systems have constant spacings and discharges. Uniformity coefficients for moving lateral systems are as high as 80 to 90 percent even in winds up to 20 mph. Because the uniformities are higher than for intermittent-move systems, the application efficiencies are also higher. The rate of lateral movement determines the depth of application, but does not affect the sprinkler discharge. For energy savings, some systems have been designed to operate at pressures as low as 20 to 40 psi. At these low pressures, the application rate is higher because the wetted area is smaller.

Center-pivot and linear-move systems are usually designed for continuous operation to supply the peak water use rate of the crop in arid regions. Where rainfall and stored soil water can provide some of the water, the design rate may be less than the peak. These systems should be well maintained, since a crop may be lost or severely damaged if the system is not able to operate during a high water-use period.

17.10   Other Moving Sprinkler Systems

Other moving sprinkler systems, such as traveling guns, may be moved by cable or self-powered units. Water is often supplied through a flexible hose that must be moved and reconnected as irrigation progresses across the field, thus increasing the labor requirement.

The traveling gun, which is moved with a power winch and cable or is self-propelled, has a high capacity sprinkler. A traveling-boom system is similar to the traveling gun, except the rotating boom usually has a gun on one arm and several smaller sprinklers on the other arm. These systems are limited to soils with a high infiltration rate, or to fields with good vegetative cover to protect the soil surface from surface sealing and erosion.

SPECIAL SPRINKLER APPLICATIONS

17.11   Sprinkler Systems for Environmental Control

Sprinkling has been successful for protecting small plants from wind damage, soil from blowing, and plants from frosts or freezing. It has also been successful in reducing high air and soil temperatures. Since the entire area usually needs protection at the same time, solid-set systems are required. The rate of application should generally be low or just enough to achieve the desired control. Small pipes and low-volume sprinklers may be desired to reduce costs, but normal sprinkler systems may be modified for dual use. Because water is applied without regard to irrigation requirements, natural drainage should be adequate or a good drainage system should first be provided.

In organic or sandy soils where onions, carrots, lettuce, and other small seed crops are grown, the soil dries out quickly and the seed may be blown away or covered too deeply for germination. When these plants are small they are also easily damaged by wind-blown soil particles. Protection for such conditions can be provided with a sprinkler system that will apply low rates up to 0.1 in./h. Operation at night when winds are usually at a minimum will provide more uniform coverage.

Low-growing plants can be protected from freezing injury, which is likely either in the early spring or late fall. Sprinkling has been most successful against radiation frosts. Water must be applied continuously at 0.1 in./h or greater (depending on the wind and temperature), until the plant is free of ice. Sprinkling should be started before the temperature reaches 32°F at the plant level. Strawberries have been protected from temperatures as low as 21°F. Tomatoes, peppers, cranberries, apples, cherries, and citrus have been successfully protected.

Tall plants, such as trees, may suffer limb breakage when ice accumulates, but in a few areas, low-level undertree sprinkling has provided some control. Rates of application may be reduced by increasing the normal sprinkler spacing. A slightly higher pressure may be desirable to increase the diameter of coverage and to give better breakup of the water droplets.

Sprinkling during the day to reduce water stress has been successful with many plants, including lettuce, potatoes, green beans, small fruits, turf, tomatoes, cucumbers, and cantaloupe. This practice is sometimes called "misting" or "air conditioning" irrigation. Maximum stress in the plant usually occurs at high temperatures, at low humidity, with rapid air movement, on bright cloudless days, and/or with rapidly growing crops on dry soils. Under these conditions, crops at a critical stage of growth, such as emergence, flowering, or fruit enlargement, may benefit greatly from the low application of water during the midday. At 81°F, water loss was reduced 80 percent, with an increase of humidity from 50 to 90 percent. Measured temperature reductions in the plant canopy of about 20°F were attained by misting in an atmosphere of 38 percent relative humidity. Green bean yields were increased 52 percent by midday misting during the bloom and pod-development period. Potatoes and corn respond to sprinkling, especially when temperatures exceed 86°F. The tasseling period is a critical time for corn. Small quantities of water applied frequently to strawberries increased the quality and the yield by as much as 55 percent. For low-growing crops, the same sprinkler system can be adapted for frost protection. Misting in a greenhouse (or under a lath house) to reduce transpiration of nursery plants for propagation increases plant growth and root development.

During periods of high incoming radiation, the soil temperatures may be 60°F greater than the ambient air temperature. Seedlings emerging through soil temperatures as high as 120°F frequently die as a result of high transpiration. Small applications of water at this critical period often ensure emergence and good stands. Sprinkling at this stage also enhances the effectiveness of herbicides applied to control weeds. For further details, see Pair et al. (1983) and other current references.

17.12   Sprinkler Systems for Fertilizer, Chemical, or Waste Applications

Fertilizer, soil amendments, and pesticides may be injected into the sprinkler line as a convenient means of application to the soil or crop. This method primarily reduces labor costs, and in some instances, may improve the effectiveness and timeliness of the application. Liquid manure and sewage wastes are applied with sprinklers. A large amount of specialized commercial equipment is on the market. Cannery wastes are usually sprinkled on wooded areas or on land in permanent grass. Good subsurface drainage is required.

Liquid and dry fertilizers have been successfully applied with sprinklers. Dry material must first be dissolved in a supply tank. The liquid may be injected on the suction side of the pump, forced under pressure into the discharge line, or injected into the discharge line by a differential pressure device such as a venturi section. The material is applied for a short time during the irrigation set. Sprinkling should be continued for at least 30 min after it has been applied to rinse the supply tank, pipes, and the crop. For center-pivot or other moving systems, the material must be applied continuously or until the field has been completely covered.

For waste disposal systems, the solids should be well mixed with the liquids and small enough not to plug the nozzles. When pumping these solids, an effective nonplugging screen on the suction side of the pump is necessary. Liquid must be stored in a lagoon or other holding pond. Equipment should be resistant to corrosion from chemicals that may be present in the water. The system may be designed to apply water during subfreezing weather, or sufficient storage should be provided during the nonoperating period. Applications are not advisable on frozen soils. Such systems are usually solid-set and are operated for long periods. The application rate should be lower than the infiltration rate so that the water does not run off the surface and cause stream pollution. The depth of application should be monitored or managed to prevent excess water from passing through the soil and carrying harmful chemicals to the groundwater supply.

17.13 Sprinkler Irrigation Systems for Turf

Sprinkler systems are generally used to irrigate turf, such as golf courses, parks, athletic fields, and lawns. These systems are usually permanent installations, and the management criteria and design are similar to those for agricultural crops. There are, however, several important differences: (1) operation is often limited to the night hours to allow human use during daylight, (2) triangular rather than rectangular sprinkler layouts may be used, (3) heads are installed flush with the soil surface so they do not interfere with use and maintenance, (4) sprinklers usually have a "pop-up" feature in which the nozzles rise up to 2 in. above the base, so spray is not affected by the grass, (5) irregular-shaped areas make designs more difficult, and (6) spray is not allowed beyond the boundaries onto roads, walkways, or adjacent property, yet the entire area must be irrigated. To conserve fresh water supplies, treated waste-effluent waters are applied to turf. Installations must meet standards and codes for safety and prevent contamination of the water supply. For example, backflow prevention devices are required so that water cannot flow back into the supply system.

MANAGEMENT AND EVALUATION

Management of sprinkler systems is based on the principles described in Chapter 15. The irrigation manager or operator should verify that adequate, but not excessive, amounts of water are being applied by the sprinkler system. Also, irrigation must be started soon enough to cover the field before plants in the last portion suffer from water stress. To determine if the system was designed properly and is being managed and operated efficiently, an evaluation of the system is required. The procedure for evaluation is partially dependent on the type of system, however, a few simple measurements are applicable to any system. Field observations of the rate and uniformity of plant growth may serve as an evaluation. For example, poor growth might indicate insufficient water is being applied, however, poor soil or low fertility might show the same signs. Better growth near the sprinklers typically indicates a poor wetting pattern, and nozzle diameters, pressure, and distribution uniformity should be determined.

Nozzle diameters should be compared to the design specifications. Intermittent-move or solid-set systems usually have the same nozzle diameters on all sprinklers. Center-pivot systems frequently have varying diameters, thus it is essential to carefully check the design specifications to ensure that the proper nozzles are installed at the correct location along the lateral.

Another important measurement involves determining the soil water content by probing as described in Chapter 16. These measurements will show if the proper amount of water is being applied. The application depth is controlled by adjusting the application time.

17.14   Evaluation of Intermittent-Move Sprinkler Systems

A simple uniformity evaluation of an intermittent-move system requires measuring the pressure and flow rate at various nozzles throughout the system. The pressure is measured with a pressure gauge having a small tube (pitot) attached. The pitot is placed about $\frac{1}{8}$ in. from the nozzle, and pressure is read from the gauge. The pressure should not have more than a 20 percent variation along a lateral. Poor uniformity of water distribution occurs when the pressure variation is too large. Excessive pressure variation can be caused by worn or oversize nozzles, incorrect pipe size, elevation differences along the lateral, or high operating pressures. High and low pressures can be detected by observation of the sprays. High pressures produce small droplets (or fogging) and irregular rotation, with most of the water falling near the sprinkler and excessive drift in windy conditions. Low pressures cause large droplets, and most of the water falls in a circle away from the sprinkler in a "doughnut" pattern.

The flow rate is determined by catching the discharge from a sprinkler in a container of known volume or by weighing. A small hose is placed over the nozzle to divert the flow to the container for a short time, as determined with a stopwatch. For example, if a 2-gal container is filled in 30 sec, the flow is $2 \times (60/30) = 4$ gpm. The variation in flow rate along a lateral should not exceed 10 percent.

Uniformity may also be evaluated by probing the soil. For example, if the soil is appreciably drier at the far end of the lateral, it could indicate excessive pressure and discharge variation.

Runoff from high areas of the field also reduces uniformity and may result in underirrigation of some areas with overirrigation of low areas, or loss of water from the entire field. Runoff can be reduced by decreasing the nozzle size, which may require reducing the sprinkler spacing. Increasing the pressure reduces the size of the water droplets, which will have a lower impact on the soil surface and maintain a higher infiltration rate. Decreasing the application time also can reduce runoff, but may result in more frequent irrigations.

A full evaluation uses most of the measurements described plus measurement of the sprinkler wetting pattern. Catch containers are placed on a grid spacing of 10 by 10 ft or less to determine the wetting pattern. The containers can be placed in a pattern as shown in Fig. 17.9. Where several laterals are operated simultaneously, as with a solid-set system, the containers should be placed in the area between four adjacent sprinklers. Each container is assumed to catch a depth representative of the square defined by the grid spacing. Measurements should be taken at more than one location to detect sprinkler and pressure differences. Containers must be placed above the plant canopy in an

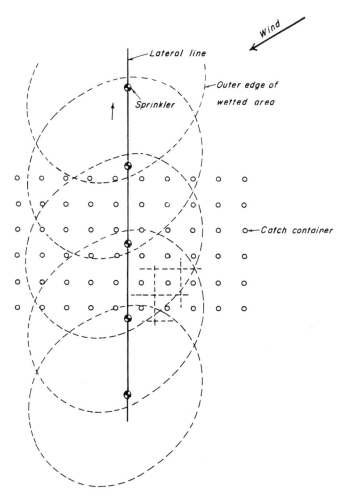

FIGURE 17.9 Layout of catch containers for testing the uniformity of distribution along a single sprinkler lateral line. (Source: SCS, 1983.)

upright position. The depth caught in the containers is best measured volumetrically, but the depth can be measured directly. The containers will indicate less water compared to the depth delivered, based on the sprinkler discharge. This is a result of evaporation from the spray and from the water in the container. A container with the amount of water expected to be caught can be placed outside the irrigated area and evaporation measured to correct for container loss. However, this will not include evaporation that can occur from the walls of the containers being sprinkled.

The Distribution Uniformity DU is calculated from the depths caught in the containers. When only a single lateral is operating, as in Fig. 17.8, the wetting pattern that would result after the next lateral setting is obtained by overlapping the container depths. Thus, depths on the left and right sides of the lateral are added as appropriate to the lateral spacing. With close lateral spacings, overlapping from more than two laterals may occur. Since winds affect the wetting pattern, the tests should be performed under typical wind conditions. A DU of 80 percent is acceptable, however, lower values are allowed where rainfall supplies a substantial portion of crop water needs. For details on evaluating sprinkler systems, see Merriam and Keller (1978).

17.15 Evaluation of Continuous-Move Irrigation Systems

The evaluation of continuous-move irrigation systems is similar to that for intermittent-move systems except for catch container layout. At least two rows of containers are placed in the path of the system. With center-pivot systems, the containers can be placed relative to the wheel tracks. Spacings should be 15 ft or less. After the system passes over the containers, the water caught is measured. A plot of depth versus distance along the lateral is helpful for presenting the distribution and to determine where problem areas occur. When calculating the uniformity of center-pivot systems, a weighting factor proportional to distance from the pivot must be used to account for the increasing area represented by each container. With traveling systems, overlapping a portion of the pattern will be necessary to include the effects of adjacent application paths.

REFERENCES

Addink, J. W., J. Keller, C. H. Pair, R. E. Sneed, and J. W. Wolfe (1983) "Design and Operation of Sprinkler Systems." In Jensen, M. E. (ed.) *Design and Operation of Farm Irrigation Systems*. ASAE, St. Joseph, Michigan.

Berge, I. O., and M. D. Groskopp (1964) *Irrigation Equipment in Wisconsin,* Univ. Wisc. Special Cir. 90.

Frost, K. R. (1963) "Factors Affecting Evapotranspiration Losses During Sprinkling," *ASAE Trans. 6,* 282–283, 287. ASAE, St. Joseph, Michigan.

Hansen, V. E., O. W. Israelsen, and G. E. Stringham (1980) *Irrigation Principles and Practices,* 4th ed. Wiley, New York.

"Irrigation Survey," *Irrigation Journal* (1994). Encino, California.

Jensen, M. E. (ed.) (1983) *Design and Operation of Farm Irrigation Systems,* Monograph 3. ASAE, St. Joseph, Michigan.

Merriam, J. L., and J. Keller (1978) *Farm Irrigation System Evaluation: A Guide for Management.* Utah State Univ., Logan, Utah.

Pair, C. H., W. W. Hinz, C. Reid, and K. R. Frost (eds.) (1983) *Irrigation,* 5th ed. The Irrigation Association, Silver Spring, Maryland.

Schwab, G. O., D. D. Fangmeier, W. J. Elliot, and R. K. Frevert (1993) *Soil and Water Conservation Engineering,* 4th ed. Wiley, New York.

U.S. Soil Conservation Service (SCS) (1983) *Sprinkler Irrigation,* 2nd ed. National Engineering Handbk., Section 15. USDA, Washington, D.C.

PROBLEMS

17.1 Determine the acreage to be irrigated per day and the depth of water to be pumped to irrigate 30 acres of corn in a hot climate. The soil has a final infiltration rate of 0.4 in./h and a water holding capacity of 4.2 in. in the root zone. Assume a water-application efficiency of 70 percent. Irrigation is to begin when 40 percent of the water has been depleted.

17.2 Determine the sprinkler capacity in gpm for a 60- by 80-ft spacing if the water application rate is 0.3 in./h.

17.3 What is the application rate in in./h for a 25 gpm sprinkler where the spacing is 60 by 80 ft?

17.4 Determine the discharge in gpm for a sprinkler operating at 45 psi and having one $\frac{3}{16}$- and one $\frac{7}{64}$- in. diameter nozzle.

17.5 Determine the required capacity of a lateral with 10 sprinklers that apply water at 0.4 in./h where the spacing is 40 by 60 ft.

17.6 For winds below 8 mph, determine the maximum sprinkler spacing for nozzles with a wetted diameter of coverage of 100 ft at 50 psi.

17.7 Determine the total head in feet to be developed by a pump to operate sprinklers at 50 psi. The elevation difference between the ends of the laterals and the main is 10 ft, and the suction lift is 8 ft. The riser height is 5 ft, and the total friction loss in the main and lateral is 20 ft.

17.8 Determine the head loss in 600 feet of 4-in. (o.d.) aluminum pipe with couplers if the flow rate is 200 gpm.

CHAPTER 18 _____
MICROIRRIGATION

Microirrigation is a method of delivering slow, frequent applications of water to the soil near the plants through a low-pressure distribution system and special flow-control outlets. Microirrigation is also referred to as drip, subsurface, bubbler, or trickle irrigation, all of which have similar design and management criteria.

These systems deliver water to individual plants or rows of plants. The outlets are generally placed at short intervals along small tubing, and unlike surface or sprinkler irrigation, only the soil near the plant is watered. The outlets include emitters, orifices, bubblers, and sprays or microsprinklers with flows ranging from 0.5 to 50 gallons per hour (gph).

According to Karmeli and Keller (1975), microirrigation research began in Germany about 1860. In the 1940s, it was introduced in England, primarily for watering and fertilizing plants in greenhouses. With the increased availability of plastic pipe and the development of emitters in Israel in the 1950s, it has since become an important method of irrigation in Australia, Europe, Israel, Japan, Mexico, South Africa, and the United States. According to the "1993 Irrigation Survey" in *Irrigation Journal* (1994), California had 1 451 000 acres and Florida 426 000 acres; the U.S. total was over 2.4 million acres.

Microirrigation has been accepted mostly in the more arid regions for watering high-value crops, such as fruit and nut trees, grapes and other vine crops, sugar cane, pineapples, strawberries, flowers, and vegetables. Although successfully used on cotton, sorghum, and sweet corn, microirrigation is not as well adapted to field crops. Microirrigation systems are very convenient for nurseries and landscape plantings. To reduce the cost, these systems often irrigate plants with different water requirements on the same lateral or valve. The water required by the various plants can be delivered by careful selection of the number and flow rate of the emitters at each plant.

18.1 Advantages and Disadvantages of Microirrigation

With microirrigation, only the root zone of the plant is supplied with water, and with proper system management deep percolation losses are minimal. Soil evaporation may be lower because only a portion of the surface area is wet. Like solid-set sprinkler systems, labor requirements are lower and the systems can be readily automated. Reduced percolation and evaporation losses result in a greater economy of water use. Weeds are more easily controlled, especially for the soil area that is not irrigated. Bacteria, fungi, and other pests and diseases that depend on a moist environment are reduced as the above-ground plant parts normally are completely dry. Because soil is kept at a high water content and the water does not contact the plant, the use of more saline water may be possible with less stress and damage (such as leaf burn) to the plant. Field edge losses and spray evaporation, such as occur with sprinklers, are reduced with these systems. Low rates of water application at lower pressures are possible, which eliminate runoff. With some crops, yields and quality are increased by the maintenance of a high temporal soil water level, adequate to meet evapotranspiration demands. Crop yield experiments have shown wide variations— from little or no difference to a 50 percent or greater increase—compared with other methods of irrigation. Crop quality may also be improved. Some fertilizers and pesticides may be injected into the system and applied in small quantities, as needed, with the water. With good system design and management, this practice can minimize chemical applications and reduce chemical movement to the ground-water supply.

The major disadvantages of microirrigation are high cost and the clogging of system components, especially emitters, by particulate, biological, and chemical matter. Emitters are not well suited to certain crops, and special problems may be caused by salinity. Salt tends to accumulate along the fringes of the wetted surface strip (Fig. 18.1). Rainfall could move the salt to the plant roots to cause injury. Since microirrigation systems normally wet only part of the potential soil-root volume, plant roots may be restricted to the soil volume near each emitter, as shown in Fig. 18.1. The dry soil area between emitter lateral

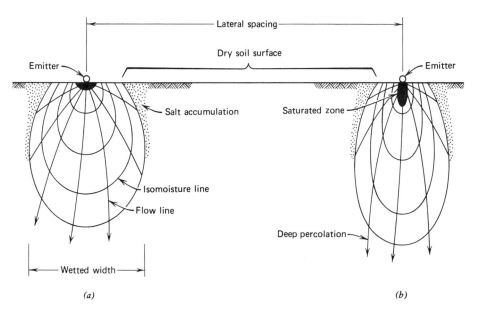

FIGURE 18.1   Soil water distribution pattern with microirrigation for (*a*) medium- and heavy-textured soil, and (*b*) sandy soils. (Adapted from Karmeli and Keller, 1975.)

lines may result in dust formation from tillage operations and subsequent wind erosion. Compared to surface irrigation systems, more highly skilled labor is required to operate and maintain the filtration equipment and other specialized components.

18.2   Components of Microirrigation Systems

Microirrigation system layouts are similar to sprinkler systems (Chapter 17), and as with sprinkler systems, many arrangements are possible. Fig. 18.2 shows a split-line operation for a 3.75-acre orchard. The well is on the south side of the field, centrally located for all areas to be irrigated. Tree rows are next to the 17 laterals at a spacing of 20 ft. The entire field is to be irrigated at the same time as a solid set system. For a larger field with longer tree rows, several submains and manifolds could be provided (Fig. 18.3).

As shown in Fig. 18.3, the primary components for a trickle system are an efficient filter, a main and submain, a manifold, and lateral lines to which the emitters are attached. The manifold is a line to which the laterals are connected. Pressure regulators, pressure gages, a water meter, flushing valves, time clocks, and automatic control devices are other desirable components. The manifold, submain, and main pipelines may be laid on the surface or buried underground. The lines are usually plastic pipe, such as polyethylene (PE) or

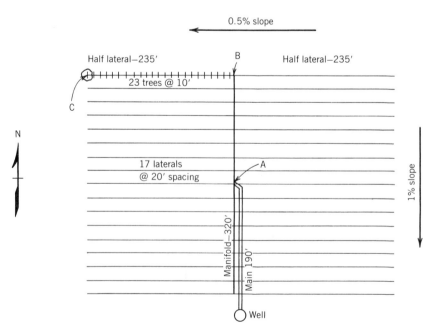

FIGURE 18.2   Microirrigation system layout for a 3.75-acre orchard with 10 by 20 ft tree spacing.

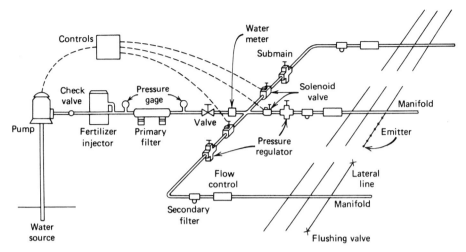

FIGURE 18.3   Components and nomenclature for a microirrigation system.

polyvinylchloride (PVC). The lateral lines that have emitters are usually flexible PVC or PE tubing. They generally range from $\frac{3}{8}$ to $1\frac{1}{4}$ in. in diameter and have emitters spaced at short intervals appropriate for the growth of the crop.

The filter is one of the most important components of a microirrigation system because of emitter clogging. Most water should be cleaner than drinking water. Microirrigation systems generally require screen, gravel, graded sand, or diatomaceous earth filters. Recommendations of the emitter manufacturer should be followed in selecting the filtration system. In the absence of such recommendations, the net opening diameter of the filter must be smaller than $\frac{1}{4}$ to $\frac{1}{10}$ of the emitter opening diameter. For clean ground water, an 80- to 200-mesh screen filter may be adequate. This filter will remove soil, sand, and debris, but it should not be used with high-algae water. For high-silt and high-algae water, a sand filter backed up with a screen filter is recommended. A sand separator ahead of the filter may be necessary if the water contains considerable sand. In-line strainers with replaceable screens and clean-out plugs may be adequate with small amounts of sand. Secondary filters may be installed at the inlet to each manifold. These are recommended as a safety precaution should accidents during cleaning or filter damage allow particles or unfiltered water to pass into the system. Filters must be cleaned and serviced regularly. Pressure loss through the filter should be monitored as an indication for maintenance.

18.3   Emitters

Many types of emitters are commercially available, some of which are shown in Fig. 18.4. The emitter controls the flow from the lateral. The pressure is greatly decreased by the emitter via small openings, long passageways, vortex chambers, manual adjustment, or other mechanical devices. Some emitters may be pressure-regulated by changing the length or cross section of passageways or

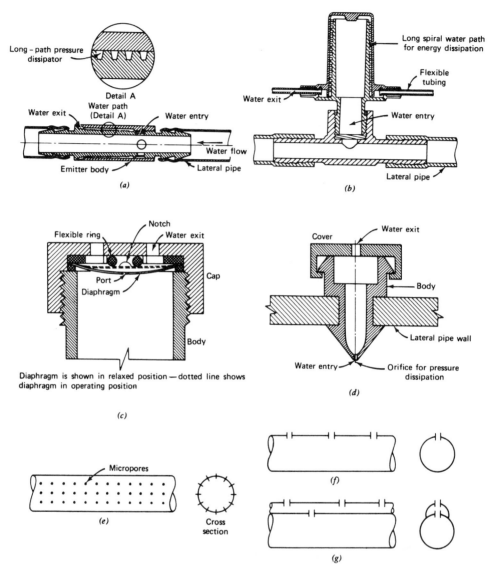

FIGURE 18.4 Types of emitters and emitter laterals showing *(a)* in-line long-path single-exit emitter, *(b)* in-line long-path multiple-exit emitter, *(c)* flushing-type emitter, *(d)* orifice-type emitter, *(e)* porous-tubing lateral, *(f)* single-tube lateral, and *(g)* double-tube or double-wall lateral. (Figs. 18.4*a*–18.4*d* redrawn from Karmeli and Keller, 1975.)

the size of orifice. These emitters (Fig. 18.4*c*) give nearly a constant discharge over a wide range of pressures. Some are self-cleaning and flush automatically. Porous pipes or tubing may have many small openings, as shown in Figs. 18.4*e* to 18.4*g*. Actual hole sizes are much smaller than indicted in the drawing. Some holes are barely visible to the naked eye. The double-tube lateral shown in Fig.

18.4*g* has more openings in the outer channel than in the main flow channel. Such tubes have thin walls and are low in cost. In Hawaii, they are often discarded after the sugar cane is harvested and replaced with new lines. Most emitters are placed on the soil surface, but they may be buried at shallow depths for protection.

The discharge of any emitter may be expressed by the power-curve equation (Karmeli and Keller, 1975) in which

$$q = Kh^x \qquad (18.1)$$

where q = emitter discharge, gph
K = discharge constant for each emitter
h = pressure head in psi
x = emitter discharge exponent

The discharges from several types of emitters are shown in Fig. 18.5. Emitter discharges usually vary from about 0.3 to 8 gph, and the pressures range from about 5 to 40 psi (10 to 90 ft). The average diameters of openings for emitters range from 0.0001 to 0.01 in. The flow characteristics of emitters are

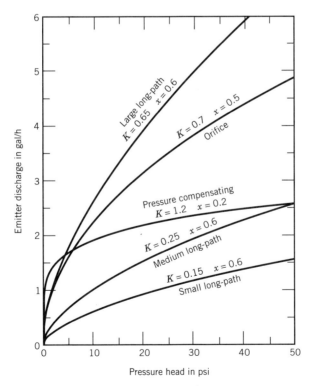

FIGURE 18.5   Discharge of various emitters versus pressure head. Values of *K* above the curves are for Eq. 18.1, with pressure in psi and flow in gph.

determined by the design. Some emitters have turbulent flow with x in Eq. 18.1 near 0.5, whereas others have laminar flow with x near 1.0. In a pressure-compensating emitter, the flow is nearly constant over a wide range of pressures and x is near zero. Because of the large number of emitters available, it may be more convenient to determine the discharge directly from manufacturer's curves or tables. Sample curves are shown in Fig. 18.5. The small, medium, and large long-path emitters are nominally rated at $\frac{1}{2}$, 1, and 2 gal/h, respectively. Double-tube laterals are typically rated as discharge per length, for example, 0.4 gal/min per 100 ft at 10 psi. Because emitters made from thermoplastic material may vary in discharge, depending on the temperature, the discharge should be corrected for temperature.

18.4 Water Distribution

Microirrigation was developed to provide more efficient application of water. An ideal trickle system should provide a uniform discharge from each emitter. Application efficiency depends on the variation of emitter discharge, pressure variation along the lateral, and seepage below the root zone or other losses, such as soil evaporation. Emitter discharge variability is greater than that for sprinkler nozzles because of smaller openings (lower flow) and lower design pressures. Such variability may result from the design of the emitter, materials, and care in manufacture. Solomon (1979) found that the statistical coefficient of variation may range from 0.02 to 0.40. The coefficients of variation (C_v = standard deviation/mean) should be available for emitters and provided by the manufacturer. ASAE guidelines for classification of emitter uniformity are shown in Table 18.1.

To obtain high efficiencies, microirrigation systems must deliver the water required to each plant with minimum losses. This is achieved by a high uniformity of water delivery by each section of the system with a separate control valve. Thus, each manifold in Fig. 18.3 should be designed with a high uniformity, since each manifold has a separate solenoid valve. The uniformity varies

TABLE 18.1   Recommended Classification of Manufacturer's Coefficient of Variation, C_v

Emitter Type	C_v Range	Classification
Point source	<0.05	Excellent
	0.05–0.07	Average
	0.07–0.11	Marginal
	0.11–0.15	Poor
	>0.15	Unacceptable
Line source	<0.10	Good
	0.10–0.20	Average
	>0.20	Marginal to unacceptable

Source: ASAE (1993a).

TABLE 18.2   Recommended Ranges of Design Emission Uniformity, (EU)

Emitter Type	Spacing (ft)	Topography	Slope (%)	EU Range (%)
Point source	>12	Uniform	<2	90–95
		Steep or undulating	>2	85–90
Point source	<12	Uniform	<2	85–90
		Steep or undulating	>2	80–90
Line source	All	Uniform	<2	80–90
		Steep or undulating	>2	70–85

Source: ASAE (1993a).

with pressure, emitter variation, and number of emitters per plant. This is defined by the Emission Uniformity as

$$EU = 100 \left(1 - \frac{1.27 C_v}{n^{0.5}} \right) \frac{q_{min}}{q_{avg}} \qquad (18.2)$$

where EU = emission uniformity

C_v = manufacturer's coefficient of variation

n = number of emitters per plant

q_{min} = minimum emitter discharge rate for the minimum pressure in the section

q_{avg} = average or design discharge rate for the section

Recommended values for *EU* are shown in Table 18.2.

Example 18.1.   Determine the emission uniformity of a microirrigation system using emitters with a manufacturer's coefficient of variation of 0.06. The system has 3 emitters per plant, the average emitter discharge is 1.0 gph, and the minimum emitter discharge is 0.92 gph.

Solution.   Substituting in Eq. 18.2,

$$EU = \left(1 - \frac{1.27 \times 0.06}{3^{0.5}} \right) \left(\frac{0.92}{1.0} \right) = 0.88$$

According to Table 18.2, an *EU* of 0.88 is acceptable for a point source and emitter spacings less than 12 feet and for spacings greater than 12 feet with slopes greater than 2 percent.

18.5   Layout of Microirrigation Systems

Microirrigation systems may be designed using the general rules and procedures for sprinkler systems. The primary differences are (1) the spacing of

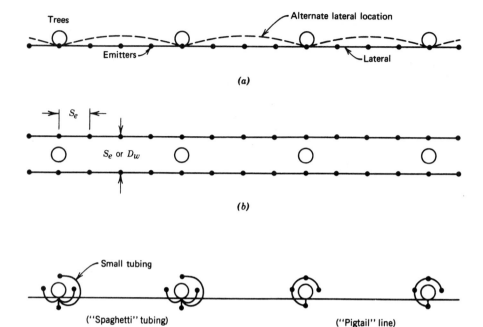

FIGURE 18.6   Lateral and emitter locations for an orchard showing (a) single lateral for each row of trees, (b) two laterals for each row of trees, and (c) multiple emitter layouts with a single lateral.

emitters is much less than the spacing required for sprinkler nozzles and (2) the water must be filtered and treated to prevent clogging of the small emitter openings. Lateral lines may be located along rows of trees, with several emitters required for each tree, as shown in Fig. 18.6. Many laterals have multiple emitters, such as the "spaghetti" tubing or "pigtail" lines shown in Fig. 18.6c. One or two laterals per row (Fig. 18.6a or 18.6b) may be provided, depending on the size of the trees. With small trees a single line is adequate.

Another major difference with microirrigation, especially for widely spaced tree crops, is that not all of the area will be irrigated. Karmeli and Keller (1975) suggest that a minimum of one-third of the potential root volume should be irrigated. For closely spaced plants, a much higher percentage may be necessary to assure sufficient water for the plants. In design, the water-use rate or the area irrigated may be decreased to account for this reduced area. Karmeli and Keller (1975) suggest the following water-use rate for microirrigation design,

$$ET_t = ET \times \frac{P_s}{85} \tag{18.3}$$

where ET_t = average evapotranspiration rate for crops under microirrigation, in./day

P_s = percent of the total area shaded by the crop

ET = conventional ET rate for the crop, in./day

For example, if a mature orchard shades 70 percent of the area and the conventional ET is 0.25 in./day, the irrigation design rate is 0.21 in./day ($0.25 \times \frac{70}{85}$). The shaded area is proportional to the leaf area that contributes to ET. For P_s greater than 85 percent, $ET_t = ET$.

The diameter of the lateral or of the manifold should be selected so that the difference in discharges between emitters operating simultaneously will not exceed 10 percent. This allowable variation is the same as for the sprinkler irrigation laterals discussed in Chapter 17. To stay within this 10 percent variation in flow, the head difference between emitters should not exceed 10 to 15 percent of the average operating head for long-path emitters, or 20 percent for in-line flow emitters. The maximum difference in pressure is the head loss between the control point at the inlet and the pressure at the emitter farthest from the inlet. The inlet is usually at the manifold where the pressure is regulated. In Fig. 18.2, the maximum difference in head loss to the farthest emitter is that for one half the lateral length plus one half the manifold length. For small systems on nearly level land, 50 percent of the allowable friction loss should be allocated to the lateral and 50 percent to the manifold. As in sprinkler laterals (Chapter 17), allowable head loss should be adjusted for elevation differences along the lateral and along the manifold. The friction loss for mains and submains (without emitters) can be obtained from the appropriate tables or with equations for calculating friction loss.

Emitter spacing and number required depend on the wetting pattern and plant spacing. Field tests should be conducted at several representative locations to obtain data on horizontal and vertical water movement. The horizontal area wetted is based on the wetted diameter 6 to 12 inches below the soil surface. If field measurements are not available, estimates may be obtained from Table 18.3 for the horizontal wetted diameter D_w from a single outlet. For a line

TABLE 18.3 Estimated Diameter of the Wetted Circle Formed by a Single Emission Outlet [a]

Soil or Root Depth and Soil Texture	Homogeneous Soil (ft)	Varying Layers [b]	
		Generally Low Density (ft)	Generally Medium Density (ft)
Depth 2.5 ft			
Coarse	1.5	2.5	3.5
Medium	3.0	4.0	5.0
Fine	3.5	5.0	6.0
Depth 5 ft			
Coarse	2.5	4.5	6.0
Medium	4.0	7.0	9.0
Fine	5.0	6.5	8.0

Source: (Adapted from SCS, 1984).

[a]Formed 6 to 12 inches below the soil surface by a single D_w outlet discharging 1 gph on various soils.

[b]Most soils are layered. As used here, "varying layers of low density" refers to relatively uniform texture but with some particle orientation and/or some compaction layering that give higher horizontal than vertical permeability. "Varying layers of medium density" refers to changes in texture with depth as well as particle orientation and moderate compaction.

source, the outlet spacing S_e should be less than or equal to $0.8D_w$ to overlap the patterns of adjacent emitters along a lateral. For double laterals, a spacing of D_w between laterals will adequately wet the area. The spacing between laterals and between outlets should be reduced to $0.8D_w$ if the water is saline. These closer spacings reduce the dry areas between emitters where salts might accumulate. The number of emitters required per plant, n, can be obtained from

$$n = \frac{p_w \times \text{area/plant}}{\text{effective area wetted per emitter}} \tag{18.4}$$

where p_w is the percentage of the total area/plant to be wetted, area/plant is based on the plant spacing, and the effective area wetted by one emitter depends on the wetted diameter, emitter layout, water quality, and the amount of water to be applied. For closely spaced plants, 100 percent of the area may be irrigated. For widely spaced crops, such as trees, the area wetted may be as low as 30 percent of the shaded area in humid regions and nearly equal to the shaded area in arid regions.

Example 18.2. Determine the number of emitters needed per tree if the tree spacing is 15 ft by 20 ft and 50 percent of the area is to be wetted. The soil is a loam to 5 ft with slight compaction, and emitters will be installed on two laterals as shown in Fig. 18.6*b*.

Solution. Assume the rooting depth is 5 ft; then from Table 18.3 D_w = 7.0 ft. The effective wetted area is assumed to be $S_e \times D_w = 0.8(7.0 \text{ ft}) \times 7.0$, which is substituted into Eq. 18.4

$$n = \frac{0.5 \times 15 \times 20}{0.8(7.0) \times 7.0} = 3.8 \text{ or round to 4 emitters per tree.}$$

18.6 Maintenance

The plugging of emitters caused by physical, chemical, or biological materials is the biggest maintenance problem. Main filters and screens should be cleaned periodically to prevent emitter clogging. The discharge of emitters should be uniform and sufficient to provide adequate water to the crop. Some emitters are designed to be self-cleaning. Secondary filters and screens on manifolds and laterals should be routinely checked. Sulfuric and hydrochloric acids are commonly used to reduce the chemical precipitation of irrigation water with a high pH and high concentrations of calcium and magnesium carbonates (Pair et al., 1983). Monthly flushing can remove accumulations of sediment, bacterial slime, and iron in the tubing and can significantly reduce plugging. Flushing can be done manually by opening the ends of the lines and using as high a velocity as possible. If lines are used for more than one season, they should be flushed at the beginning and end of each season and regularly thereafter. Chlorine or copper sulfate are common chemicals to kill bacteria and algae.

18.7   Chemical Applications Through Microirrigation Systems

As with sprinkler irrigation systems, soluble chemicals such as fertilizers, pesticides, and other chemicals can be applied with the water in microirrigation systems. The application of chemicals in the water reduces labor, energy, and equipment costs compared with conventional methods. Less chemicals are required because they are added at the point of use where they are most effective. Any material that causes precipitation and clogging of the emitters should be avoided. Some fertilizers are made specifically for this application. The fertilizer injection rate depends on the concentration of liquid fertilizer and the quantity needed. This concentration can range from 4 to 10 ppm (parts per million) (Howell et al., 1980). Nitrogen injection will usually not significantly increase clogging. With some nitrogen sources, aqua ammonia and anhydrous ammonia will increase pH, which can result in precipitation of insoluble calcium and magnesium carbonates. Phosphorus, potassium, micronutrients, and herbicides have been successfully applied through microirrigation systems.

Chemicals should be injected into microirrigation systems before the filters, in case chemical reactions (which could cause plugging) occur. Some chemicals should be injected into the system at such a rate as to maintain a specific concentration, whereas other chemicals should be applied at a rate to apply a given amount. Thus, the injection method and equipment are important for chemical management. A simple method is to use the suction on the inlet side of the pump, however, corrosion of pump materials may occur. Special injector pumps are available to meter and inject precise quantities of chemicals. The chemicals are injected at a higher pressure than in the irrigation system. This method has the disadvantage that the concentrated chemical is under pressure, with a greater potential for hazardous leaks. A venturi injector creates a suction that mixes the chemical with the water in the body of the injector. In small systems, the venturi can be installed in the main line before the filter. In large systems, a small centrifugal pump takes water from the main line through the venturi and back to the main line. The concentrated chemical is at less than atmospheric pressure and is partially diluted in the injector, which is safer than injecting under pressure. Special valves and flow meters are available to assist in controlling the amount and concentration of the chemical in the irrigation water.

MANAGEMENT AND EVALUATION

Management of microirrigation systems is based on the principles described in Chapter 15. The irrigation manager or operator should verify that adequate, but not excessive, amounts of water are being applied. This is crucial for microirrigation systems to attain the high efficiencies expected and because only a limited soil volume is wetted and a limited volume of water is stored and available for plant use. If water is inadequate, plants excessively deplete the soil water reservoir before the next irrigation and yield is reduced. If this occurs, the quantity of water applied may be too small and/or the number or spacing of emitters is inadequate. In the latter case, the wetted soil volume may be insufficient to adequately store the water needed to meet the crop requirement between irrigations. To determine if the system was designed properly and is being managed and operated efficiently, an evaluation of the system is required.

The procedure for evaluation depends on the system, however, a few measurements are applicable to any system. Information should be obtained from the operator on the duration and frequency of irrigation, area irrigated, and flow rate to that area. The resulting volumes can be compared to estimated or calculated values for *ET* as a first approximation regarding the adequacy of irrigation. Eq. 18.3 is helpful for determining the *ET* requirement.

As with other types of systems, observations of rate and uniformity of plant growth can serve as a simple evaluation. Poor growth may indicate insufficient water is being applied, however, poor soils, low fertility, and diseases may have similar symptoms. Irregular growth patterns probably indicate some emitters or outlets are plugged. A consistent decrease in growth with distance along the lateral suggests a decrease in pressure, and hence flow rate, with distance. This can occur if the flow rate is too high, which causes excessive friction loss, or if elevation increases. High flow rates are the result of poor design, or the operator has added emitters and/or used replacements with higher flow rates. Laterals with flow moving uphill also lose pressure due to elevation and must be much shorter than those on level fields or with a downhill flow.

Soil water measurements before and after irrigation, by probing the soil, can show if adequate water is being applied and provide some indication of uniformity. However, this is more difficult with microirrigation than with surface and sprinkler irrigation, because only a portion of the soil is wetted. Therefore, numerous locations around a plant must be probed to determine the wetting pattern. If the flow rates indicate that sufficient water is being delivered, but the soil is dry and plant growth is poor, the volume of soil wetted is too small. This can be corrected by adding more emitters per plant, and Eq. 18.4 can be used to verify emitter spacing and number required. Adding emitters can increase the flow rate in a lateral and decrease uniformity because of high pressure losses.

Since filters are important components of any microirrigation system, they should be checked for excessive pressure losses. Typically, the pressure difference between the inlet and outlet should not exceed 5 to 10 psi or the manufacturer's recommended value. High pressure losses reduce the flow and indicate the filter needs cleaning. Sand filters can be cleaned by back flushing, whereas the cleaning element in screen and cartridge filters should be removed and washed or replaced.

The hydraulics of the system should be checked. For a simple procedure, select four laterals operating on a manifold. One lateral should be near the inlet and another near the far end. The other two laterals should be spaced near $\frac{1}{3}$ and $\frac{2}{3}$ of the manifold length. Measure the pressure at the inlet and outlet ends of the four laterals. The difference between the average pressure and the lowest pressure should not exceed 10 to 15 percent of the average pressure.

The variation in discharge from individual emitters is the main indicator of uniformity of a microirrigation system. Emitter discharge is affected by the design hydraulics, emitter type and variability, emitter plugging, and water temperature. The statistical uniformity is recommended as a measure of emitter discharge variation. The statistical uniformity is given by

$$U_s = 100(1 - Sd/\bar{q}) \qquad (18.5)$$

where U_s = the statistical uniformity of emitter discharge

Sd = the standard deviation of the measured emitter discharges, and

$\bar{q}$ = the average of the measured emitter discharges

At least 36 measurements should be taken, however, the confidence level of the measurements increases as the number of measurements increases (ASAE, 1989b). The area to be sampled, usually the area of a submain or the system, should be divided into a number of subareas equal to the number of measurements to be taken. An emitter should be randomly selected from each subarea and the discharge determined with a stopwatch and a small container (such as a 250-ml graduate cylinder). The statistical uniformity can be related to the emission uniformity from Table 18.4.

Flows from perforated or porous tubing can be measured by using a trough to collect the flow from several outlets or the flow can be caught from individual outlets. If the tubing is buried, a pit will need to be dug beside the tubing to collect the flow from individual outlets. If plugging is observed, the filters should be checked, the system should be flushed, acids injected to dissolve precipitates and control pH, emitters manually flushed or replaced, and the system retested. Chlorine can be injected on a regular basis as a preventive measure to control algae.

TABLE 18.4   Relationship Between Statistical Uniformity and Emission Uniformity

Acceptability	Statistical Uniformity U_s, *percent*	Emission Uniformity EU, *percent*
Excellent	100–95	100–94
Good	90–85	87–81
Fair	80–75	75–68
Poor	70–65	62–56
Unacceptable	<60	<50

Source: ASAE (1993b).

REFERENCES

American Society of Agricultural Engineers (ASAE) (1993a) *Design, Installation, and Performance of Trickle Irrigation Systems,* EP 405.1. ASAE, St. Joseph, Michigan.

American Society of Agricultural Engineers (ASAE) (1993b) *Field Evaluation of Microirrigation Systems,* EP458. ASAE, St. Joseph, Michigan.

Hoffman, G. J., T. A. Howell, and K. H. Solomon (eds.) (1990) *Management of Farm Irrigation Systems,* ASAE Monograph. ASAE, St. Joseph, Michigan.

Howell, T. A., D. S. Stevenson, F. K. Aljibary, H. M. Gitlin, I. Wu, A. W. Warrick, and P. A. C. Raats. (1980) "Design and Operation of Trickle (Drip) Systems," Chapter 16. In Jensen, M. E. (1980) *Design and Operation of Farm Irrigation Systems,* Monograph No. 3. ASAE, St. Joseph, Michigan.

"1993 Irrigation Survey," *Irrigation Journal* (1994), Encino, California.

Karmeli, D., and J. Keller (1975) *Trickle Irrigation Design.* Rain Bird Sprinkler Manufacturing Corporation, Glendora, California.

Merriam, J. L., and J. Keller (1978) *Farm Irrigation System Evaluation: A Guide for Management.* Utah State University, Logan, Utah.

Nakayama, F. S., and D. A. Bucks (eds.) (1986) *Trickle Irrigation for Crop Production.* Elsevier, Amsterdam.

Northeast Regional Agricultural Engineering Service (NRAES) (1981) *Trickle Irrigation in United States.* Cornell Univ., Ithaca, New York.

Pair, C.H., W. W. Hinz, K. R. Frost, R. E. Sneed, and T. J. Schiltz (eds.) (1983) *Irrigation,* 5th ed. The Irrigation Association, Silver Springs, Maryland.

Solomon, K. (1979) "Manufacturing Variation of Trickle Emitters," *ASAE Trans.* 22(5):1034–1038, 1043, ASAE, St. Joseph, Michigan.

U.S. Soil Conservation Service (SCS) (1984) "Trickle Irrigation," Chapter 7, *National Engineering Handbook,* Section 15, Washington, D.C.

Walker, W. R. (1979) *Sprinkler and Trickle Irrigation.* 3rd ed. (mimeo.). Dept. of Agr. and Chem. Engr. Colorado State Univ., Fort Collins, Colorado.

PROBLEMS

18.1 Determine the daily ET rate for a young orchard in which the tree canopies shade only 40 percent of the total land area. The conventional ET rate for a mature orchard with 90 percent cover is 0.30 in./day.

18.2 How many gallons of water should be delivered per irrigation to each tree if conventional ET is 0.25 in./day, 40 percent of the area is shaded, irrigation occurs every 2 days, and the tree spacing is 10 ft by 12 ft? (1 cubic ft of water = 7.48 gallons)

18.3 Determine the emission uniformity for a system with medium long-path emitters if the manufacturer's coefficient of variation is 0.04. There are four emitters per plant and the average operating pressure is 25 psi and the minimum pressure is 20 psi.

18.4 Estimate the spacing between emitters along a line source in a fine-textured homogeneous soil, assuming plants with a 2.5-ft root zone.

18.5 Determine the number of emitters required per tree for a tree spacing of 12 ft by 16 ft if the soil is coarse textured, layered of low density. Assume a 5-ft root zone, a double lateral layout, and 60 percent of the area is to be wetted.

CHAPTER 19

WATER
MEASUREMENT

Effective use and management of water require that flow rates and volumes be measured and expressed quantitatively. Water measurement is based on application of the formula

$$q = av \tag{19.1}$$

where q = flow rate, volume per unit time in cubic feet per second or cubic meters per second

a = cross-sectional area of flow in square feet or square meters

v = mean velocity of flow in feet per second or meters per second

Flow measurement thus involves determination of mean velocity and area of flow. Some techniques make each of these determinations separately and use them directly in Eq. 19.1. In others, the calibration of the measurement device gives the flow directly. Calculation of flow volume requires integration of the flow rate over the time period involved.

Units of volume and volume per unit time are required in water measurement. In agriculture, the common units of volume are gallons, cubic feet, and acre-feet or liters and cubic meters. Common units of rate of flow are gallons per minute (gpm), cubic feet per second (cfs), Miner's inches, liters per second (L/s), and cubic meters per hour (m³/h) or per second (m³/s). The Miner's inch, which is defined by state legislation and varies from state to state, is about $\frac{1}{40}$ cfs.

The type of flow measurement device selected depends on whether the flow is in a pipe or open channel. In addition, such factors as cost, accuracy, variability of flow, level of automation, as well as whether volumetric flow rate and/or total volume is needed, affect selection of a flow measurement device.

WATER MEASUREMENT IN PIPES

Orifices and propeller meters are commonly used to measure pipe flow. Pipe flow measurement devices require that the pipe flow full. If the pipe does not flow full, the device will not operate or will produce erroneous values.

19.1   Orifices

An orifice is usually a circular opening (in a plate) through which water flows. When the orifice plate is thick, the orifice becomes a short tube. The orifice is a good measuring device, since the velocity is a function of the water head (Fig. 19.1a). As water flows through the orifice, head is lost by friction and the contraction of the jet with free outflow reduces the cross-sectional area of the jet to less than the orifice. The head lost by friction slightly reduces the flow or slightly increases the pumping head. The discharge of an orifice with a thin plate and free flow is

$$q = Ca(2gh)^{1/2} \tag{19.2}$$

where q = discharge in cubic feet per second or cubic meters per second
   C = discharge coefficient, usually 0.6
   a = cross-sectional area of the orifice in square feet or square meters

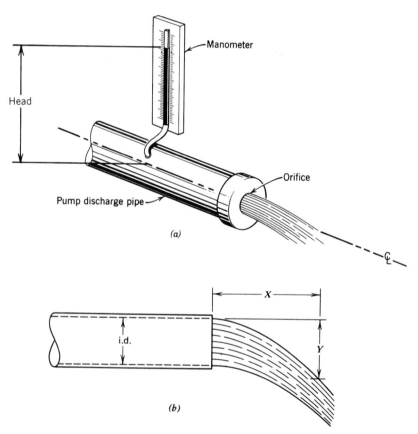

FIGURE 19.1   (a) End-cap orifice, and (b) coordinate method for measurement of flow from a pump discharge.

g = acceleration of gravity, 32.2 ft/sec² or 9.81 m/sec²

h = static water head above the center of the orifice in feet or meters

The coefficient C depends on the orifice configuration and must be determined for each orifice design; however, the value is 0.6 for most orifices. Figure 19.1*a* illustrates an end-cap orifice for measuring the discharge from a horizontal pipe (such as an irrigation pump). The head producing the flow through an orifice can be measured with a single tube manometer, as shown in Fig. 19.1*a*. The height of water in the tube above the center of the orifice is the head required in Eq. 19.2.

19.2   Impeller Meters

Impeller meters are commonly used to measure the flow in pipes (Hayward, 1979). In this application, the meters are calibrated for a given pipe size to read directly in cumulative volume. Meters are available that will also indicate the instantaneous flow rate. Figures 19.2 and 19.3 show typical installations of impeller meters in closed conduits. These meters are convenient to use and are accurate if properly maintained.

19.3   Other Devices

Flow from pipes can be estimated by measuring the trajectory of the flow (Fig. 19.1*b*). The pipe should be horizontal and long enough to produce smooth flow. The flow rate is obtained from appropriate calibration tables using the measured vertical distance Y from the water surface at the end of the pipe to the upper surface of the water trajectory at a measured horizontal distance X from the end of the pipe.

Pipe elbows offer an opportunity for measurement of flow. Water flowing through an elbow exerts different centrifugal pressures at the inside and outside radii of the elbow. The resulting pressure difference can be measured with a differential manometer. Some commercial meters are calibrated to record the

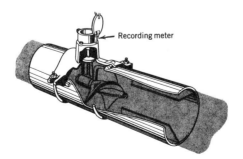

FIGURE 19.2   Impeller meter for a low-pressure pipeline. (Courtesy Sparling Instruments Company, Inc., El Monte, California.)

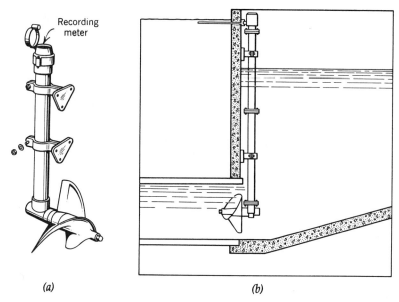

FIGURE 19.3   Impeller meter for open channels. *(a)* Basic meter assembly. *(b)* Installation at an inverted siphon. (Courtesy Sparling Instruments Company, Inc., El Monte, California.)

accumulated volume with an additional small meter connected across the inside and outside radii of the elbow.

Electronic devices using electrical signals at specific frequencies are used for flow measurement in pipes. The signals are distorted by fluid movement and converted to velocities. For each pipe size, a second component integrates the velocity with time and provides a digital readout of both volumetric flow rate and accumulated volume. In remote locations, these can be powered by solar panels. Also, radio transmitters can be added to facilitate data collection.

Many other types of commercial meters are available, usually with calibration tables for determining the flow rate.

WATER MEASUREMENT IN OPEN CHANNELS

Measurement of water flow in open channels is more difficult than measurement of water flow in pipes, because the flow area in open channels changes with the flow rate, whereas the flow area in pipes is constant.

Simple methods are available for estimating the flow rate, but they are not accurate. Accurate measurement of flow in open channels requires structures of known hydraulic characteristics. These structures, such as weirs and flumes (Ackers et al., 1978; USDI, 1975), have consistent relationships between head and discharge.

19.4   Floats

A crude estimate of the velocity of a stream may be made by determining the velocity of an object floating with the current. A straight uniform section of stream several hundred feet long should be selected and marked by stakes or range poles on the bank. The time required for an object floating on the surface to traverse the marked course is measured and the velocity calculated. The average surface velocity is determined by averaging float velocities measured at several distances from the bank. Mean velocity of the stream is often taken as 0.7 to 0.9 of the average surface velocity.

Floats consisting of a weight attached to a floating buoy can estimate the mean velocity. The weight is submerged to the depth of the mean velocity, and the buoy marks its travel downstream. The float method has the advantage of estimating velocity with a minimum of equipment; however, it lacks precision.

Example 19.1.   Determine the discharge in a 10-ft-wide irrigation canal with an average depth of 2.4 ft. Time trial runs for surface floats to travel 250 ft were 96, 91, 93, 95, and 90 sec. Assume the average velocity of flow is 80 percent of the surface (float) velocity.

Solution.   The average time of flow for the five trials is 93 sec. The cross-sectional area of the canal is $2.4 \times 10 = 24$ sq ft. Substituting into Eq. 19.1,

$$q = 0.8 \times 24 \text{ ft}^2 \times 250 \text{ ft}/93 \text{ s} = 51.6 \text{ cfs } (1.46 \text{ m}^3/\text{s})$$

19.5   Slope Area

The basic equations for velocity of open-channel flow may be applied to stream flow measurements. The equation most commonly accepted is the Manning formula, discussed in Chapter 8. A nomograph for calculating the velocity from the Manning formula is also given in Chapter 8. The formula for open-channel flow requires the slope of the water surface and properties of the cross section of the stream. The length of channel selected should be uniform in size and slope and, if possible, as long as 1000 ft. The roughness coefficient n (given in Chapter 8) must be estimated, which is difficult to do accurately. Values of n helpful in arriving at such estimates are given in Chapter 8.

The slope area method is sometimes used in estimating the discharge of past flood peaks. The cross-sectional area and flow gradient are measured from high-water marks along the channel. This method gives only a rough approximation of the peak flow.

Example 19.2.   Determine the discharge rate of a concrete-lined 2-ft wide irrigation canal, rectangular in cross section with a uniform depth of flow of 1.1 ft. By instrument survey, the difference in elevation of the water surface at points 300 ft apart was 0.21 ft.

Solution.   The cross-sectional area of flow, $a = 1.1 \times 2.0 = 2.2$ sq. ft and the wetted perimeter, $p = 1.1 + 2.0 + 1.1 = 4.2$ ft. The slope of the water surface, $s = (0.21/300)$

= 0.0007. From Chapter 8, select a roughness coefficient, $n = 0.015$ for concrete. Substituting into Eq. 8.1 (the Manning formula),

$$v = (1.49/0.015) \times (2.2/4.2)^{2/3} \times 0.0007^{1/2}$$
$$= 99.33 \times 0.65 \times 0.0265 = 1.71 \text{ ft/s (Fig 8.2)}$$

Substituting into Eq. 19.1,

$$q = 2.2 \text{ ft}^2 \times 1.71 \text{ ft/s} = 3.76 \text{ cfs } (0.11 \text{ m}^3/\text{s})$$

19.6 Current Meters

Current meters employing an impeller that rotates at speeds proportional to the velocity of flow are often used for velocity determinations in open channels. Figure 19.4 shows a typical current meter with accessories. The essential part of the meter is the impeller that revolves when suspended in flowing water. An electrical circuit indicates the speed of revolution of the impeller. The meter is suspended by a cable for deep streams or attached to a rod in shallow streams. When supported by a cable, a streamlined weight holds the meter against the current, and a vane attached to the rear of the meter keeps the impeller headed into the stream. Some meters determine the velocity by measuring the disturbance in an electrical field. Current meters are used in rivers and streams (where it is impractical to install a permanent measuring device) when calibrating a gaging station or when infrequent measurements are needed.

When the discharge of a stream is determined with a current meter, the

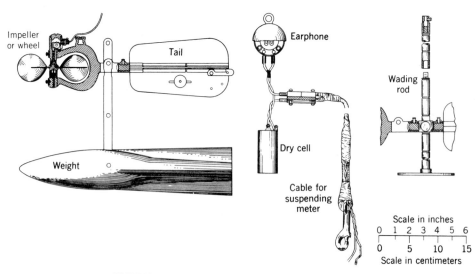

FIGURE 19.4 Price current meter and attachments.

cross section is divided into a number of subareas. Width of the subareas depends on the size and shape of the stream and precision desired. The subareas may be indicated by marks on a tape or cable stretched across the stream or by marks on a bridge railing or other convenient structure. The average velocity in each subarea is multiplied by the area in each subarea to obtain the flow rate. The average velocity of each subarea is determined by readings at 0.2 and 0.8 of the depth. When the stream is too shallow, the velocity at 0.6 of the depth below the surface is taken for the average velocity.

19.7 Weirs

Weirs consist of a barrier or plate (with a sharp crest) placed in a stream to both constrict the water flow and direct it over the crest (Fig. 19.5). Weirs provide accurate measurements of flow rate, are easy to construct, allow passage of floating materials, and are durable; however, sediment deposition in front of the weir reduces the accuracy, and they require a drop in water surface. To achieve this drop in water surface, the downstream water level should be at least 0.2 ft below the weir crest. Figure 19.6 shows the flow profile over a sharp-crested weir and the recommended location for measuring the head h. Fig. 19.6 also shows that the nappe is free at the crest and that air must pass freely under the nappe for accurate readings. The most common weir shapes are rectangular, trapezoidal, and triangular.

SUPPRESSED RECTANGULAR WEIRS Suppressed rectangular weirs have crests the same width as the channel structure in which they are installed (Fig. 19.7). An open-

FIGURE 19.5 Rectangular weir for the measurement of flow in a small stream.

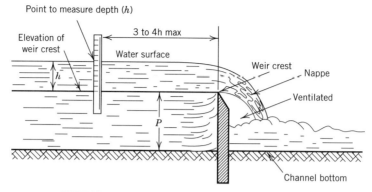

FIGURE 19.6   Profile of flow over a sharp-crested weir.

ing must be provided to ventilate the underside of the nappe with suppressed weirs. The flow through these weirs is determined by the equation

$$q = CLh^{3/2} \tag{19.3}$$

where q = flow rate in cfs
C = discharge coefficient
L = length of the weir crest in feet
h = head on the weir in feet

The discharge coefficient C is a function of the ratio h/P, where P is the height of the weir crest above the channel bottom. Values of C range from 3.3 for $h/P = 0.2$ to 3.6 for $h/P = 1.0$. Typically, h/P of 0.5 is recommended, for which $C = 3.4$.

CONTRACTED RECTANGULAR WEIRS   The crest length of a contracted rectangular weir is less than the channel width, which causes the flow to contract on the sides of

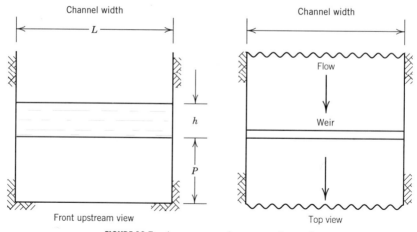

FIGURE 19.7   A suppressed rectangular weir.

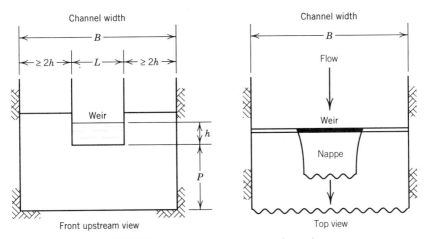

FIGURE 19.8   A contracted rectangular weir.

the nappe. To be fully contracted, the distance from the side of the weir to the sides of the approach channel must be at least twice the head but not less than 1 ft (Fig. 19.8). The flow is determined from Eq. 19.3. For contracted weirs, C is a function of both h/P and L/B, where B is the channel width upstream from the weir. If L/B is 0.3 and h/P is 0.5, the value of C is 3.16. For more specific values of C, see hydraulic handbooks or Bos (1989).

TRAPEZOIDAL OR CIPOLETTI WEIRS   The sides of the trapezoidal weir slope outward at a 1 horizontal to 4 vertical ratio to compensate for the side contractions (Fig. 19.9). For the configuration shown in Fig. 19.9, C is 3.37.

TRIANGULAR (V-NOTCH) WEIRS   Triangular weirs generally have a 90-degree V-notch, but any angle may be used (Fig. 19.10). Their main advantage is that they have more accuracy at low flows than the rectangular or trapezoidal weirs. The flow over a 90-degree V-notch weir is determined by the equation

$$q = Ch^{5/2} \tag{19.4}$$

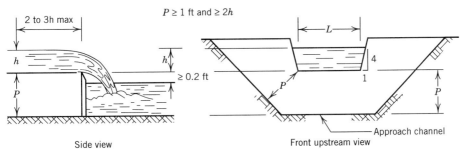

FIGURE 19.9   A trapezoidal or Cipoletti weir.

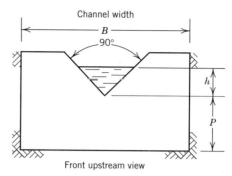

FIGURE 19.10 A 90-degree triangular weir.

This weir is fully contracted if $h/P < 0.4$, $h/B < 0.2$, $P > 1.5$ft, and $B > 3$ft, then C is 2.47. However, C can range from 2.46 to over 2.61 for other configurations. Values of C can be obtained from hydraulic handbooks or the manufacturer.

19.8 Flumes

Flumes are specially shaped, stabilized channel sections installed to measure flows. They are less inclined to catch sediment and floating debris than weirs, and thus are particularly suited for measurement of runoff. Some flumes are difficult to construct, but flumes require only a small difference in elevation between the upstream and downstream water surfaces (head loss).

PARSHALL FLUMES The Parshall flume was one of the first flumes standardized for flow measurement (Fig. 19.11). Discharge tables for all sizes of flumes are available in Parshall (1950) or hydraulic handbooks. When the head at H_b is less than 0.7 H_a, the equation for flow through Parshall flumes as revised by Blaisdell (1994) is

$$q = 4.1 \ WH_a^{1.584} \tag{19.5}$$

where q = flow rate in cfs

W = throat width in feet

H_a = head in the converging section in feet

When H_b is greater than 0.7 H_a, both H_a and H_b must be considered in determining the discharge, and reference should be made to calibration tables.

LONG-THROATED FLUMES The long-throated flume (or broad-crested weir) is formed by installing a ramp and sill in a straight, uniform channel (Fig. 19.12). The structure is easily added to existing concrete channels, or prefabricated,

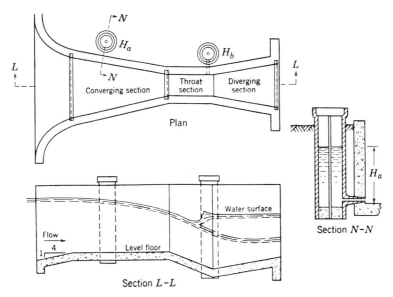

FIGURE 19.11   Parshall measuring flume. (Redrawn from Parshall, 1950.)

portable sections can be placed in unlined channels. This flume is easy to construct, but a discharge equation or table of discharge versus head values must be determined for each flume (Bos et al., 1993). Table 19.1 shows discharges and heads for two specific flumes. Although trapezoidal and rectangular shapes are most common, ratings are available for other shapes such as circular, triangular, and parabolic.

OTHER FLUMES   Other flumes such as Cutthroat flumes (Skogerboe et al., 1972), WSC V-notch flumes (SCS, 1962), and H-flumes (Holtan et al., 1962) are available. The shapes of these flumes vary, but their characteristics are similar to those previously described.

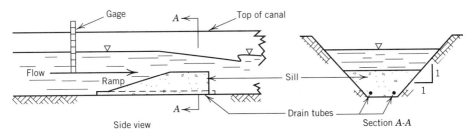

FIGURE 19.12   Long-throated flume in a trapezoidal channel. (From Clemmens and Replogle, 1980.)

TABLE 19.1 Typical Heads Versus Discharges for Long-Throated Flumes[a]

	Discharge	
Head Reading[b] feet	Trapezoidal[c] Flume ft^3/s	Rectangular[d] Flume ft^3/s
0.10		0.09
0.20	0.72	0.27
0.30	1.38	0.52
0.40	2.23	0.81
0.50	3.26	1.15
0.60	4.48	1.53
0.70	5.88	
0.80	7.48	

Source: Clemmens et al. (1987).

[a]Long-throated trapezoidal flume installed in a trapezoidal channel and a rectangular flume installed in a rectangular channel.

[b]Head readings are referenced to the top of the sill.

[c]The trapezoidal channel has a 1-ft bottom width and 1:1 side slopes. The sill height is 0.75 ft and sill length is 2 ft.

[d]The rectangular channel and flume are 1 ft wide. The sill height is 0.5 ft and sill length is 1 ft.

REFERENCES

Ackers, P., W. R. White, J. A. Perkins, and A. J. M. Harrison (1978) *Weirs and Flumes for Flow Measurement.* Wiley, New York.

Blaisdell, F. W. (1994) "Results of Parshall Flume Tests," *J. Irrig. and Drain. Engr.* ASCE, 120(IR2):278–291.

Bos, M. G. (ed.) (1989) *Discharge Measurement Structures,* 3rd ed. ILRI Pub. 20, International Inst. for Land Reclamation and Improvement, Wageningen, Netherlands.

Bos, M. G., J. A. Replogle, and A. J. Clemmens (1993) *Flow Measuring Flumes for Open Channels.* ASAE, St. Joseph, Michigan.

Clemmens, A. J., and J. A. Replogle (1980) *Constructing Simple Measuring Flumes for Irrigation Canals,* USDA-SEA, Farmer's Bull. 2268. U.S. Government Printing Office, Washington, D.C.

Clemmens, A. J., J. A. Replogle, and M. G. Bos (1987) *FLUME: A Computer Model for Estimating Flow Through Long-Throated Measuring Flumes,* ARS-57, USDA-ARS. National Tech. Infor. Ser., Springfield, Virginia.

Hayward, A.T. J. (1979) *Flowmeters: A Basic Guide and Source-Book for Users.* Wiley, New York.

Holtan, H. N., N. E. Minshall, and J. J. Harrold (1962) *Field Manual for Research in Agricultural Hydrology,* Agr. Handb. 224. USDA, Washington, D.C.

Parshall, R. L. (1950) *Measuring Water in Irrigation Channels with Parshall Flumes and Small Weirs,* U.S. Dept. of Agr. Cir. 843. USDA, Washington, D.C.

Skogerboe, G. V., R. S. Bennett, and W. R. Walker (1972) "Generalized Discharge Relations for Cutthroat Flumes," *J. Irrig. and Drain. Div.,* ASCE, 98(IR4):569–583.

U.S. Department of Interior (USDI) (1975) *Water Measurement Manual,* U.S. Bureau of Reclamation. U.S. Government Printing Office, Washington, D.C.

U.S. Soil Conservation Service (SCS) (1962) "Measurement of Irrigation Water," Chapt. 9, Sec. 15, *National Engineering Handbook*. USDA, Washington, D.C.

PROBLEMS

19.1 Determine the discharge of a 4-in. diameter end-cap orifice on a 6-in. diameter pipe if $C = 0.7$ and the head on the orifice is 12 in.

19.2 Discuss the advantages and disadvantages of impeller meters for flow measurement in pipes.

19.3 Determine the discharge in a trapezoidal concrete-lined canal using the float method. The canal has a 1-ft bottom width and $1:1$ side slopes with a flow depth of 1.3 ft. The times for a float to travel 200 feet were 72, 79, 74, 76, and 79 sec.

19.4 Determine the discharge of a stream having a cross-sectional area of 50 ft^2 and a wetted perimeter of 16 ft, using the slope area method. The difference in elevation of the water surface 500 ft apart is 0.3 ft. The channel has straight banks with some weeds.

19.5 A velocity of 2 ft/s is measured with a current meter placed 0.4 ft above the bottom of a small stream. The estimated cross-sectional area of the stream is 2 ft^2 and the flow depth is 1 ft. What is the estimated flow rate?

19.6 Determine the flow rate over a suppressed weir if the weir crest is 24 in. long and the head 30 in. upstream is 6 in. Determine the flow rate over a contracted weir having the same dimensions.

19.7 A trapezoidal weir has crest length of 18 in. Determine the discharge over the weir if the head is 5 in. at a point 2 ft upstream.

19.8 Determine the discharge over a 90-degree V-notch weir if the head is 0.6 ft.

19.9 Determine the discharge through a Parshall flume having a throat width of 1.5 ft for $H_a = 1.1$ ft and $H_b = 0.75$ ft.

19.10 A long-throated flume is installed in a trapezoidal canal having a 1-ft bottom width and $1:1$ side slopes. The sill height is 0.75 ft and length is 2.0 ft. Determine the discharge in the canal if the head over the flume is 0.3 ft.

GLOSSARY OF _____
SPECIAL TERMS

_____

Definitions of words and phrases found in this text are given below. In general, they do not include units of measurement nor terms normally found in a collegiate dictionary. Some definitions are simplified and reduced in length, compared to more complete and technically correct ones, most of which are included in American Society of Agricultural Engineers, S 526, *Soil and Water Engineering Terminology* (1993) or Soil and Water Conservation Society of America, *Resource Conservation Glossary*, 2nd ed. (1976). Definitions may change with time and may differ by state or country.

Abney level. A small hand level for leveling or measuring slope in percent or degrees (Chapter 3).

Acid rain. Precipitation that has a low pH (less than 5.6, which is normal) and has a harmful effect on some plants, aquatic organisms, soils, and buildings.

Alfalfa valve. An outlet valve attached to the top of a pipeline riser, with an adjustable lid to control irrigation water flow.

Aquifer. A geologic formation that holds and yields useable amounts of water.

Available soil water. Amount of soil water, released between *in situ* field capacity and the permanent wilting point, that can be absorbed by the plant.

Backfurrow. A small ridge formed when soil plowed out of two adjacent furrows is thrown together.

Backsight. A level rod reading on a point of known elevation that is always added to the elevation to obtain the height of a surveying instrument, also called a plus sight.

Backslope. Land area on the downhill side of a terrace ridge or earth embankment.

Base height. The vertical distance from the sight bar or laser detector on a plow or trenching machine to the bottom of the digging mechanism (grade line).

Base line. A reference line from which measurements are taken or a parallel of latitude for public land surveys.

Bedding. A method of surface drainage consisting of narrow-width plow lands in which deadfurrows are parallel to the prevailing land slope and serve as field drains, also called crowning or ridging.

Bedding angle. The angle of a V-groove in the bottom of a trench in which pipe drains are laid.

Bench mark. A temporary or permanent surveying reference point, the elevation of which is assumed or known (often referenced to mean sea level).

Berm. Strip or area of land, usually level, between the edge of the spoil bank and the edge of the ditch or canal (Chapter 12).

Best management practice (BMP). Structural, nonstructural, and managerial techniques recognized to be the most effective and practical means to reduce surface- and ground-water contamination.

Break tape. A procedure for measuring horizontal distances in short increments with a tape on sloping land (Chapter 2).

Chute spillway. A steep stable (lined) channel for conveying water to a lower level without erosion.

Claypan. A dense, compact layer in the subsoil having a much higher clay content than the overlying material, separated by a sharply defined boundary. Claypans are usually hard when dry, and plastic and sticky when wet.

Conservancy District (Natural Resources District). A legally organized enterprise for the purpose of soil and water management and flood control.

Conservation practice factor. The ratio of soil loss for contouring, strip cropping, or terracing, to that for up and down the slope farming, as used in the soil loss equation.

Convective storm. Rainfall caused by condensation of heated air that moves upward, being cooled both by the surrounding air and by expansion.

Cover management factor. The ratio of soil loss for a given condition to that from cultivated continuous fallow land, as used in the soil loss equation.

Creep distance. The longitudinal length along the outer surface of a pipe or conduit plus the length along antiseep collars within an earth embankment.

Crest. Top of a dam, dike, spillway, weir, or wave and the peak of a flood.

Curve number. An index of the runoff potential that is related to the soil and vegetative conditions of the watershed.

Cut-fill ratio. The ratio of the cut volume to the fill volume in relation to land grading and earthwork operations.

Cutoff. Wall, collar, or other structure (such as a trench) filled with relatively impervious material intended to reduce water flow.

Cutslope. The uphill side slope of a broadbase terrace channel.

Deadfurrow. Empty furrow(s) left when plowing; a single furrow when plowing in one direction or a double furrow when plowing in opposite directions.

Differential leveling. A method of leveling in which the difference in elevation between two or more points is determined.

Diversion. A channel or dam constructed across the slope for intercepting surface runoff and diverting it to a safe or convenient discharge point.

Division box. A canal or ditch structure for dividing the water flow into predetermined portions.

Drain. Any closed conduit (perforated tubing or tile) or open channel for removal of surplus ground or surface water.

Drainage area (Drainage basin or watershed). The land area from which runoff is collected and delivered to an outlet.

Drainage coefficient. The rate of removal of water from the drainage area, usually expressed as the depth to be removed in one day (24 h) or the flow rate per unit area.

Drainpipe. (See Pipe drain).

Drain plow. A machine with a vertical blade, chisel point, and shield or boot to install corrugated plastic tubing or drain tile.

Drain tile. Short length of pipe made of burned clay, concrete, or similar material, usually laid with open joints, to collect and remove subsurface water.

Drop-inlet spillway. Overfall hydraulic structure in which water is discharged through a vertical riser conduit (Chapters 8 and 11).

Drop spillway. Overfall hydraulic structure in which water drops over a vertical wall onto an apron (Chapter 8).

Dumpy level. A surveying level in which the telescope is rigidly attached to the instrument frame.

Electrical conductivity. A measure of the salt content of irrigation or drainage water or solution extract of a soil.

Emergency spillway. An auxiliary channel for a dam to carry flood runoff exceeding the capacity of the principal spillway.

Emitter. A small microirrigation dispensing device to dissipate pressure and discharge a small uniform flow of water at a constant discharge rate.

Envelope filter. Granular material or geotextile sheet fabric that surrounds subsurface pipe drains to prevent soil inflow and enhance water entry.

Ephemeral gully. Small channels eroded by runoff that can be easily filled and removed by normal tillage, only to reform again in the same location.

Erosion. The wearing away of the land surface by running water, wave action, glacial scour, wind, and other physical processes.

Evapotranspiration. The combination of water transpired from vegetation and evaporated from the soil and plant surfaces.

Eyepiece. The small end of a telescope at which the eye is placed to read the level rod.

Fallow. The practice of allowing cropland to lie idle, either tilled or untilled, during the whole year or the greater part of the growing season.

Farm planning. An organized and systematic scheduling of all soil and water management practices integrated with a well-balanced farming program.

Field capacity. Water content of a soil after it has been saturated and allowed to drain freely, usually expressed as a percentage of its oven-dry weight or volume.

Field ditch. An open channel constructed within a field either for irrigation or drainage.

Field drain. A shallow-graded channel, usually having relatively flat side slopes, that collects surface water within a field.

Flood spillway. (See Emergency spillway).

Flood storage depth. The vertical distance between the normal water level in a storage reservoir and the bottom of the emergency spillway.

Flume. An open conduit for conveying water across obstructions; an elevated canal above natural ground level (an aquaduct); or a specially calibrated structure for measuring open channel flow.

Foresight. The level rod reading on a turning point or on some other point having an unknown elevation.

Freeboard. The vertical distance from the top of the embankment to the maximum expected water level in a reservoir or channel.

Frisco rod. A type of telescoping survey rod, usually of three short sections.

Frontslope. The upslope or channel side of a terrace ridge.

Gage height. The vertical distance from the sight bar, batter board, or laser beam to the bottom of the finished cut.

Gated pipe. Portable pipe with small gates installed along one side for distributing irrigation water to corrugations or furrows (Chapter 16).

Geographic Information System (GIS). A computer data-based management system, which includes remote sensing, mapping, cartography, and photogrammetry for conducting spatial searches and making map overlays.

Geotextile. A fabric or synthetic material placed at the boundary between granular material or an impermeable surface and the soil to enhance water movement and retard sediment inflow.

Global Positioning System (GPS). A computer-based satellite system for establishing precise horizontal and vertical positions on the earth's surface.

Grade. The degree of slope expressed as a percent and equal numerically to the rise or fall in a horizontal distance of 100 feet.

Gravel envelope (Gravel filter). Graded sand and gravel aggregates placed around a subsurface drain or around a well screen to prevent the infiltration of fine material.

Guard stake. A flag, lath, or stake placed beside a hub stake (for elevation) for location and identification of the point.

Gully erosion. A type of soil removal by water that produces well-defined channels that usually cannot be obliterated by normal tillage.

Hardpan. A hardened or cemented soil layer in the lower A or in the B soil horizon.

Head. The energy in the liquid system expressed as the equivalent height of a water column above a given datum.

Height of instrument. The elevation of the line of sight of a surveying instrument.

Horizontal interval. The horizontal distance between consecutive terraces.

Hub stake. A short stake driven nearly flush with the ground surface for an elevation reference in surveying.

Hydraulic conductivity. The rate at which water will move through soil under a unit hydraulic gradient.

Hydraulic gradient. Change in the hydraulic head per unit distance.

Hydraulic radius. The cross-sectional area of a channel or conduit divided by the wetted perimeter.

Hydrograph. A graphical or tabular representation of the flow rate of a stream with respect to time.

Hydrologic soil group. A special grouping of soil types into four major categories for estimating runoff.

Impeller meter. A rotating mechanical device for measuring water flow rates in a pipe or open channel.

Infiltration rate (Intake rate). The downward rate that water can enter the soil.

Interrill erosion. Water erosion due to raindrop splash and shallow overland flow.

Irrigation district. A cooperative, self-governing corporation set up as a subdivision of the state and organized primarily to provide irrigation water to several farms in a local area.

Irrigation interval (Frequency). The average time between the commencement of successive irrigations for a given field.

Key terrace. Terrace that is selected as a reference for laying out other terraces.

Land capability classification. The designation of soil units for showing their suitability for specific uses, such as cropping, grazing, woodland, wildlife, or others, usually divided into eight classes.

Land grading (Land forming, leveling, or shaping). The operation of shaping the land surface to predetermined grades for drainage or irrigation.

Land plane. A machine for land smoothing operations.

Land smoothing. The process of removing minor differences in elevations without changing the general contours of the land (the finishing operation after land grading).

Laser beam. A collimated beam of light from a laser transmitter.

Laser grade control. A method for controlling the grade of an earth-moving machine from a laser plane or beam.

Lateral. A secondary or side channel, ditch, or conduit that removes water or delivers irrigation water.

Lathe box (Spile). A device placed through a ditch bank for transferring irrigation water from a ditch to the field.

Lenker rod. A survey level rod with a moveable loop or ribbon graduated with an inverted scale to read elevations directly.

Level. A surveying instrument for establishing a horizontal line.

Level rod (Frisco or Philadelphia). A surveying rod from which elevations can be determined.

Manifold. Pipeline that supplies water to irrigation laterals.

Metes and bounds. A method of public land survey, by which land is described by direction and distance.

Multistage pump. A pump having more than one impeller mounted on a single shaft.

Net positive suction head. Head required to move water into the eye of a pump impeller without causing cavitation (Chapter 10).

Nonpoint source pollution (NPS). Water pollution from diverse land or air sources, such as a field or a watershed.

Orifice. An opening with a closed perimeter and of regular form through which water flows.

Orographic storm. A weather pattern in which precipitation is caused by the rising and cooling of air masses as they are forced upward by higher topography.

Phreatophyte. A nonbeneficial water-loving plant that derives its water from subsurface sources.

Pipe drain (Subsurface or tile drain). Any subsurface drain made from clay, concrete, plastic, or other material.

Pipe spillway (Culvert). A structure for carrying water through an earth embankment.

Point row. A short crop row that forms an acute angle with another row or with the field boundary.

Prime land. The best quality land capable of producing high sustainable yields of crops and fiber, economically, when properly treated and managed.

Profile leveling. A method of surveying to secure the elevation of a series of points located along a line.

Radius of influence. Maximum distance from a well at which drawdown is significant.

Rainfall factor. A relative value for comparing rainfall intensities at different geographical locations.

Rainfall frequency (Return period). Frequency of occurrence of a rainfall event whose intensity and duration can be expected to be equalled or exceeded.

Range lines. True meridian lines that are the east and west boundaries of townships in the rectangular system of public land survey.

Range pole. A sight stake (red and white) for marking points to be seen from a long distance.

Ratio of error. The difference of two measurements between the same points divided by the average distance.

Relief pipe. A vertical riser from a pipe drain to the ground surface or above to relieve hydrostatic pressure.

Retardance class. A characterization of vegetation with respect to its resistance to flow, used primarily for designing grassed waterways.

Return period (Recurrence interval). Time in years in which a given event can be equalled or exceeded, once on the average.

Rill (Ephemeral gully). Small channels eroded into the soil surface by runoff that can be filled easily and removed by normal tillage.

Riser. A vertical conduit for air or water flow extending from a subsurface drain to the ground surface (or above) or a pipe attached to a sprinkler head.

Roughness coefficient (Resistance coefficient). A constant in the Manning velocity equation representing channel resistance to flow.

Runoff. The portion of precipitation, snowmelt, or irrigation water that eventually makes its way into a surface stream.

Saltation. Soil movement by wind or water in which particles skip or bounce along the soil surface or streambed (Chapter 9).

Sheet erosion. Removal of a fairly uniform layer of soil from the land surface by runoff.

Shelterbelt. A long barrier of living trees and shrubs for the protection of farmland or buildings from erosion or damage.

Side slopes. Slope of the sides of a channel or embankment, horizontal to vertical (written 2:1).

Slope (Grade or gradient). Rate of rise or fall from the horizontal, expressed as percent or in degrees.

Slope length factor. A relative number for evaluating the length of slope in the soil loss equation.

Slope stake. A survey stake to mark the edge of a dam or ditch.

Slope steepness factor. A relative number for evaluating the land slope in the soil loss equation.

Soil and Water Conservation District. Normally, a county-size area organized under state laws to conserve natural resources.

Soil erodibility factor. A numerical value by soil type for estimating the tendency of a soil to be eroded in the soil loss equation.

Soil permeability (Hydraulic conductivity). A soil characteristic indicating the rate water moves through the soil.

Soil series. A group of soils having horizons similar in differentiating characteristics in the soil profile, except for the texture of the surface soil.

Soil structure. The arrangement of individual soil particles into larger crumbs, granules, or aggregates.

Soil texture. A term related to the size of primary mineral particles in the soil, such as sand, silt, and clay.

Soil type. A subdivision of a soil series based on surface soil texture.

Specific yield. The fraction of a unit volume of porous media or soil that drains water by gravity.

Spile (Lathe box). A device placed through a ditch bank for transferring irrigation water from the ditch to the field.

Spillway. A channel or structure for conveying runoff through or around an earth embankment.

Splash erosion. Soil movement caused by impact of raindrops on the soil.

Sprinkler head. A device for distributing water under pressure, including a rotating mechanism and nozzles.

Sprinkler riser. (See Riser).

Stadia. A method of measuring distance by reading the rod interval between stadia hairs in a telescope.

Steel tape (Chain). A narrow strip of metal or other material for measuring distances in the field.

Subirrigation (Reverse drainage). Application of water to the root zone by raising the water table through a pipe or ditch distribution system.

Subsoil. That part of the soil beneath the topsoil.

Subsurface drain (Tile, concrete, plastic, pipe, or underdrain). (See Pipe drain).

Suction lift. Vertical distance between the elevation of the surface of the water source and the center of the pump impeller.

Summer fallow. The practice of tilling uncropped land during the summer to control weeds and store soil water for a later crop.

Surface creep. Coarse sediment that moves in almost continuous contact with the soil surface during wind erosion.

Surface inlet. A structure for conveying surface water into an open ditch, pipe, or subsurface drain (Chapter 12).

Suspended sediment. Material that remains in suspension in flowing water or in air for a considerable period of time.

Taping. The practice of measuring distances with a surveying tape.

Terrace. A broad channel, bench, or embankment constructed across the slope to intercept runoff and detain or channel it to a protected outlet.

Terrace inlet riser. (See Riser).

Terrace spacing (Interval). The vertical or horizontal distance between adjacent terraces, except for the top terrace.

Tile. A subsurface conduit made from clay or concrete. (See Pipe drain).

Tile depth (Pipe depth). Vertical distance from the soil surface to the grade line (bottom) of a tile drain.

Toe drain. A subsurface drain located at the downstream toe of an earth embankment.

Topographic factor. Product of the slope factor and the slope length factor in the soil loss equation.

Topographic map. A map showing contour lines, stream channels, and other surface features of the land.

Topsoil. A nonspecific term for the top surface layer of the soil, normally high in organic matter.

Total station. A surveying instrument similar to a transit, but capable of storing and displaying data electronically.

Township lines. Parallels of latitude at 6-mile intervals that are the north and south boundaries of townships in the rectangular system of public land survey.

Transit. A surveying instrument with visual scales capable of measuring elevations, distances, and vertical and horizontal angles.

Tripod. A three-legged stand to which a surveying instrument is attached.

Turning point (Temporary bench mark). An identifiable temporary point whose elevation is or can be determined by leveling.

Vertical interval. (See Terrace spacing).

Water harvesting. Any practice that increases runoff, such as covering the surface with plastic, applying sealants, paving, etc.

Water holding capacity (Available soil water capacity). Amount of soil water available to plants, usually field capacity less the wilting point percentage.

Watershed (Catchment). Land and water area that contributes runoff to a given point along a stream.

Watershed gradient. Average slope in a watershed, measured from a given point along the path of water flow to the most remote point upstream.

Watershed planning. Formulation of a plan to use and treat water and land resources.

Water table. The maximum height to which water rises in a vertical hole in the soil.

Water yield. Volume of surface or ground water from a watershed (not flood volume), usually minimum annual flow.

W-Drain (Ditch). Two closely spaced parallel field drains between which the spoil from construction is placed (Chapter 12).

Weir. A structure across a stream to control, divert, or measure the water flow.

Wetland. Area of wet soil that has a predominance of hydric soils, is inundated or saturated by surface or ground water, and under normal circumstances would support a prevalence of hydrophytic plants (Chapter 12).

Wetted perimeter. Length of the wetted contact line between water and its containing conduit, measured on a plane at right angles to the direction of flow.

Wilting point (Permanent wilting point, or permanent wilting percentage). Soil water content at or below which plants permanently wilt.

Windbreak. Any type of barrier for protection from winds, especially for buildings, gardens, orchards, and feed lots.

INDEX